D0535428

## About Island Press

Island Press is the only nonprofit organization in the United States whose principal purpose is the publication of books on environmental issues and natural resource management. We provide solutions-oriented information to professionals, public officials, business and community leaders, and concerned citizens who are shaping responses to environmental problems.

In 2005, Island Press celebrates its twenty-first anniversary as the leading provider of timely and practical books that take a multidisciplinary approach to critical environmental concerns. Our growing list of titles reflects our commitment to bringing the best of an expanding body of literature to the environmental community throughout North America and the world.

Support for Island Press is provided by the Agua Fund, Brainerd Foundation, Geraldine R. Dodge Foundation, Doris Duke Charitable Foundation, Educational Foundation of America, The Ford Foundation, The George Gund Foundation, The William and Flora Hewlett Foundation, Henry Luce Foundation, The John D. and Catherine T. MacArthur Foundation, The Andrew W. Mellon Foundation, The Curtis and Edith Munson Foundation, National Environmental Trust, The New-Land Foundation, Oak Foundation, The Overbrook Foundation, The David and Lucile Packard Foundation, The Pew Charitable Trusts, The Rockefeller Foundation, The Winslow Foundation and other generous donors.

The opinions expressed in this book are those of the author(s) and do not necessarily reflect the views of these foundations.

## About *Defying Ocean's End*

The concept of *Defying Ocean's End* was inspired by an urgent need—to address the sharp decline in ocean wildlife, a disturbing increase in ocean pollution, the neglect of policies and resources directed to solve these problems, and a widespread complacency in the general public borne of a profound lack of awareness of how much the health of the ocean affects their daily lives and their future.

An invited group of international ocean scientists, economists, lawyers and conservationists—and representatives from governments, corporations and the media—gathered in Los Cabos, Mexico, from May 29th to June 3rd 2003, to build a global action plan for ocean conservation. A *DOE* business team assisted the experts in identifying near-term priorities and costs. The resulting plan includes high-level recommendations for global ocean governance, fisheries reform, communications, marine protected areas/large marine ecosystems (MPAs/LMEs) and ocean science.

Since the Conference a coalition of major conservation NGOs—including Conservation International, the Natural Resources Defense Council, The Nature Conservancy, The Ocean Conservancy, Wildlife Conservation Society, World Wildlife Fund and IUCN–The World Conservation Union, and U.S. and other national governments—have worked together to begin putting the agenda into concerted action.

**Preceding page**
Juvenile lionfish in Suruga Bay, Japan.
DAVID DOUBILET

*Defying Ocean's End*

# *Defying Ocean's End*

## *An Agenda for Action*

**Edited by Linda K. Glover and Sylvia A. Earle**

*Assistant Editor, Arlo H. Hemphill*
*Foreword by Graeme Kelleher*
*Maps by Debra Fischman*

Island Press / WASHINGTON • COVELO • LONDON

Copyright © 2004 Island Press
Copyright © 2004 Sylvia A. Earle, *Preface* and *Time for a Sea Change*

All rights reserved under International and Pan-American Copyright Conventions. No part of this book may be reproduced in any form or by any means without permission in writing from the publisher: Island Press, 1718 Connecticut Ave., Suite 300, NW, Washington, DC 20009.

Island Press is a trademark of The Center for Resource Economics.

*Library of Congress Cataloging-in-Publication Data*

Defying ocean's end : an agenda for action / edited by Linda K. Glover and Sylvia A. Earle ; assistant editor, Arlo H. Hemphill ; foreword by Graeme Kelleher ; maps by Debra Fischman.
p. cm.
Includes bibliographical references (p. ).
ISBN 1-55963-753-6 (cloth : alk. paper)
ISBN 1-55963-755-2 (pbk. : alk. paper)
1. Marine ecology—Congresses. 2. Aquatic resources conservation—Congresses. I. Glover, Linda K. II. Earle, Sylvia A., 1935–
QH541.5.S3D37 2004
333.95'616—dc22 2004024488

Printed on recycled, acid-free paper using soy-based inks

*British Cataloguing-in-Publication Data available*

Book design by Creative Project Management, Inc., and Kim Bieler Graphic Design

Manufactured in the United States of America
10 9 8 7 6 5 4 3 2 1

**Preceding page**
Golden cownose rays glide through waters surrounding the Galápagos Archipelago.
STERLING ZUMBRUNN, CONSERVATION INTERNATIONAL

## About the Editors

**Linda K. Glover** is a marine geologist with research and ocean policy experience. While working for the Navy, NOAA, the National Advisory Committee on Oceans and Atmosphere and the National Reconnaissance Office, she authored numerous scientific papers and national ocean policy reports and has edited several books on ocean issues. She is known for presenting complex information in easily understandable language. Currently Sole Proprietor of GloverWorks Consulting and Senior Marine Policy Advisor to Conservation International, Ms. Glover coordinated *Defying Ocean's End (DOE)* efforts before, during and since the *DOE* Conference held from May 29th to June 3rd 2003, in Los Cabos, Mexico.

**Sylvia A. Earle** is Executive Director of Conservation International's Global Marine Program, Explorer-in-Residence at the National Geographic, Program Coordinator for the Harte Institute for Gulf of Mexico Research and Chairman of Deep Ocean Exploration and Research, Inc. She has served on the boards of numerous corporate, scientific and conservation institutions, is author of more than 150 publications on ocean science, technology and conservation, leader of more than 50 ocean expeditions and recipient of numerous honors including the Netherlands Order of the Golden Ark, Explorers Club Medal, *Time* magazine's first Hero for the Planet and the Library of Congress Living Legend. Dr. Earle conceived the rationale for the *Defying Ocean's End* Conference and co-convened the event with Dr. Gordon Moore, Founder of Intel Corporation.

## About the Authors

All of the senior authors participated in preparations for—and deliberations during—the *Defying Ocean's End* Conference. Most of the co-authors were members of theme groups at the conference, but a number of ocean experts who did not attend have joined the effort as co-authors as interest in the *DOE* agenda for action has spread.

# Contents



**Facing page**
Olive ridley sea turtle hatchling headed for the Pacific Ocean off Mexico.
RODERIC MAST, CONSERVATION INTERNATIONAL

# Acknowledgements

Profound thanks first to Gordon Moore and Peter Seligmann for their leadership, vision and deep conservation ethic and to the Gordon and Betty Moore Foundation for the vital grant to Conservation International that made possible the *Defying Ocean's End* Conference, as well as a series of follow-on meetings in Washington, D.C. Valuable contributions were also made by the BP Corporation, Environmental Systems Research Institute (ESRI), Royal Caribbean Cruise Lines, the International SeaKeepers Society, an anonymous donor, and by the Edith and Henry Munson Foundation for preparation and printing of the *DOE* Executive Summaries.

Special thanks are due to Cinco Hermanos for supporting us in implementing the *DOE* vision and preparing this book; to Stone Gossard for supporting this volume; and to the leadership and staff of Conservation International for their above-and-beyond efforts to make this book a reality: to Roger McManus for valuable oversight, Arlo Hemphill for tireless, good-humored scientific and technical support, Sterling Zumbrunn for contributing his skillful photographic expertise, Debra Fischman for knowledgeable and timely building of the maps, Nicole Manning and Santi Duewel for assistance near the end and Tim Werner for his critical role in crafting the beginning of *DOE*. Cyndi Wood of Creative Project Management and her team—Kim Bieler of Kim Bieler Graphic Design and Mac Hobbs and Steve McKenzie of the McArdle Printing Company—deserve special credit for design, editing and printing above and beyond normal standards. Special thanks also to the many photographers who donated their images for this volume.

The powerful movement arising from the *DOE* Conference stems directly from support of the more than seventy organizations involved and the dedication of the many authors who brought the distillation of years of thoughtful research and concern to the issues addressed here. May their vision become reality.

# *Invitation to Action*

In recent years, human actions have had an unprecedented impact on the health of the ocean. A crisis is growing that threatens not only coastal communities that depend directly on living marine resources, but also all people everywhere. The Earth's ocean is a unique feature in our solar system and is essential for maintaining life on this planet.

As never before, we are seeing the consequences of abuse in collapsing fish populations, biodiversity loss and physical and chemical changes that are leading to the decline of entire ecosystems. As never again, we have an opportunity now to respond to this crisis, learn from centuries of experience in resource management on land and move beyond localized and ad hoc initiatives—however good they may be—to coordinated global action.

Our collective challenge is nothing less than the creation of a framework and steps for a practical agenda of global action—including costs and impacts—to safeguard the ocean for generations to come.

Sincerely,

Gordon E. Moore    Sylvia A. Earle

*Original invitation letter to participants of the* Defying Ocean's End *Conference from the co-conveners—Gordon Moore, the founder of Intel Corporation, and Sylvia A. Earle of Conservation International.*

**Facing page**
A reef off Grand Cayman Island in the Caribbean.
© JENNIFER JEFFERS

# Foreword

Graeme Kelleher

The beginning of this new millennium is a critical time for humanity, for our fellow animals, for plants and for the biosphere. For the first time, as far as we know, a single species has become so numerous and so powerful that its activities have already caused significant changes in the biosphere and exhibit the potential to have much more serious, even catastrophic, effects on our planet.

The purpose of the *Defying Ocean's End* Conference in Los Cabos, Mexico, in May/ June 2003 was to bring the consciousness and commitment of an ocean-experienced cross-section of international experts to focus on how to make humanity more of a friend to our living Earth.

Fortunately, we were not starting from scratch. Over the past decade in particular, there has been a series of developments that provided a foundation on which to build. To mention just a few:

- The United Nations Convention on the Law of the Sea in 1992 (UNCLOS)
- The Convention on Biological Diversity (CBD)
- The World Summit on Sustainable Development (WSSD)
- Recent United Nations General Assembly resolutions
- Implementation of the Kyoto Protocol on Climate Change
- The establishment of a High Seas Marine Protected Area (MPA) Executive Committee, which brings together IUCN–the World Conservation Union, the World Wildlife Fund (WWF), the World Commission on Protected Areas (WCPA) and some governments in the cause of high seas MPAs.

Since the *DOE* event, its themes have also been addressed at:

- The Workshop in Cairns, Queensland, Australia on 17–21 June 2003, focusing on high seas biodiversity—an expression of the Australian Government's commitment to this issue.
- The World Parks Congress in Durban, South Africa, on 8–17 September 2003.
- The 7th Meeting of the Conference of the Parties (COP7) to the Convention on Biological Diversity in Kuala Lumpur, Malaysia, in February 2004.

The future offers good opportunities for a continuum in the process. The third IUCN World Conservation Congress in Bangkok, November 2004 and the first

**Facing page**
Australia's Middleton Reef in the Tasman Sea—an atoll atop a volcanic seamount—is one of the world's southernmost coral reefs.
DAVID DOUBILET

International MPA Congress (IMPAC1) in Geelong, Australia, scheduled for October 2005 are two examples.

We used to say that the world is littered with paper parks. That's still true, but referring to the ocean it might be fair to say we are awash in wish lists. We don't want another one. The *Defying Ocean's End* Conference was organized into seven theme groups and five regional case study groups, with a great deal of preparatory work done before we convened in Los Cabos. The chairs of the Case Study and Theme Groups had a major responsibility in defining realistic Action Plans for three target times: one, three and ten years. The expectations during the Conference were to create an Action Plan, defining specific:

- Outcomes—targets
- Target groups
- Actors
- Timelines
- Costs
- Sources of funding

Results of this work before, during and since the Conference are presented in the chapters of this book.

There were four cross-cutting attributes of the Conference in general, and of the theme groups in particular, that should be recognized and applauded. The first was the enthusiasm. The enthusiasm that was shown by everybody involved was wonderful. Most of us are familiar with those terrible conferences where people preach from the podium and people from the audience sneak out to talk to each other rather than listen and, as a consequence, they don't take part in the creative function of the event.

ROGER STEENE

Yellow crinoids, a starfish relative, living among sponges on a reef in the Philippines.

Fortunately, that was not the case here—*Defying Ocean's End* was a conference characterized by commitment, communication and collaboration.

The second element we should identify is thoughtfulness. The *DOE* Conference was unusual in the deliberate abandonment of mental boundaries. The thoughtfulness demonstrated by the speakers, non-speakers, the theme groups and the case study teams was quite remarkable.

Third, the professionalism was outstanding. All the participants, in my view, demonstrated impressive professionalism in explaining and sharing their views.

Fourth, and just as important, was cross-professionalism. The language used was generally understandable even by people from other professions. I think that's going to be one of the most important things to foster as we continue along the path we are following together. Cross-professional communication will be absolutely fundamental and will lead to increased collaboration. We have to continue to collaborate, not just among professions, but also among nations and among communities and organizations around the world.

There are another four things that I should like to share with you as you use this volume. There are four lessons we can all apply personally and collectively:

First, follow your dream. All of us involved with *Defying Ocean's End* have a dream of a world where humans are not the enemy of the biosphere, but the friend. I should like to quote from Polonius in Shakespeare's *Hamlet,* who said, "and this above all, to thine own self be true, and it must follow, as the night the day, thou canst not then be false to any man." Following our dream is going to be an absolutely essential path to the future we hope for.

The second lesson is based on a quotation from the owl in the Pogo comic strip: "I have met the enemy, and it is us!" We need to realize that most, if not all, of us have inherited or developed attitudes and instincts that threaten our species and the entire biosphere. They include selfishness, hostility, prejudice against others and the desire for revenge. These attitudes were markedly absent during the *DOE* Conference. We need to be increasingly conscious of such attitudes and instincts in others and ourselves so we can act cooperatively and creatively.

The third quotation, from the book of Genesis, is probably well known to most of you: "Be fruitful and multiply and subdue the earth." If there is one lesson that the problems of our environment should have taught us, it is that this injunction is disastrous. Humanity must live in harmony with the Earth and not try to subdue it. If human populations continue to increase, we will eventually overwhelm the biosphere.

The fourth lesson I should like to mention is: Never give up! The quotation for that is: "There is a tide in the affairs of men, which taken at its flood leads on to fortune." (Brutus from Shakespeare's *Julius Caesar*). Now, if you never give up, you will eventually encounter the incoming tide and can ride it to the fulfillment of your dream. Our dream is for humanity to live in harmony with the biosphere. If there was ever a tide in humanity's relation to the sea, it is now. It is demonstrated in conferences such as this one, in international agreements, in research into marine ecosystems and processes and their relations to the entire biosphere, and in the gradually increasing understanding by human communities of these relations.

Let us take this tide in its flood for the future of humanity and our wonderful planet!

# Preface

Sylvia A. Earle

*Eventually man . . . found his way back to the sea . . . And yet he has returned to his mother sea only on her own terms. He cannot control or change the ocean as, in his brief tenancy on earth, he has subdued and plundered the continents.*

—Rachel Carson, "The Sea Around Us," 1951

The 20th century proved to be a time of unprecedented scientific discovery, with more new insights gained about ourselves, the nature of the Earth and the universe beyond than during all preceding human history. Fifty years ago many scoffed at the notions that continents move around, that oceans come and go and that life can exist in the deepest sea. In the 1950s most people believed the ocean was so vast and so resilient that there wasn't much humans could to do alter its nature. Half a century later, two great truths have emerged based on new knowledge about the sea. First, the ocean is the cornerstone of Earth's life support system and therefore is vital to human survival and well-being. We have come to understand that the sea generates most of the oxygen in the atmosphere, absorbs huge quantities of carbon dioxide, governs our climate and weather, regulates temperature, drives planetary chemistry, harbors most of the water and contains the greatest abundance and diversity of life on Earth. As the sea changes, so do the circumstances that shape the way our world works.

The second great truth that emerged in the latter half of the 20th century was that humans *do* have the capacity to alter basic global processes. In fact, changes are occurring as a consequence of what we are putting into the ocean and atmosphere and onto the land, as well as what we are removing from the mantle of living systems on land and in the sea that are vital to maintaining the planet as a place hospitable for humankind. We learned the hard way that the ocean is not infinite in its capacity to accommodate what we put in or take out. Already, destruction of coastal and deepsea habitats, the annual removal of millions of tons of ocean wildlife and the addition of millions of tons of noxious wastes have set in motion processes that are not likely to be in our best interests. In the 20th century, we learned more about the ocean than during all preceding human history, but during the same time, we lost more.

Those who participated in the *Defying Ocean's End* Conference in Los Cabos, Mexico, in May/June 2003 came with this knowledge as a starting point, and with a com-

**Facing page**
School of jacks near Cocos Island in the eastern tropical Pacific.
STERLING ZUMBRUNN, CONSERVATION INTERNATIONAL

mitment to identify actions that realistically could be taken to reverse the worrisome global trends. The year before, in June 2002, Conservation International convened a small group of scientists and conservation experts also in Cabo San Lucas, Mexico, to consider what could be done not just to slow the decline of ocean health, but to succeed in turning the trends in a positive direction. The major outcome of that planning session was the vision of assembling the best international experts available, after a year devoted to developing background data about specific issues and actions that could "defy ocean's end" on a global scale. Thus the *DOE* Conference began to take form.

That a meeting of minds prepared to develop an action plan could actually help drive positive, international initiatives was inspired by the successful outcomes of the *Defying Nature's End* Conference held in 2000 at the California Institute of Technology. At that event, a select group of scientists, economists, leaders in business and other disciplines focused attention on actions needed to stem the loss of global biodiversity. That conference gave rise to strategies aimed at protecting biodiversity, with an emphasis on "hotspots"—areas of high diversity and high threat—as well as intact wilderness areas. Innovative programs were recommended to preserve the economic, societal and security advantages of maintaining global biodiversity, while underlining the problems that arise when biodiversity is diminished or lost.

Optimism prevailed because of the realization, as reported in the journal *Science* in May 2001, that ". . . greatly increasing the areas where biodiversity is protected is a clear and achievable goal, one potentially attainable by using funds raised in the private sector and leveraged through governments." Critical needs were identified: synthesis and distribution of available knowledge, investment in targeted research, development of research and management centers in countries where biodiversity actions are focused, and demonstrating the links among biodiversity, ecosystems, their services and people—all of these leading to benefits for humankind through increased and enduring protection for hotspots and wilderness areas worldwide.

ROGER STEENE

An improbably colorful nudibranch in the Philippines.

GETTY IMAGES®

Goldenstriped cardinalfish and a sea anemone in the waters off Japan.

Some consideration was given to coastal and ocean matters at the *Defying Nature's End* Conference, but both the recommendations and subsequent implementation focused largely on terrestrial issues. By 2002 it was time to take a hard look at the blue part of the planet—largely neglected by conservation organizations, national governments and international policymakers, despite alarming trends that closely parallel declines of natural systems and species on land, and with similar devastating consequences to the interests of humankind.

The challenge was and still is daunting. Many seemingly intractable ocean problems were identified in 1998, the International Year of the Ocean, at various national and international conferences. Over and over the need for accelerated research and exploration were expressed, recognizing that less than five percent of the ocean has been seen, let alone explored. Repeatedly, the consequences were discussed of overfishing and the use of gear and techniques that extract an enormous toll in terms of habitat destruction and many millions of tons of non-targeted marine life destroyed incidental to fishing operations. Problems relating to the downstream consequences of upstream pollution, as well as the deliberate dumping of wastes into the sea, were the subject of numerous meetings and reports. Most puzzling and damaging of all was the widespread indifference to ocean issues, a complacency borne of the lack of awareness among decision-makers and the public at large that the ocean is in trouble, and a profound lack of understanding that there is a direct connection between the declining state of the ocean and human affairs.

In response to these grave concerns, two ocean commissions were established in the United States in 2001: the Pew Oceans Commission supported privately by the Pew Foundation, and the National Ocean Commission appointed by the President and Congress to review national policies and recommend actions. Both of these Commissions, however, focused largely on U.S. national issues and actions.

A "chocolate chip" starfish from the Galápagos.

*Defying Ocean's End* took a broader view, deliberately looking for solutions that could be implemented globally over broad areas involving a wide range of participants: the public and private sectors of local communities, states, nations and international institutions. Several reports issued just before the Conference highlighted the urgency of the deliberations. Two Canadian scientists, Drs. Boris Worm and Ransom Meyers, published the results of their ten-year study of global fisheries data with the conclusion that 90% of the big fish—tuna, swordfish, marlin, sharks, grouper, snapper and others—have been extracted from the ocean in the past fifty years. New data on bycatch revealed the destruction of more than 300,000 marine mammals, tens of thousands of birds and turtles and more than 20 million tons of animals killed and discarded in industrial fishing operations every year. Another report cited a dramatic increase in the number of "dead zones" in coastal waters, created by unnaturally high levels of nitrates and phosphates flowing down rivers from fields, lawns and farms—an increase from 50 such areas a decade ago to more than 150 now. Many other recent studies have documented dramatic declines in coral reefs, seagrass meadows, kelp stands, mangrove forests and other vital systems known to have thrived half a century ago. Special concern was raised at *DOE* about the swift destruction of seamount habitats and their associated ocean wildlife by bottom-trawl fishing, coincident with news that seamounts may be likened to the Galápagos Islands on a grand scale, with each rocky, underwater "island" hosting high biodiversity and high levels of species found nowhere else on Earth.

There was also good news. The high, hidden costs of industrial fishing have not gone unnoticed and a fundamental change in thinking about commercial fishing is underway, starting with a recognition that fish and other marine life are not commodities, they are wild animals. Fishing involves hunting and capturing, not cultivating and harvesting. Some policy and regulatory revisions for commercial fishing are beginning, with a growing focus on "sustainable use" of living marine populations within national Exclusive Economic Zones and in the 60% of the ocean called high seas that lacks coherent governance. Commitment to research, education and new technologies to improve access to the ocean depths are gradually improving in some countries, including Australia, China, Japan, France and the United States.

The mood of participants at the *Defying Ocean's End* Conference was electric. It was driven in some measure by a sense of urgency, but also by a conviction that it is not too late to do something about protecting ocean health and restoring some of the damage. Ten percent of the big fish are still out there in the ocean; they're not all gone yet. Half of the coral reefs are still in good shape. Despite pollution issues, much of the ocean is still healthy and resilient. A strong sense of hope prevailed that actions could be taken in time to avoid further loss of marine species and irreplaceable ocean ecosystems, and resources could be raised and mechanisms identified to achieve the desired, successful outcomes.

Whether we act on what we know or allow the present opportunities for concerted action to slip away, the next decade may prove to be the most significant for ocean conservation—and for the future of humankind—in the next thousand years.

**Facing page**
Giant octopus with anemone off British Columbia, Canada.
DAVID DOUBILET

# *Overview*

Linda K. Glover and Arlo H. Hemphill

There are many national and international conferences held every year around the world that focus on aspects of marine conservation, but *Defying Ocean's End* in Los Cabos, Mexico, May 30th to June 3rd of 2003 had several unusual characteristics.

First, its objective was more than just communicating ideas. *DOE*'s goal was to map out a global agenda for action in marine conservation, with priorities and costs identified. Although we heard from a few invited speakers, whose remarks were very useful input to the global plan, most of the time at *DOE* was spent in working sessions and in presentations and critiques of the working groups' results. Second, most of the work was begun many months in advance. Some twenty grants and contracts were written with various groups to prepare the foundation for discussions well before the Conference began. These efforts were of two types—regional case studies and global thematic working groups.

Case studies were commissioned from five regions of the world ocean chosen for their differences—in their physical environments, marine biological assemblages, levels of threat, political and legal regimes and ongoing conservation management efforts. The five case study regions are:

- The Caribbean
- Seamounts
- Antarctic Waters
- The Coral Triangle
- The Gulf of California

The wider Caribbean, presented in Chapter 1, is a very large semi-enclosed oceanic basin with tropical and subtropical environments. Its land areas include the margins of two continents and hundreds of islands belonging to over twenty nations, so complex multinational regional efforts are required to address the problems of marine conservation in this area. The Caribbean basin is close to highly developed land areas in some places, and continuing development for tourism and other commercial uses are an increasing influence in the region. Fishing is both local and industrial-scale, but its impacts seem reversible through increasing establishment of "no-take" zones and a variety of innovative fishery management approaches like individual and community quota schemes. And the basin harbors surprisingly high biodiversity and numbers of species found only there. The Caribbean chapter thus raises most of the issues

**Facing page**
Anthias fish and chalice coral in Milne Bay, Papua New Guinea.
DAVID DOUBILET

addressed throughout the book and sets the stage for the discussions that follow.

Seamounts are large underwater mountains that exist throughout the world ocean and host unique deep-water biological communities. Deep coral reefs on the cold, dark flanks of many seamounts (Chapter 2) yield species new to science every time they are sampled and likely hold important but not yet discovered information about the workings of the ocean, the origins of life on Earth and indeed the Earth's environment as a whole. Despite their remoteness these richly diverse and intriguing environments are being destroyed by industrial-scale bottom-trawl fishing on the high seas. The seamounts are being fished faster than they can be studied, and a significant portion of the Earth's biodiversity may be lost before it can be discovered.

Antarctic waters (Chapter 3) present another environment in great contrast to the tropical coral reefs—frigid waters with their own unique biological communities. This region is also unusual politically in that all nations have stepped away from any national claims or jurisdictional controls over the land and water areas or their resources. The primary vehicle for management and protection of its biologic resources is the Convention for the Conservation of Antarctic Marine Living Resources (CCAMLR)—the first international convention to implement the "precautionary principle" and ecosystem-wide approaches, like considering the needs of natural predators before setting catch limits for the fishing industry. Its successes and limitations provide useful lessons.

The "Coral Triangle" spanning the equator in the far western Pacific (Chapter 4) has historically been, and still is, an area of high biodiversity in marine life. The

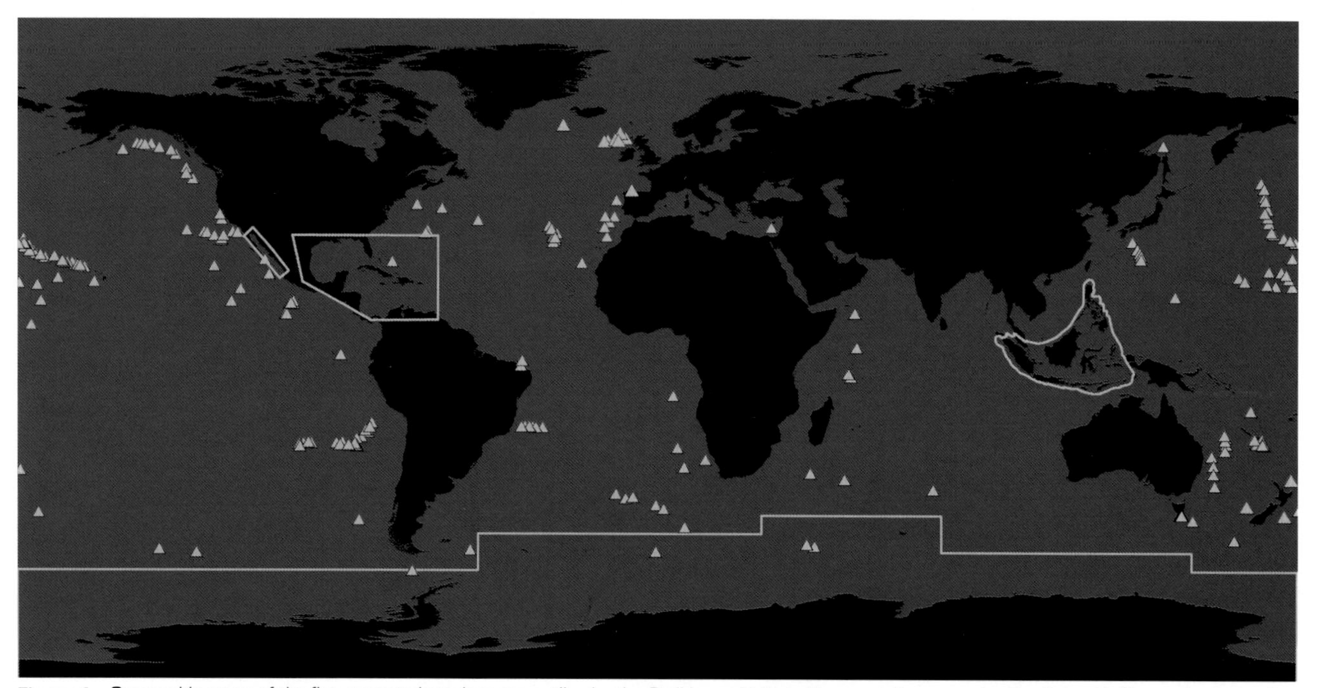

**Figure 1** Geographic areas of the five case study regions are outlined—the Caribbean, biologically sampled seamounts, the Antarctic Convention area, the Coral Triangle and the Gulf of California.

remoteness of this region has preserved some pristine environments, but unmanaged practices like clear-cutting of forests and blast fishing in many areas threaten to destroy some of the high biodiversity in this region.

The Gulf of California is another unique ocean region (Chapter 5). A deep, narrow basin caused by movement of land and ocean crust along the immense fault line on the western coast of North America, the Gulf sits in a subtropical setting and is surrounded by only one country. The successes in this region from engaging the users, conservationists, the business community and state and national governments are instructive.

The charge to these case study groups was to review threats and marine conservation activities in their regions (Figure 1), assess what is working and what needs improvement and make recommendations for improved action. Results of the case studies were reported to everyone on the first day of the *DOE* Conference to provide "lessons learned" and recommendations from each region that might be considered while creating a global action plan.

ROGER STEENE

Brightly colored crinoid in the Philippines.

Meanwhile a number of "theme groups" also began their work well before the Conference. The themes were chosen to cut across typical threats, uses and interest areas —such as fisheries or invasive species—and take a broader look at large-scale solutions for multiple environmental problems. Results of the seven global theme groups for *DOE* are also presented as chapters in this volume:

- Ocean-Use Planning & Marine Protected Areas (Chapter 6)
- Economic Incentives & Disincentives (Chapter 7)
- Land-Ocean Interface (Chapter 8)
- Maintaining & Restoring Functional Marine Ecosystems (Chapter 9)
- Communications (Chapter 10)
- Ocean Governance (Chapter 11)
- The Unknown Ocean (Chapter 12)

GREG FOSTER

Seaweed blenny emerging from its coral reef hiding place off Florida, U.S.

The charge to each of the theme groups was to:

- Follow a consistent approach to assist in integrating the results of all groups into a final agenda for action
- Make use of a map-based approach wherever possible
- Prepare a very brief background of the key issues associated with your theme that sets the stage for the recommendations
- Identify the most important and urgent issues that will affect future action
- Focus on practical and achievable actions
- Determine what each component of your strategy will cost to implement
- Propose measures for assessing success

Theme group members were chosen to include a broad representation of international experts—in ocean science, conservation, law, economics and communications—mostly from academia. In many cases the working group members had never met, but began their collaboration through email and the internet to have a draft plan ready when they did meet at the Conference.

We threw another element into the mix at the *DOE* Conference itself—almost forty additional experts who had not participated in the preparatory work joined the theme groups and brought a fresh perspective to their proceedings during *DOE*. We had a small number of very informative invited talks, and these speakers also joined the theme groups. Most of the *DOE* Conference was spent in the small theme groups working out actions, timing, priorities and costs for their pieces of a global action plan.

Another element of the Conference was the addition of a "business team"—a small group of business planning professionals who assisted the theme groups in organizing their recommended actions into 1-, 3- and 10-year outcomes and estimating total program costs; their approach is described in Chapter 13.

An impressive IT (information technology) team "wired" the entire Conference with email and computing capacity for pulling in new datasets during discussions, and for displaying interim plans and presentations in realtime. We also had a mapping and GIS (global information system) team that made maps before and during the Conference to illustrate, assess and compare threats and conservation planning priorities. A decision was made well before the Conference to use standard mapbases for the world and for each of the regions, so that data layers of different information could easily be shared and compared among the groups. The standard maps used for the Conference are visible throughout this volume, with all worldwide data shown on the same global

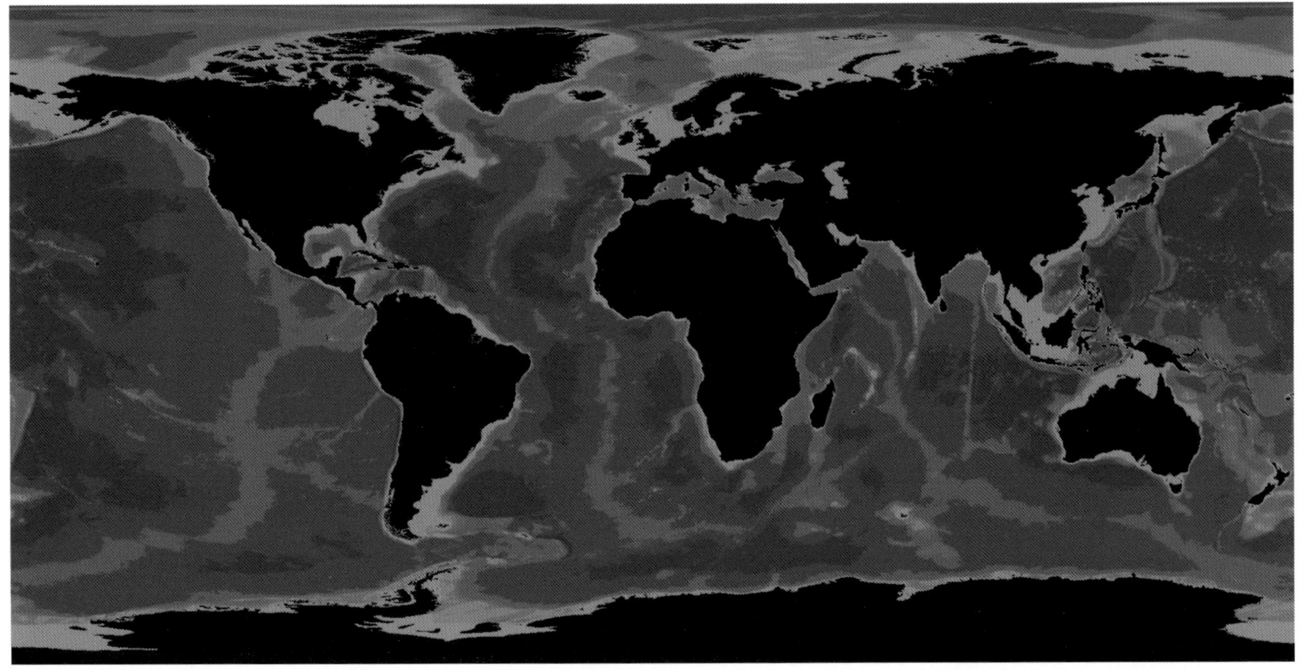


**Figure 2** Global bathymetry (water depth) clearly shows underwater mountain ranges, deep trenches and broad shallow continental margins. The same bathymetry scale is used for all maps throughout this volume.

mapbase (Figure 2). The *DOE* maps are clearly designed to focus on information in the ocean. Our IT and mapping leads at the Conference prepared a piece on technology support to conservation which is included in this volume as Chapter 14.

At the Conference, we had almost 150 experts from more than 20 countries and representing more than 70 universities and other organizations. All were energized by the urgency of the problems facing the ocean, the broad sharing of information across disciplines and the focus on trying to forge an effective set of global solutions. Strange, ancient creatures in caves and on deep seamounts captured the group's imagination. A growing sense of urgency for global reform in fisheries management, high seas legal regimes and communication of ocean concerns to the public were widely endorsed. Some theme groups worked late into the night. Some were seen in discussions on the beach at dawn. The results of their efforts and *DOE* follow-on activities since are presented throughout these pages. We have made every effort to minimize the use of, or define, technical terms so the concepts can be understood by the widest possible audience of ocean professionals, students from all fields, global decision-makers and the public.

For further information about *DOE* follow-on, contact:
*http://www.defyingoceansend.org*
DOE Staff/Global Marine Division
1919 M Street NW, Suite 600
Washington, D.C. 20036

CHAPTER 1

# The Caribbean

Mark Spalding, *University of Cambridge*
Philip Kramer, *The Nature Conservancy*

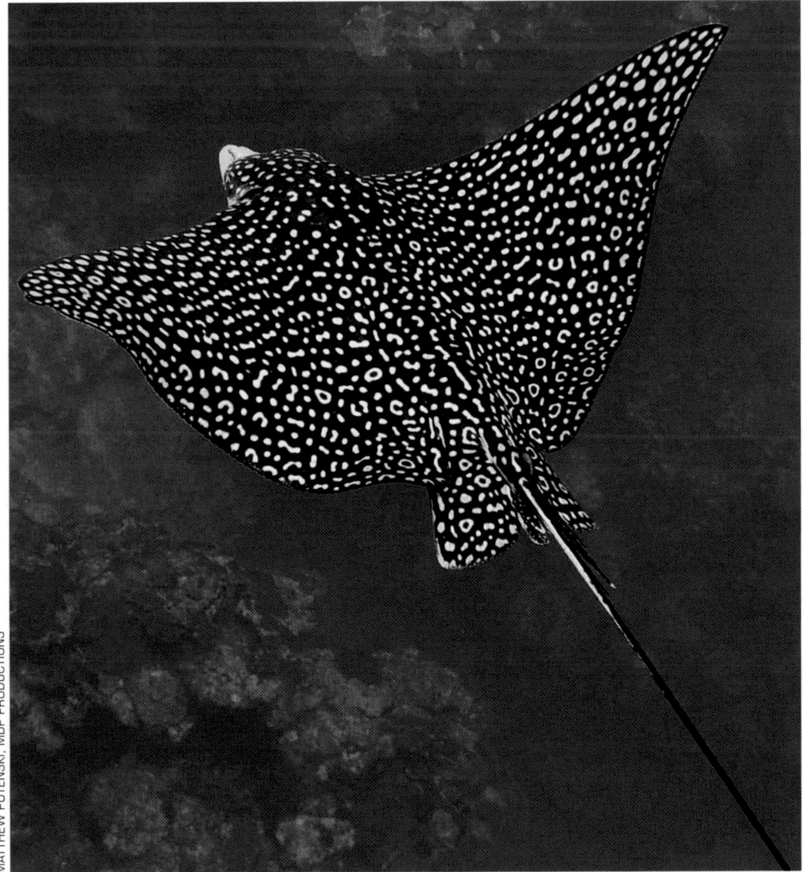

MATTHEW POTENSKI, MDP PRODUCTIONS

## Introduction

The Caribbean Region, for the purposes of this study, is an area comprising the Caribbean Sea, the Gulf of Mexico and the adjacent areas of the Bahamas and Turks and Caicos. This is a region of considerable geological, biological and political complexity (Figure 1) but also one of internal integrity, a partially closed series of basins having relatively little interplay with adjacent regions. The region is often referred to as the Wider Caribbean, the Intra-Americas Sea or the American Mediterranean. It provides an excellent case study for considering both human impacts in the marine realm and the measures required for their mitigation. Three features underpin this value:

- *Biological complexity.* The region encompasses the heart of Atlantic species biodiversity and is host to considerable numbers of "endemic" species—those found only in a geographically restricted area and nowhere else in the world. This is well documented for shallow coastal communities (Tomlinson 1986, Veron 1995, Spalding, Blasco & Field 1997, Spalding, Ravilious & Green 2001, Spalding *et al* 2003), and there is now growing evidence for similar patterns in deeper water species (Smith, Carpenter & Waller 2002).
- *Socio-political complexity.* This small region includes 35 countries and territories covering a complete spectrum of political systems and economic regimes, including the world's richest nation alongside some of the world's poorest (Schumacher, Hoagland & Gaines 1996). The history of human activity in the region has had a major role in the current status of the marine environment.
- *Pressure.* Humans have had some influence over parts of the marine environment in the Caribbean for millennia (Jackson *et al* 2001), but there is no doubt that current pressures are unparalleled (Burke & Maidens 2004).

This case study was prepared for the *Defying Ocean's End* Conference in Los Cabos, Mexico, in May/June 2003. We commence with an overview of the physical and biological geography of the region, then focus on the range of human stresses affecting the marine environment and finally consider the options for reversing the current trends of rapid environmental decline.

## Natural History

### Physical geography

The region under consideration consists of two semi-enclosed basins—the Caribbean Sea and the Gulf of Mexico—as well as the westernmost embayment of the

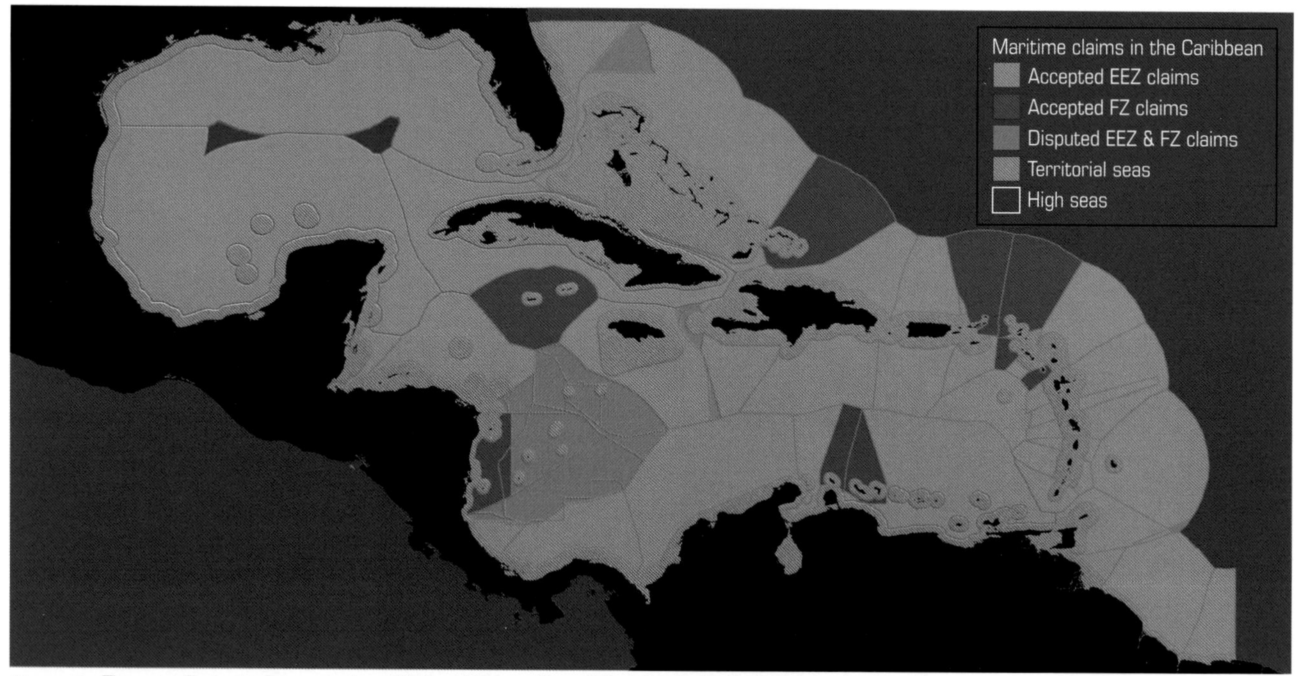


**Figure 1** The many Exclusive Economic Zone (EEZ) and Fishery Zone (FZ) claims in the Caribbean region complicate regional conservation schemes.

Atlantic Ocean. Bounded to the west and south by Central and South America, the Caribbean is bounded to the north by the Greater Antilles (Cuba to Puerto Rico) and to the east by the great arc of small islands called the Lesser Antilles.

Most of the Caribbean Sea is centered on the Caribbean Plate, a relatively small tectonic plate (section of Earth's crust) that formed between 75 and 90 million years ago. The movements of this plate form subduction zones to the east and west (where one plate slides beneath another) and transform boundaries (where two tectonic plates slide past each other) to the north and south. These tectonic interactions have created the island chains of the Greater and Lesser Antilles as well as the mountainous perimeters across Central America and the southern Caribbean Sea. Most of the remainder of the region lies on the North American Plate (Pindell 1994, Meschede & Frisch 1998).

The Caribbean Sea is itself divided into four smaller basins separated by shallower ridges, while the Gulf of Mexico can be considered a separate fifth deep basin. Deep trenches are found to the north of Puerto Rico and to the south of Cuba and the Cayman Islands. The basins, as well as the deep trenches, are isolated from each other by shallow underwater sills and, although there is some water exchange, each may contain unique and important features (Tomczak & Godfrey 1994, Smith, Carpenter & Waller 2002). With the exception of the Bahamas Banks, the only extensive shallow areas are on the continental shelves, notably around the southern U.S. coast, the northern coast of the Yucatan Peninsula and the Nicaraguan Rise.

Surface water movement is driven by the North Equatorial Current and the North Brazil (or Guiana) Current. These combine to form the Antilles Current, which sweeps up the outer edge of the Lesser Antilles and the eastern side of the Bahamas. There is also considerable flow between the Lesser Antilles islands, and to a lesser degree through the channels of the eastern Greater Antilles, creating a broad westward-flowing Caribbean Current across the Caribbean Sea. This current is weak and there are surface eddies creating a relatively complex pattern of surface flow. Further west, the water flow is concentrated through the Yucatan Channel and becomes much stronger. Once in the Gulf of Mexico, this current flows northwest, then in a broad clockwise circle —the Loop Current—which flows back southward along Florida's west coast, shedding eddies westward into the Gulf before leaving the system east of Florida. The Florida Current, enhanced by flow from the Antilles Current, forms a broad stream of warm water—the Gulf Stream—which flows northeast across the Atlantic (Tomczak & Godfrey 1994, Longhurst 1998).

Rivers have a considerable influence in the southeast Caribbean, where outflows of the Amazon and Orinoco Rivers create slightly lower salinities and more suspended sediments over wide areas (Tomczak & Godfrey 1994).

Deep water flows into the Caribbean Sea through a limited number of deep passes: through the Jungfern Passage east of Puerto Rico into the Venezuela and Colombian Basins, through the Windward Passage east of Cuba into the Cayman and Yucatan Basins and through a number of eastern passages into the Aves and Grenada Basins. The deeper water areas have remarkably consistent temperatures. Little is known about the turnover of deep water in the Caribbean basins, but most estimates suggest periods of hundreds of years (Tomczak & Godfrey 1994, Tyler 2003).

**Biodiversity**

The Caribbean is perhaps best known for its tropical shallow marine ecosystems. Its coral reefs cover about 20,000 square kilometers (Spalding, Ravilious & Green 2001). This is only about 7% of the world's shallow reefs, but they are of enormous biological and human importance. Biologically they are unique, with very little overlap in species with other parts of the world—except for Brazil and to a small degree West Africa, both of which have related assemblages but fewer species (Cortés 2003). The Caribbean is also a center of diversity and endemism for some mangrove and seagrass species, although a higher proportion of these extends into the Eastern Pacific and across to West Africa. The Caribbean has roughly 22,000 square kilometers of mangrove (amended from Spalding, Blasco & Field 1997), which is about 12% of the global total. No accurate estimate of the extent of seagrass area is available, although the Caribbean is likely to be a highly significant region for these habitats (Onuf *et al* 2003, Creed, Phillips & Van Tussenbroek 2003). Unpublished estimates from work at the United Nations Environment Programme (UNEP) World Conservation Monitoring Centre suggest as much as 33,000 square kilometers of Caribbean seagrass beds (Spalding *et al* 2003), which would be about 18 to 19% of the global total.

An explanation for this diversity and endemism in the Caribbean is at least partly linked to the geological history (Table 1) of the region. During the Triassic, about 200 million years ago, all of the world's land areas were connected

Table 1

**The geologic time scale. Epochs are shown only for the more recent Cenozoic Periods. Many of the Era and Period boundaries are based on major episodes of extinction or other biological factors.**

| Period | Epoch | Years ago |
| --- | --- | --- |
| **Cenozoic Era** | | |
| Quaternary | Holocene | 11 thousand |
| | Pleistocene (glacial) | 2 million |
| Tertiary | Pliocene | 7 million |
| | Miocene | 23 million |
| | Oligocene | 38 million |
| | Eocene | 53 million |
| | Paleocene | 65 million |
| **Mesozoic Era** | | |
| Cretaceous | | 135 million |
| Jurassic | | 195 million |
| Triassic | | 230 million |
| **Paleozoic Era** | | |
| Permian | | 280 million |
| Pennsylvanian (Upper Carboniferous) | | 310 million |
| Mississippian (Lower Carboniferous) | | 345 million |
| Devonian | | 395 million |
| Silurian | | 435 million |
| Ordovician | | 500 million |
| Cambrian | | 600 million |
| **Precambrian Era** | | |
| | | 600 million to 4 billion |

**Source:** Modified from the American Heritage Dictionary 2000.

as one supercontinent known as Pangea. Pangea was then divided into northern and southern continents by seafloor spreading, and the growing gap between them was filled by a globe-encircling equatorial sea known as the Tethys (Wikipedia 2004). Without land masses to disrupt biological migration and genetic flow, much of the subtropical and tropical marine regions of the world shared a similar array of animals, referred to as a pan-Tethyan fauna, until the Miocene (Veron 1995, Saenger & Luker 1997). Since that time, the open-sea connection between the Caribbean and the Pacific became increasingly restricted as the Central American land bridge rose above sealevel, which finally separated the two bodies of water about three million years ago (Guzmán 2003). For the corals, and likely for many other species, considerable changes were then caused by the Pliocene/Pleistocene glaciation periods, including the extinction of many marine species. This was followed by significant biological "radiation events" (the rapid evolution of new species to fill new habitats) in isolation from the rest of the world's tropical seas (Rosen 1984, Veron 1995), which led to the present levels of Caribbean coral diversity and endemism.

The boundaries and even the definitions of marine habitats other than coral reefs in the Caribbean remain poor. "Proxy" measures (scientifically assumed from the extension of known data) can be developed from knowledge of physical features such as surface circulation patterns, surface chlorophyll distributions or bathymetry. The location of upwelling areas—where deep, cold, nutrient-rich waters come to the surface—is extremely important for biodiversity and ocean productivity. Some sites are documented, such as the Yucatan Shelf edge (Merino 1997), the eastern coast of Venezuela and parts of the Gulf of Mexico (Longhurst 1998), but it may be possible to improve our understanding of the location, timing and frequency of these and other upwellings by looking at patterns of surface chlorophyll.

Dividing the Caribbean region into zones based on bathymetry (water depth) indicates that about 10% of the seafloor falls within areas less than 30 meters deep—where we find the highest light penetration and primary productivity—the transfer of sunlight into food through photosynthesis. Waters on the continental shelf edges (to 200 meters depth) cover a similar area. Waters between 2,000 and 5,000 meter depths dominate the region (Figure 2). The basins and deep trenches are clearly separated below 4,000 meters depth, however. Existing studies suggest that levels of biodiversity below the continental shelf edge may be very high and that, even in the deep trenches where relatively fewer types of organisms exist, endemism is a key feature, with important examples of species unchanged since ancient times (Zezina 1997). The few studies undertaken in the Puerto Rico Trench suggest that it has lower than expected diversity, however (Vinogradova 1997, Tyler 2003).

Some of the most important features of the deep seafloor include rocky outcrops with no sediment cover, deep shelf bioherms (limestone rock structures derived from

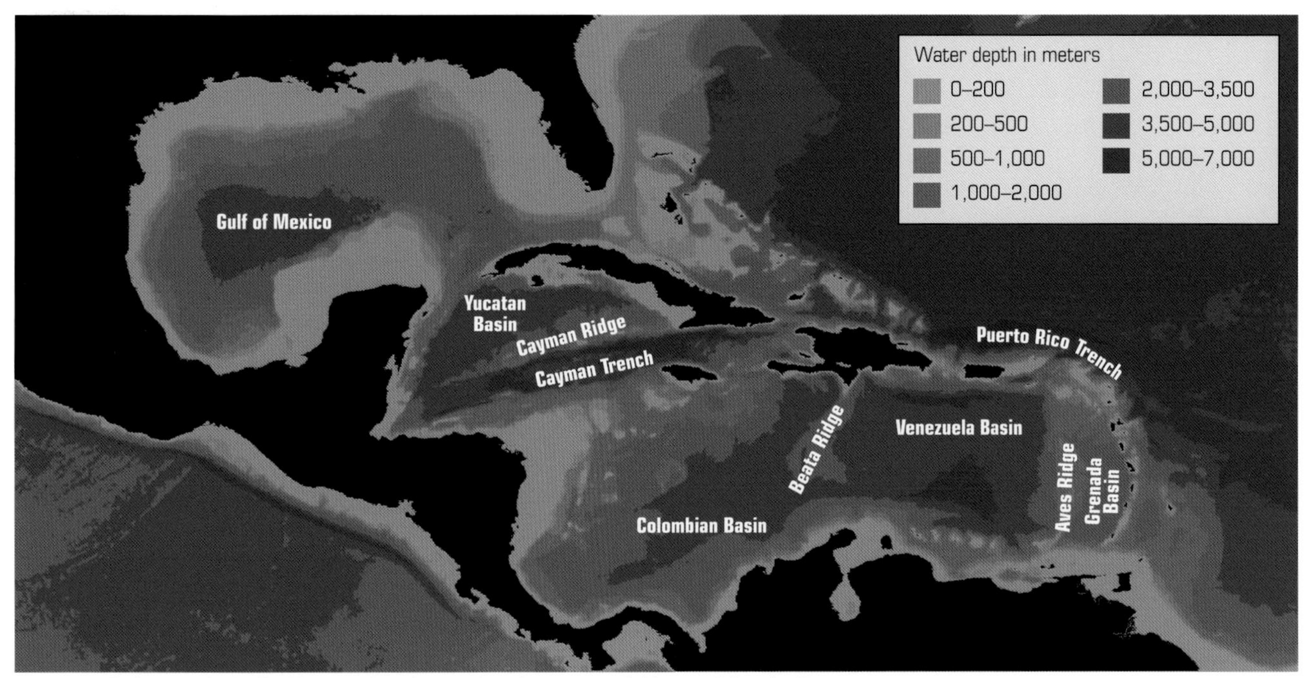


**Figure 2** Bathymetry in the Caribbean region shows a complex pattern of ridges and deep basins.

the shells of living organism) and hydrothermal vent areas where hot minerals spew up from cracks in the seafloor. There has been no systematic attempt to map these, but it might be possible to produce maps through a combined literature survey and expert interpretation of features from high-resolution seafloor mapping data.

### Species diversity patterns

Increasing scientific efforts are rapidly expanding our knowledge of the historical and present-day distribution of species. Shallow-water species have tended to dominate in several global studies, notably coral reef species (Roberts *et al* 2002), mangroves (Spalding, Blasco & Field 1997) and seagrasses (Spalding *et al* 2003). Smith, Carpenter and Waller (2002) have developed detailed maps of species ranges across the Caribbean including deeper waters (Figure 3). This work has shown previously unrecorded levels of endemism, with 227 of all fish species studied—fully 23% of them—being restricted to the Caribbean. Maps of these endemic species show new patterns of endemism concentration (Figure 4), especially in the Straits of Florida, the northern coast of South America, the Caribbean coast of Central America and the northern Gulf of Mexico. Location of all of these centers of highest endemism on continental shelf margins suggests that geological history may have a far more important role in both speciation (development of new species) and dispersal events than had previously been assumed (Smith, Carpenter & Waller 2002).

## Human History and Environmental Change

### Pre-Columbian period (9000 BC to 1492)

The Central and South American coasts have probably been inhabited since about 9000 BC, originally by nomadic hunter-gatherer tribes from North America. Early migration to Cuba apparently occurred about 6000 BC, but many of the other islands were not inhabited until 4000 BC. Many of these early peoples relied heavily on the sea as a major source of protein, but archaeological evidence suggests that extinction rates of terrestrial species within the Caribbean doubled after humans migrated into the area, despite low human populations in many areas (Wilson 1997).

When the first European explorers arrived near the end of the 15th century, two groups of people dominated the Caribbean Islands—the Arawaks in the Bahamas and Greater Antilles and the Caribs in the Lesser Antilles. In addition, small numbers of descendants from an earlier wave of settlement (the Ciboney people) were found in parts of Cuba and Hispaniola. A much broader range of ethnic

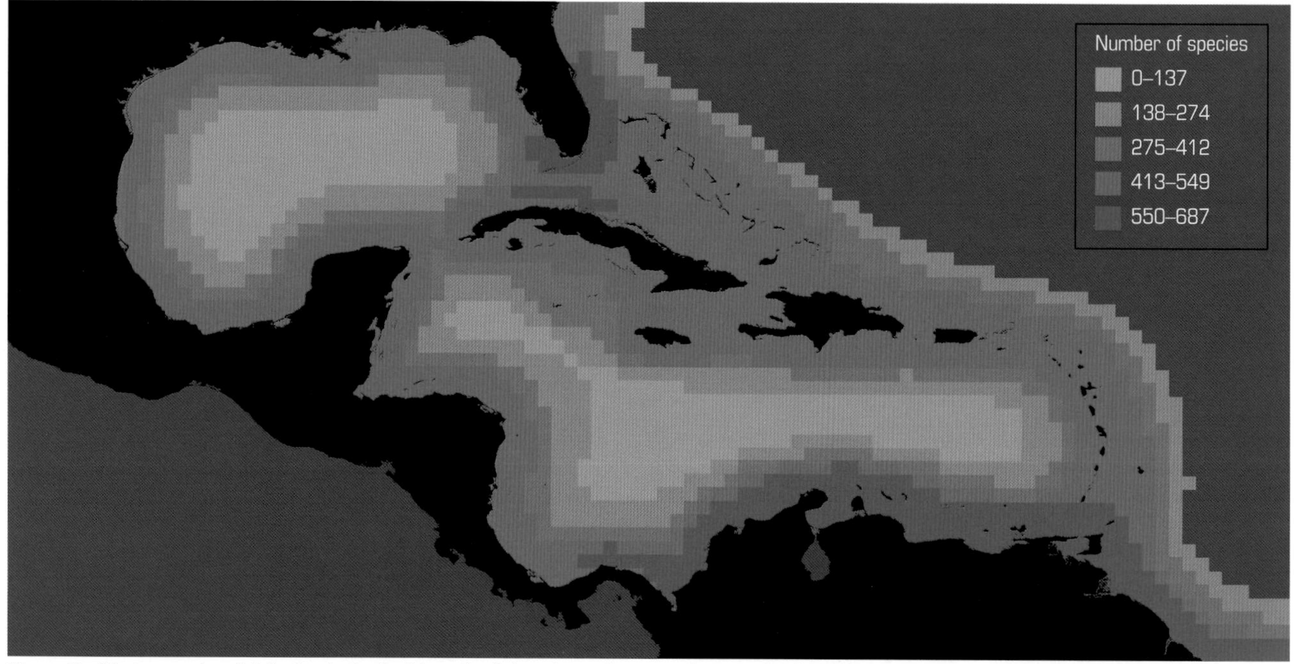


**Figure 3** Marine species distribution in the Caribbean for fishes, invertebrates and tetrapods—including sea turtles, whales, dolphins, seals and sealions (adapted from Smith, Carpenter & Waller 2002).

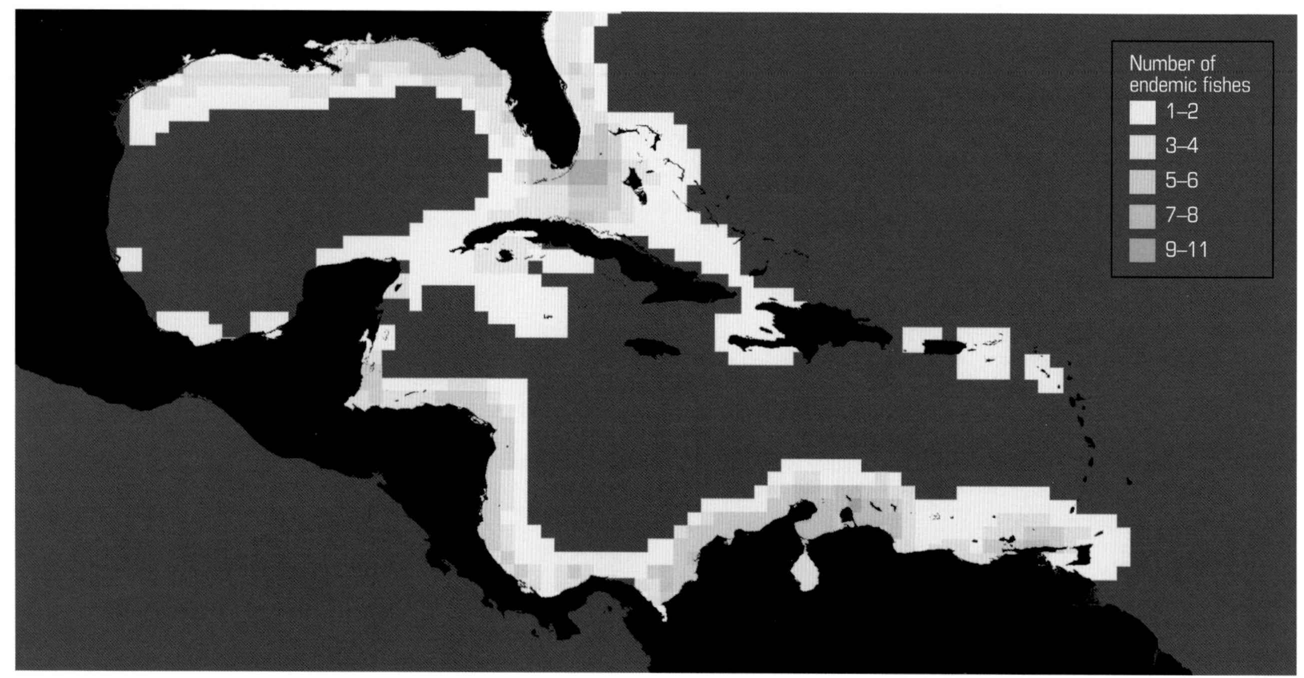


**Figure 4** Distribution of endemic fish species in the Caribbean (adapted from Smith, Carpenter & Waller 2002). Note the surprising areas of high endemism in the northern Gulf of Mexico and in the deep waters of the Florida Straits.

groups were found living along the continental coasts. Total numbers of people in the region at that time remains unknown, but estimates range from 0.25 to 6 million on the Caribbean Islands and 15 to 30 million in the bordering countries of Central and South America (Denevan 1992).

As agricultural techniques improved, population densities increased, environmental impacts may have grown, and there is evidence of overfishing on several islands of the Lesser Antilles (Wing & Wing 2001). There may also have been some clearing of mangroves for cassava cultivation, and large animals (megafauna) such as manatees, monk seals and turtles may have been particular targets for hunters (Wilson 1997).

**Colonial period (1492 to 1850)**

Within 20 years of Columbus's arrival, Spain established colonies on most islands of the Greater Antilles. During this time, enslavement, famine and exposure to new diseases decimated the native peoples. Following a series of military conquests between 1520 and 1550, and the discovery of precious mineral resources (notably gold), Spanish attention was transferred to the mainland. The native population there also suffered devastating losses, such that within a century they numbered less than three million.

Other European countries (Britain, France and the Netherlands) began to take an interest in the region, not only seeking to plunder the Spanish treasure fleets but also to establish colonies on the smaller Caribbean Islands throughout the 17th century. Sugar export, first established in Barbados in 1640, quickly transformed many of the island colonies into major national assets. A plantation economy developed which, like the mining economy on the Central and South American mainland, was heavily dependent on imported slave labor. By the early 1800s, there were an estimated 4.65 million slaves in the Caribbean region (Rogozinski 1999).

From an environmental perspective, this was a period of great change. At first, the decimation of the indigenous populations probably led to a reduction in fishing pressure, a hiatus which may have enabled recovery of fish and megafauna to near natural levels in all areas for a period of perhaps 100 years (Jackson 1997). The demise of these indigenous cultures also meant the loss of traditional knowledge, and we do not know whether any of them ever exercised controls to protect marine resources. The growth of European colonies, and particularly the plantation culture, also brought dramatic changes. As soil fertility declined, more slave labor was required to fertilize the cane fields. Vast areas of forest were cut and steeper marginal areas were logged for timber and fuel for the sugar cane boilers. Massive soil erosion resulted along with climate changes on many of the islands, making them more susceptible to drought.

In many places the colonies were too large to be fed from the surrounding natural resources, so fish and other foodstuffs were imported from Europe and North America. The introduction of new species may have exacerbated the demise of many terrestrial species (Wilson 1997). Industrial whaling briefly touched the region in the early 1700s, quickly exhausting whale populations in the Silver Banks calving area near the island of Hispaniola. The demise and eventual extinction of the Caribbean monk seal also began during this period. Turtles were another major source of protein, with one of the most remarkable and devastating fisheries of the period being the virtual mining of the green turtle nesting colonies. From 1688 to 1730, a dedicated fleet took some 13,000 turtles per year from Grand Cayman Island to help feed the large British colony in Jamaica. In the latter half of the 17th century the fishery collapsed, and by 1800 the turtle population of Grand Cayman had been decimated (Jackson 1997).

**Industrial period (1850 to 1950)**

The abolition of slavery during the 1800s paved the way for a gradual transition away from the plantation way of life—with everyone supported through income from exports—toward a semi-subsistence economy. Former slaves had to make their own living from the land and sea. Sugar remained the principal economic engine for many islands with the most suitable land and climate, while bananas replaced sugar as the primary export crop on others. Marginal land not used for sugar, such as mountain slopes, was quickly cleared and cultivated with subsistence crops or livestock, which further increased soil erosion. Reliance on the sea increased and both subsistence and commercial fishing expanded greatly.

Independence from colonial rule first began on the continents but soon followed on the larger islands in the Caribbean. The United States emerged as the principal economic and military power and began to exert its influence on many newly established governments. Huge American-owned banana plantations were established in Honduras during the early 1900s and, through U.S. investments, Cuba was developed into the principal sugar-producing country in the world.

The taking of sea turtles continued across the Caribbean (Rebel 1974), with British Honduras (now Belize) forming a new centre for green turtle export to England toward the end of the 19th century (Craig 1966). In the Florida Keys, shark leather factories were built and an estimated 100 sharks, many in excess of 45 kilograms, were harvested daily for decades (Viles 1996). Other marine-related resource extraction also began to occur. In the Bahamas and south Florida, sponge gathering flourished after the American Civil War. As technologies improved, first steam, then diesel engines on ships supported the increasing industrialisation of fisheries and gave access to many more remote areas.

### Recent (1950 to present)

Independence from colonial rule continued to occur on the smaller Caribbean islands into the early 1970s, although a number of smaller territories chose to remain linked to the United Kingdom, the Netherlands, France and the U.S. Agricultural production of a broad range of export crops (notably sugar, bananas, tobacco and cotton) has remained of primary importance, but this period has also seen a boom in tourism. Tourism is now the principal industry for many island nations and an important part of the economy region-wide.

Through this period, the population has continued to grow very rapidly in many Caribbean countries and there has been a significant migration of population to the coastal areas in the U.S. and some of the South American countries. Human impacts have been higher during this period than in any other. The following section considers these pressures in more detail.

## Driving Forces and Future Concerns

Despite a long history of human interactions with the natural environment in the Caribbean, there is no doubt that the past 50 years have wrought changes across the region at an altogether different scale. It is critically important that we understand the processes driving these changes and their underlying causes before recommending management actions. In future, such changes will continue or accelerate —and new concerns can be anticipated—unless there are shifts in policy, attitudes and management approaches. In this section, we consider the forces driving change in the marine environment of the Caribbean, both directly and indirectly through socio-economic forces.

### Direct human impacts

Human impacts that directly impinge on specific localities or marine communities are the most easily identified and often present the simplest targets for management efforts.

#### *Overfishing*

The most important commercial fishery species are the Gulf menhaden *Brevoortia patronus* and the American cupped oyster *Crassostrea virginica,* both predominantly taken from the Gulf of Mexico. Additionally, yellowfin tuna is an important part of the catch throughout the Caribbean (Caddy & Garibaldi 2000, FAO data). Small-scale coastal fisheries may be considerably more important for the region in terms of employment and socio-economic function, however, providing food, income and employment to coastal communities, including many areas where there are few alternative occupations.

The United Nations Fisheries and Agriculture Organisation (FAO) estimates that in 2002 only about 25% of global fish stocks were under-exploited or moderately exploited. The remainder were "fully exploited" (giving stable, but probably non-sustainable yields), over-exploited or significantly depleted. These trends are mirrored in the Caribbean. Catches from the western central Atlantic, as reported to FAO, increased steadily to a peak in 1984 of some 2.2 million tonnes followed by an 18% decline to 1.8 million tonnes. Although figures are unavailable for the region, catch per unit effort is likely to have declined even more dramatically over this period.

Myers and Worm (2003) report that fishery declines appear to be particularly severe for large predatory fish, with decreases of 90% from levels reported in all oceans before industrial-scale fishing began. This finding receives further support from a recent study of shark fisheries in the Northwest Atlantic, including the Caribbean and Gulf of Mexico. Among the most depleted species were oceanic hammerheads and white sharks with 89 and 79% declines respectively since 1986. Coastal sharks (genus *Carcharhinus)* have suffered declines of 61% since 1992 (Figure 5). These collapses are linked to both targeted shark fisheries and to shark bycatch—non-targeted species caught unintentionally and usually thrown back dead—in longline fisheries. It should be noted also that the size of these declines is being measured against an already reduced baseline, as sharks were already being fished prior to the starting point of the data (Baum *et al* 2003). Although these studies are largely focused on pelagic species (those living in offshore open water), Caddy and Garibaldi (2000) have shown similar declines in shallow benthic (bottom-dwelling) species of predatory fish like groupers and snappers.

Inshore fisheries are particularly poorly documented and there is no region-wide information. Compared to many other countries, Jamaica has been relatively well studied and presents an important regional case study. There is evidence of overfishing in Jamaica for several decades. The inshore fisheries are complex, with over 200 species being landed by Jamaican fishers. Fish biomass is thought to have been reduced by up to 80% by 1960. Although levels of catch remained stable between 1968 and 1981—in spite of a doubling in the number of fishers (Hughes 1994, Gustavson 2002)—such a reduction in catch per unit effort indicates a continuing decline in the populations of fish. Even for the wealthy countries, a lack of information is a hindrance to active management. Of the 179 fishery stocks falling under the jurisdiction of the U.S. Caribbean Fisheries Management Council, the status of 175 (98%) is reported as unknown or undefined (Cochrane 2001 in Smith, Carpenter & Waller 2002).

One fishery approach of particular concern is the targeting of spawning aggregations, where benthic species gather at particular times of the year for mass spawning events. Fishermen targeting such aggregations have caught a remarkably large number of fish such as large snappers and groupers, but this large-scale interruption of fish reproduction has stark long-term consequences, with localised extinctions of such species as the Nassau grouper *Epinephelus striatus* in many parts of the region. Luckhurst (2002) lists almost 200 spawning aggregations in the Wider Caribbean, of which only ten are described as stable or protected. It seems likely that many, perhaps hundreds, of spawning aggregations have been lost throughout the Caribbean. In some countries where fishing pressure is particularly high—in Jamaica, Dominican Republic and much of the Lesser Antilles—no remaining sites are known at all.

Another highly problematic fishery is bottom trawling, notably for shrimp. In addition to the destructive impacts of trawling described below, there are considerable ecological impacts from bycatch. Bycatch is particularly high in tropical shrimp fisheries—in the Gulf of Mexico eleven kilograms of bycatch are discarded for every one kilogram of shrimp caught, while one study of Venezuelan shrimp trawling recorded a bycatch ratio of 40 to 1 (EJF 2003). Many marine turtles are still caught although "Turtle Excluder Devices" (TEDs) are mandatory for U.S. trawlers and all countries importing shrimp to the U.S. (Figure 6). One study estimated, however, that 41% of trawlers from the U.S. state of Texas were violating this law (EJF 2003). Trawling also impacts other fisheries. In 1995 it was estimated that Gulf of Mexico shrimp trawlers caught and discarded some 34 million red snappers *Lutjanus campechanus*—mostly young or immature individuals. The red snapper fishery itself typically only catches about three million fish per year and reports rapidly declining yields (EJF 2003).

DAVID BRYAN

**Figure 5** Coastal sharks, like this Caribbean reef shark off Eleuthera, are top predators on coral reefs and in other coastal marine environments. Their drastic declines negatively impact the entire ecosystem.

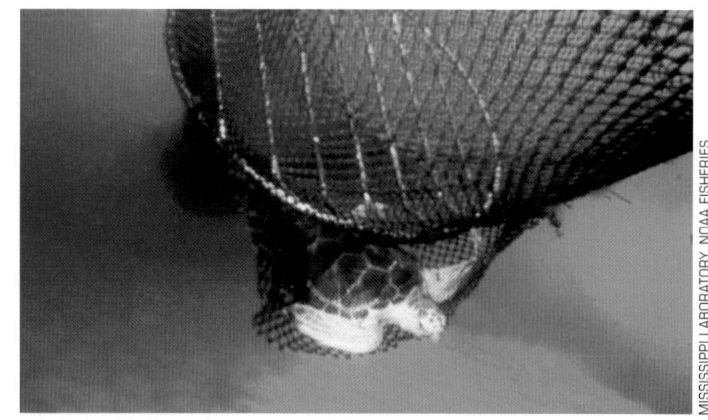

MISSISSIPPI LABORATORY, NOAA FISHERIES

**Figure 6** Turtle excluder devices (TEDs) are a simple grid of bars with an opening at one end that allow sea turtles to slip out of shrimp trawls and avoid dying as bycatch.

*Destructive impacts on the marine environment*

Direct physical interference with coastal and marine environments is mostly related to coastal development. Problems include the drainage of coastal mangrove areas, dredging for marinas, clearance of beach vegetation and the dumping of sand for "beach enhancement." The destruction of mangrove areas for coastal development has been cited as one of the major coastal problems in a number of Latin American countries, with adverse effects on coastal processes, productivity and fisheries. The clearance of mangroves and other coastal wetlands in the development of shrimp aquaculture has not been as widespread in the Caribbean as in other tropical areas but has occurred in Mexico, Venezuela, Colombia and the Dominican Republic (EJF 2004).

Coastal engineering, such as the building of seawalls and jetties, has had adverse consequences for many miles of coastline. As wave and coastal current patterns are altered, unpredictable levels of erosion or deposition (build-up of sand or soils) result. In the Caribbean region, tourism is often seen as a particular cause for concern, often leading to sprawl along coastlines. Turtle nesting beaches have suffered in many areas from direct interference but also from the light and noise of developed areas that disturb their nesting behavior. At a smaller scale, SCUBA diving has been responsible for direct damage to corals and reef communities through the touching and trampling of corals and other organisms, while dive and pleasure boats damage the seafloor with their anchors.

Bottom-trawl fishing also has a heavy impact on marine life. There is considerable damage to the ocean floor, typically on the continental shelf but increasingly moving into deeper waters. Trawls dragged across the seafloor destroy a large proportion of the bottom-dwelling plants and animals. Where this includes long-lived, slow-growing species such as corals and sponges, recovery may take years or decades (Morgan & Chuenpagdee 2003, EJF 2003). McAllister *et al* (1999) estimate that 321,000 of the 360,000 square kilometers of the U.S. continental shelf in the Gulf of Mexico are used as trawling grounds. Remaining areas may include those "protected" by offshore oil drilling and production platforms.

### Indirect human impacts

The highly dynamic, liquid nature of the ocean allows movement of organisms and materials among locations in various nations' waters and onto the high seas. This movement cannot be stopped by legal or administrative means, or even by physical interventions. Some 75% of pollutants in the sea originate on land, often far from the marine communities they impact (Schumacher, Hoagland & Gaines 1996). In addition to these, ocean dumping of land-derived wastes and the release of ballast water from ships add solid waste, nutrients, toxins and invasive species that rapidly spread across boundaries.

*Toxic waste*

Agrichemicals, pesticides, herbicides and fertilizers used on crops such as cotton, bananas, sugar, rice and coffee are probably the major source of toxic pollutants in the Caribbean. Rawlins *et al* (1998) have shown increases of pesticide imports between 1961 and 1995 that vary from doubling (U.S. Virgin Islands) to increases of hundreds of times as much (Belize). Some of the highest rates of pesticide application in the world have been reported for Caribbean cotton crops (up to 80 kilograms per hectare). Shrimp aquaculture can also lead to high levels of both toxic and nutrient pollution (UNEP 1999). Unfortunately there have been few measures of such pollutants in the soils or the surface waters of the Caribbean, and there have been no major studies of their impacts on the marine environment.

The Caribbean has a relatively low industrial base, but there are problems caused by cyanide and mercury release from gold mining in countries such as Colombia. Hydrocarbon pollution is also a problem in parts of Venezuela and around major port areas such as Havana Bay, Veracruz and Cartagena Bay (UNEP 1999).

*Nutrient pollution*

In 1993, only 10% of the waste in the region was treated (Colmenares & Escobar 2002). Direct impacts of this—on both humans and the marine environment—are considerable. A recent study suggested that, at the global level, there are some 250 million cases of gastroenteritis and upper respiratory disease worldwide per year related to bathing in polluted seas, with a global cost to society of US $1.6 billion (GESAMP 2001). Pathogenic disease-causing bacteria such as hepatitis survive well in the ocean—as well as in fish or shellfish that ingest them—and there are an estimated 2.5 million cases per year associated with sewage-contaminated shellfish (GESAMP 2001). Caribbean statistics are unavailable.

Untreated sewage leads to high levels of nutrients in

marine waters, and this often leads to very high levels of primary productivity (such as "algal blooms") in surface waters. This can lead to short-term increases in fisheries productivity, but the additional organic matter falls to the seafloor where decay processes remove oxygen from the water. This can lead to conditions of hypoxia or anoxia (low or no oxygen) in deeper waters, which most marine organisms cannot survive. The resultant permanent or seasonal "dead zones" are being reported worldwide in increasing numbers. One of the largest and best studied occurs during the summer months off the Mississippi River in the Gulf of Mexico and covers an area of more than 20,000 square kilometers. Several other seasonal dead zones are observed in the northern Gulf of Mexico, and a persistent dead zone also occurs off the Caribbean coast of Venezuela (GESAMP 2001, Diaz 2001, UNEP 2004).

*Sedimentation and solid waste*

Sediment inputs from the Mississippi River (320 million tonnes per year) and the Magdalena River (235 million tonnes per year) dominate sediment discharges in the region, with other rivers contributing an additional 350 million tonnes per year. Sediments produce shading effects, reducing sunlight penetration down through the water column. They can also have direct physical effects by smothering benthic organisms. Corals are particularly sensitive to high levels of sediment loading. Where rivers have continued to deliver sediments over long time periods, marine communities are usually well adapted, but recent human impacts have changed the patterns of sediment delivery from many watersheds in the region. Forest cover on many Caribbean islands has halved since the 1960s and this, combined with poor agricultural practices, has increased soil erosion to the detriment of both terrestrial productivity and the health of coral reefs and other marine ecosystems (UNEP 1999) (Figure 7).

Sediments can also have severe localised impacts around marine and coastal engineering projects. Some coral reefs in Florida have been devastated by offshore sand-mining projects and the associated dumping of sand in certain areas for beach replenishment. Solid waste is also a widespread problem for many countries, and particularly for the small island nations in this region. Although dumping at sea is largely prohibited there is considerable illegal dumping, and inadequate facilities on land cause high volumes of waste products to enter coastal waters (UNEP 1999).

*Deep ocean impacts*

Deep ocean waters, particularly far from land, have long been regarded as unaffected by human activities, but increasingly this view is changing. Long-lived pollutants, including persistent organic pollutants (POPs), are now increasing in these areas. The major source of deepsea pollution comes from ocean dumping of wastes. Trawling in bathyl or abyssal zones (200 to 6,000 meters deep) regularly yields large quantities of solid waste (plastics, glass, metal and masonry). One of the most common waste products is "clinker" (burned coal) from the steamship era. Sewage sludge is also widely dumped in the ocean. Between 1949 and 1967 the U.S. disposed of over 32,000 drums of radio-

GREG FOSTER

**Figure 7** The mountainsides of Haiti have suffered widespread deforestation. Associated soil erosion leads to unhealthy levels of sedimentation and nutrients on nearby coral reefs.

active waste in the western Atlantic. Concerns about the build-up of these materials in the deep ocean are heightened by our general lack of knowledge of ecosystem processes in these waters and by the generally low rates of biological and physical turnover which might otherwise help in the metabolizing or removal of these materials. Concerns are now also growing about the possible use of deep oceans as a dumping ground for carbon dioxide in an effort to control climate change (Thiel 2003).

*Species introduction*

The introduction of "exotic" species to marine environments is a widespread problem, though the Caribbean has probably suffered less than other regions from these impacts. It has been estimated that there are some 3,000 marine plant and animal species being carried alive in the ballast water tanks of the world's ocean fleets (Bright 1999). These tanks are filled with "raw," or untreated, seawater to improve ship stability when carrying a light cargo load. The picking up of ballast water in one area and releasing it and its organisms in another port area is the greatest single source of exotic and invasive species in marine environments. There is some speculation that the pathogen that killed off the majority of the long-spined sea urchin *Diadema antillarum* in the Caribbean may have come through the Panama Canal in ships' ballast water, though there remains no firm evidence (Sapp 1999).

*Climate change*

Human-induced changes in global climate are now widely feared as one of the most pressing environmental impacts on the natural environment (Gitay *et al* 2002). Significant, measureable changes have begun with rises in sea surface temperatures of 0.6 to 0.8°C and rises in sealevels of some 18 centimeters in the last hundred years. These changes have already begun to affect ocean chemistry, with a higher concentration of carbon dioxide ($CO_2$) in the water column affecting various chemical processes. They may also begin to influence patterns of wind and current flows, as well as the intensity and location of extreme weather events.

Corals exhibit a stress response known as "coral bleaching" during even relatively minor high temperature perturbations. There have been a number of mass-bleaching events in the region, but the most extensive was probably in 1998, particularly in the Mesoamerican reef complex and in the eastern Caribbean. Unlike other parts of the globe, this event did not lead to mass coral deaths in the Caribbean except in places where there were other negative effects compounding the stress of bleaching (Wilkinson 2002). Global climate models predict the Caribbean region will experience a 2° to 4°C increase in atmospheric temperatures by 2070, with the largest changes occurring in the northern Caribbean and around the continental margins (IPCC 2001). Assuming that corals and other marine organisms are already at their limit of high temperature thresholds, it is expected that bleaching events will become more common. The long-term survival of shallow-water corals will depend on how well they are able to adapt to the changing environmental temperatures.

Mangrove ecosystems may be threatened by rising sealevels. Although these ecosystems are highly opportunistic and would be likely to migrate along with changing sealevels, intensive coastal development in many areas is likely to lead to a squeeze in the available space for mangroves to grow.

The impacts of climate change on other offshore biological communities are difficult to predict, but many pelagic and migratory species use water temperatures as cues for their movements, and changes could have profound effects not only on population dynamics of the fish, but also the livelihoods of fishers using these stocks.

*Hurricanes*

Although human impacts are a powerful shaping force in modern natural ecosystems, it remains important to consider natural forces as well, and particularly those associated with extreme events.

Most of the Caribbean is subject to violent tropical storms and hurricanes between June and October/November of each year (Figure 8). The frequency and intensity of hurricanes is driven largely by climatological conditions in the northern Atlantic, being more frequent during periods when the North Atlantic is warmer and atmospheric high pressure systems are less prevalent in the southern Atlantic (Goldenberg *et al* 2001). There appear to be decadal-scale variations in hurricane frequency and intensity, with a period of higher hurricane frequency from 1940 to 1970, followed by significantly fewer hurricanes from 1971 to 1994. Since 1995, hurricane frequency has risen again, associated with warmer water in the North Atlantic.

The impact of these storms on all shallow marine ecosystems and adjacent coastlines is often considerable, including wind and wave damage to mangroves, coastal erosion, the building and removal of coral rubble islands, and the alteration of patterns of coastal currents and water flow through channels and lagoon systems (Stoddart 1971, Smith *et al* 1994).

**Symptoms of change**

Many of the impacts discussed above are relatively easy to conceive, with clear-cut, directly measurable effects such as loss of habitats or species. Other changes have been observed in marine ecosystems across the Caribbean, however, that cannot be directly attributed to particular stresses. These include the growth and spread of disease, coral bleaching and blooms of toxic algae. Some seem to be attributable to particular causes, such as coral bleaching coincident with high temperature events, but may also arise under different conditions.

*Disease*

The occurrence of diseases in wild populations has been recorded in numerous marine taxa (types of plants and animals) and ecosystems of the Caribbean. The shallow tropical environments of the Caribbean appear to have become the global epicenter for diseases affecting corals and also bear the scars of other diseases. Widespread sponge deaths have been reported since the mid-1800s, including the 1930s "blackwater" event which led to the collapse of the sponge fishery in Florida and the Bahamas. In 1983/84 a pathogen struck the entire population of the long-spined sea urchin *Diadema* and in this short period removed 93% of all individuals from the population. The results were startling, with massive increases in benthic algae beginning within days of the urchins' demise (Carpenter 1997). The populations of this urchin have still not recovered, and there have been profound and far-reaching effects on the region's ecology, compounded by the effects of other stresses.

Since the 1930s, a seagrass wasting disease has been reported from the northern Atlantic, and the marine slime-mold *Labyrinthula* has been implicated as the pathogen. More recent "die-offs" of seagrass in the Florida Bay from 1987 to 1994 have been linked to the same pathogen, probably enhanced by adverse conditions including algal blooms and high turbidity—the cloudiness of water, which diminishes sunlight penetration (Peters 1997, Onuf *et al* 2003).

The first reports of disease in corals arose from the 1970s, with black-band disease reported in the reefs of Bermuda and Belize, and white-band disease reported from Acroporid corals in the U.S. Virgin Islands (Peters 1997). The impact of these and other diseases has been far-reaching. The Acroporid corals responsible for reef-building and coastal protection, once dominant across the Caribbean (Figure 9), are now rare in most areas. The United Nations Environment Programme's World Conservation Monitoring Centre (UNEP-WCMC) has developed a global dataset of coral disease events and, although it has been difficult to link coral disease to specific human activities, Green and Bruckner (2000) found that 97% of observed cases in the Caribbean were in areas of medium to high human impacts. More recently, African dust blown across the Atlantic has been recognized as a possible mechanism for transporting fungi and other microorganisms into the Caribbean basin (Griffin *et al* 2001). Other diseases less widely reported include fibropapillomas in sea turtles, first observed in Key West in 1938 and in the Caymans in 1980 and now widespread, and a range of fungal and viral diseases widely reported in fish populations in the Gulf of Mexico (HEED 1998).

JACQUES DESCLOITRES, MODIS RAPID RESPONSE TEAM, NASA/GSFC

**Figure 8** Hurricane Isabel headed for the Bahamas and the eastern U.S. in September 2003.

DAVID BRYAN

**Figure 9** Two signature species of the Caribbean marine environment, elkhorn *Acropora* coral and the long-spined *Diadema* sea urchin, are seen here in the British Virgin Islands. Both have suffered widespread declines attributable to disease and other possibly human-related factors.

*Harmful algal blooms*

The Health and Economic Dimensions of Global Climate Change Program (HEED 1998) considers "major marine ecological disturbances" to include both diseases and harmful algal blooms (HABs). Such blooms are typically made up of planktonic (drifting) algae, notably dinoflagellates, but can also be linked to benthic species like the dinoflagellate *Gambierdiscus*, which causes ciguatera (a poisoning from fish consumption) in humans. These organisms produce potent biotoxins that accumulate in certain species (particularly filter-feeding molluscs and high-level predators) until they reach highly toxic levels. They have also been implicated in the deaths of marine mammals and humans. Even non-toxic algal blooms can cause ecosystem damage by blocking sunlight penetration to benthic communities. There is now evidence that HABs, reported from numerous localities along the northern Gulf of Mexico, may be linked to high levels of nitrogen and phosphorus in the water and to sea surface warming (HEED 1998).

*Coral bleaching*

Under certain stress conditions, corals undergo a response known as bleaching. The process involves the loss (or expulsion) of zooxanthellae, an algae corals depend on for nutrients. Without the algae, corals lose their colour and become transparent, which makes their white skeletons visible (hence the term "bleaching"). Shallow tropical corals normally receive a considerable proportion of their nutrition from these algae, and their loss considerably weakens the corals. Various conditions lead to bleaching responses, including unusually warm or cold conditions. A sharp increase since the 1980s in the number of coral bleaching observations worldwide has been linked almost exclusively to warm water events.

**Synergistic effects**

Many phenomena—HABs, coral diseases and seagrass die-off events—may well occur at background levels in natural, undisturbed conditions. There is growing evidence, however, that their increased occurrence and geographic distribution is often linked to human impacts on the natural environment (HEED 1998, Green & Bruckner 2000). Such a compounding effect was recently observed in Belize, where corals weakened by bleaching appeared to be more susceptible to coral disease (Gibson & Carter 2003).

In some cases more complex sequences have been inferred, with Jamaica offering a classic case. Overfishing had denuded herbivorous (plant-eating) fish stocks there for many years. After coral damage from a hurricane in 1980, a slow recovery began. But in 1983, the only remaining major herbivore, *Diadema,* was wiped out by disease. Since then, algae have become the dominant benthic organisms. The algae obviously benefited from the lack of herbivores and may have been further encouraged by raised nutrient levels coming off the land. A subsequent hurricane in 1988 further entrenched this "phase-shift" in biological community structure from coral reefs to algal ecosystems (Hughes 1994, Hughes & Connell 1999).

**Ultimate drivers**

In order to establish effective management responses, it is necessary to understand the proximate causes of problems, their underlying mechanisms and the ultimate drivers of negative changes. Legal and administrative mechanisms may succeed in controlling pressures over short to medium timeframes, but if ultimate drivers are not understood and addressed, such pressures will continue to mount.

Population growth and increasing wealth (combining both per capita consumption and level of industrialisation) are two of the greatest drivers of change in most environments, and while both may occur with minimal environmental impact, this is not usually the case. Economically the Caribbean region includes some of the world's wealthiest communities along the U.S. coast and some of the poorest areas in the Americas, such as Haiti and some coastal communities in Central America.

*Population growth*

Human population in the Caribbean has more than doubled in the last 40 years and now exceeds 40 million for the Caribbean islands alone. Many of the economically poorer countries have experienced significant emigration rates while wealthier countries have experienced net immigration, but populations are increasing overall in the region. Birth rates tend to be highest in poor countries such as Haiti (3.2% per year), although these countries also have the highest death rates—1.51% per year in Haiti (Sealey 1994). The poorer countries with fast population growth are most likely to be places where poor land-use practices and natural resource over-exploitation are major influences on the natural environment.

*Increasing wealth*

Industrialisation and the growth of personal wealth have caused dramatic environmental changes across the Caribbean. The production of both solid waste and greenhouse gas emissions are almost entirely driven by the wealthier nations. The development of industrial-scale farming—such as citrus and other crops in Florida and the U.S.-owned banana plantations in Central America—has released large quantities of agrichemicals into the marine environment. The mechanisation and industrialisation of fisheries has led to rapid declines in fish stocks: small-scale artisanal fishing has been widely replaced by highly efficient vessels and gear; and at-sea storage systems ("factory" ships) have allowed industrial fishing in remote locations. The rapid growth of international markets for fishery products has pushed up values, and over-exploitation has been reported for most high-value species (lobster, conch, snapper, grouper, tuna and swordfish) in nearly all areas of the Caribbean (FAO 2002).

*Tourism*

Tourism is one of the largest and fastest growing industries in the region. In the Caribbean islands, it provided about 2.4 million jobs and roughly 25% of Gross Domestic Product (GDP) in 1996 and these numbers have undoubtedly increased (UNEP 1999). The Caribbean attracts 57% of the world's international SCUBA divers, and dive tourism (Figure 10) alone was predicted to generate US $1.2 billion by 2005 (UNEP 1999). During the 1990s, the wider Caribbean region continued to experience significant increases in tourism with an estimated 16 million stopover and 10 million cruise tourists visiting the region in 1998. The most tourists are from the United States (48% in 1998), followed by Europe and Canada. The cruiseship industry has continued to expand and the region currently has 50% of the global cruiseship passengers, generating about US $3 billion per year (UNEP 1999). Cruise-related tourism accounted for slightly less than 30% of all tourists to the region in 1998.

The development of infrastructure to support increased tourism, and the supply of materials to support this industry, are among the major drivers behind coastal development, natural habitat clearance and the fishing of key species such as lobster, snapper and conch. The tourism industry has encouraged the movement of people from rural

© WOLCOTT HENRY 2001

**Figure 10** Recreational diving, the levels of which are predicted to increase over the next decades, is a major source of revenue in the Caribbean.

to urban areas and out of traditional industry into service industries (WTTC *et al* 2002). Certain sectors of this industry, notably the ecotourism sector and some SCUBA diving and bonefishing operations, have been recognised as providing very high economic benefits to small rural coastal areas with low environmental impact and are being promoted in several countries.

*Political systems and education*

In many cases, there are simple, cost-effective solutions to environmental issues that, if properly applied, will lead to net societal and economic benefits. Unfortunately many of these solutions are poorly understood among policymakers, managers and the wider populace. Better information and education at all levels may help greatly in addressing many environmental problems and in establishing an electorate or public that both understands the problems and presses for change.

Some of the solutions may require changes not only in understanding, but also in political systems or long-standing traditions. Over-exploitation of marine resources, for example, is often linked to uncontrolled open access. Cases where access has been restricted and ownership given to adjacent local communities have been highly effective in reducing these problems. Other problems require coordination of activities among government departments and communities—and even among countries—to establish more integrated management.

Education and political change must work together. In quite a number of cases across the Caribbean region, efforts to mitigate change by the establishment of management regimes have been thwarted by non-compliance or active resistance from communities who were unable to perceive the benefits to them of proposed changes.

## Management Processes

Given the vast array of threats—direct and indirect, local to global—facing the world's ocean, there can be no simple management solution. Some problems, such as damage to corals from boat anchors, can be solved at the site level by simple intervention like the placement of mooring buoys. Other issues, like local overfishing, may require a more comprehensive involvement of politics and communities (establishing local cooperation and setting up a "no-take" zone). But many issues require coordination of effort across broad spatial scales and across societal sectors. The demise of turtles and tuna, for example, can only be addressed through international collaboration. The dying of mangroves may be linked to upstream freshwater diversion, requiring collaboration of several government departments and multiple stakeholders. The threat of global climate change is strongly linked to energy consumption in wealthy countries worldwide, and the spread of coral disease in the Caribbean has even been linked to deforestation in Africa, causing more pathogen-carrying dust to be blown across the Atlantic.

We review some of the existing management processes that have been used across the Caribbean region. Many offer viable and highly acceptable solutions to current problems, but few have been fully or effectively put into practice in more than a few sites.

### Fisheries controls

In almost every Caribbean country, declines in fish stocks have led to efforts to control and manage fisheries on behalf of, and usually in discussion with, fishers and in consultation with fisheries scientists. The success of these controls has typically been hampered by the fishers themselves; despite their concerns, they often continue to maintain unsustainably high fishing levels.

One of the underlying problems facing many fisheries has been that of open access, allowing free competition to take catches beyond economic or sustainable yields. These problems have been exacerbated by government subsidies, including tax benefits, fuel subsidies, income support and payment of foreign license fees. Global fishery subsidies of these types have begun to decline in recent years but still remain high. A broad array of fishery management tools has been used in various parts of the world with varying levels of success.

*Output controls*

These are limits placed on the size of the catch. The total allowable catch (TAC) system, widely used in Europe, sets a maximum overall catch allowed for a fish stock but does not restrict the activities of individual vessels (an unallocated quota), leading to a "race to fish" for the stock when the quota is announced, as well as problems of non-compliance. Once the TAC is reached, any additional fish of that type caught unintentionally in other fishing operations must be discarded to avoid punitive measures. TAC systems are the main forms of control for the Atlantic tuna fishery, but it is widely agreed these have been ineffective in stalling the decline of most stocks (Singh-Renton, Mahon & McConney 2003). There is growing interest in the establishment of allocated quotas. A system of Individual Transferable Quotas (ITQs)—where each member of a fishing community is allotted a quota that he/she may catch, but quotas may be bought or sold among fishers—may offer a more sustainable solution in industrial scale fisheries. Such systems can also lead to self-policing within the industry, as fishers do not want to see others taking more than their quota. Bycatch, and the problem of "high-grading" (discarding smaller-sized or lower-valued fish so that the quota is reached only with top quality individuals) still remain with this approach however.

For multi-species fisheries, such as many of the near-shore, demersal (near-bottom) fisheries of the Caribbean, single stock quotas are not appropriate. There are also considerable difficulties with stock assessments—and management approaches based on them—in such fisheries because of a lack of understanding of life-history characteristics, interactions between species and interactions between communities.

*Input controls*

These are management approaches that seek to reduce fishing effort, for example by restricting the total numbers of fishers allowed through licensing systems or informal agreements or by controlling the numbers of days at sea.

The provision of property rights to fish stocks or fishing areas is increasingly used as a tool to manage fisheries. This approach immediately places a general restriction on the access, creating "owners." These owners have a strong incentive to manage the resource in a sustainable manner and tend to rapidly establish their own system for restricting access. Similar systems have developed in many traditional cultures. Sometimes called Territorial Use Rights in Fisheries (TURFs), these have typically involved community-level ownership of marine areas and their stocks (Lutchman & Hoggarth 1999). Such ownership may be traditional, informal or covered by modern legal systems. The trap fishery of Jamaica, which still makes up about 60% of all nearshore catches in that country, was operated on an informal TURF system for many years (Figure 11). There was close community regulation of who could fish and a number of informal sanctions to deter newcomers from fishing in community "owned" areas. This system unfortunately broke

**Figure 11** Trap fisheries, using traps like this one off Virgin Gorda, British Virgin Islands, are common throughout the Caribbean.

down as a result of government intervention and subsidies in the 1950s (Brown & Pomeroy 1999). Another TURF-type system is widely operated in the Lesser Antilles amongst beach-seine fishers who have developed a complex series of rules outside of the law that divide up the access to this multi-species fishery in both time and area. Recently there have been concerns that modern developments—including both tourism and increased use of motorised fishing vessels—may be undermining this system.

*Technical measures*

Quite a number of countries impose restrictions on the methods of fishing or types of fishing gears that may be used. In nearshore fishing areas these include bans on the use of spear guns by SCUBA divers or the use of SCUBA gear in the harvesting of shellfish. A number of Caribbean island nations now require that all fish traps include biodegradable panels so they do not continue to catch fish—called ghost fishing—if the traps are lost (Spalding 2004). Other controls include restrictions on net sizes or on mesh sizes of the nets. The U.S. forbids the use of shrimp trawls that are not equipped with Turtle Excluder Devices and indeed will not accept imports of any shrimps unless the trawlers catching them use these devices. Even in international waters there are moves to restrict gears, notably the United Nations General Assembly's moratorium on the use of driftnets on the high seas.

Another suite of technical measures looks at ecological parameters. For example, the capture of conch and lobster in the Caribbean is generally restricted by minimum sizes, and the capture of juvenile or gravid (egg-bearing) individuals is prohibited. The use of closed seasons is also growing in many fisheries, typically to prevent capture of breeding individuals.

One widely used technical measure for the protection of coastal fisheries in the Caribbean is the spatial closure of fishing grounds in protected areas or "no-take" zones. These have proved highly effective in supporting fisheries at local scales and have had considerable side benefits in terms of tourism growth and provision of alternative incomes. They are mostly concentrated in nearshore ecosystems and particularly on coral reefs. In wider shelf areas, and in pelagic fisheries, the target species may be highly mobile and the value of protected areas is more complex. There is a danger that poorly planned protection of areas in these cases may lead to re-distribution of fishing effort, placing greater strain on other areas or even damaging formerly pristine areas.

*Economic measures*

Fishery subsidies have been one of the most destructive measures in terms of supporting overfishing and the destruction of fish stocks. The removal of grants for new vessels or new gear, tax breaks and cheap fuel are some of the first steps required to reduce overfishing. Many countries have gone further than this, establishing alternative incomes within fishing communities to draw people away from fishing. In a few cases, governments have sought to buy back fishing vessels, pay fishers subsidies not to fish or provide grants for them to set up alternative businesses.

It can be seen from the above that a broad range of tools is available for fisheries management, but policing of many of these is difficult and compliance is often poor. Among the most successful approaches are those that establish a degree of ownership of resources by the fishing communities themselves. This has been variously achieved by providing ownership of stocks (through ITQs), ownership of marine areas (TURFs) or at least an active role in the management of the fisheries. Such co-management measures range from "instructive"—where the government instructs the users—to "informative" where decision-making authority is handed over to the users, but they must inform the government of status (Sen & Nielson 1996). At present the levels of community involvement are low in much of the Caribbean. Part of this relates to the relatively recent history of fishing in the Caribbean and a generally lower level of social cohesion among communities than may be found in many older cultures. Exceptions to this include long-established fishing cooperatives in areas such as Belize, St. Lucia and Jamaica (Brown & Pomeroy 1999).

Fishing controls in coastal demersal fisheries must ultimately be driven by the specific needs and vagaries of local cultures and politics. However, offshore fisheries, particularly pelagic fisheries, will require considerable cooperation among countries (Lutchman & Hoggarth 1999). The interconnectedness of fish stocks is a particularly Caribbean issue, given the very large number of small countries with adjacent Exclusive Economic Zones (EEZs), and efforts to work in collaboration are taken through a number of regional organisations.

**Protected areas**

The setting aside of areas of the ocean and coastline as protected areas has been widely undertaken for two related purposes: wildlife conservation and fisheries management. In certain regions, especially the Caribbean, tourism development has been a major beneficiary and in some cases may represent a third driving force behind protected area site designation.

*Fisheries protection*

There is a rapidly growing body of examples showing that the strict closure of selected areas to all fishing (often termed marine reserves or no-take zones) can produce dramatic improvements in fish stocks (Mosquera *et al* 2000, Roberts & Hawkins 2000, Roberts *et al* 2001, Ward, Heinemann & Evans 2001, Gell & Roberts 2002). Most sites show significant increases in abundance, size, biomass or fecundity (fertility or reproductive rate) of protected stocks, as well as "spillover" of stocks to surrounding areas. High levels of larval "export" from protected areas have been more difficult to quantify, though they are likely to be occurring.

Positive changes are typically noted in reserve areas within one or two years. Bohnsack (2000) noted a 15-fold increase in yellowtail snapper *Ocyrus chrysurus* in the no-take zones of the Florida Keys between 1997 and 2001. It has also been widely observed that benefits from reserves continue to accrue over relatively long timescales, particularly where there are indirect effects of no-take, such as habitat recovery following an end to trawling (Gell & Roberts 2002).

The primary benefits of marine reserves are for non-migratory demersal species whose home range is covered by the reserve area. Another critical group that benefits are those species protected during a critical phase of their life history, which may require protection of spawning grounds or nursery areas (Gell & Roberts 2002). The protection of spawning aggregations is critical, not only for species conservation but for the economic protection of fishers. Very large area-based restrictions would possibly also benefit highly mobile pelagic species, but the closure of sites without further controls to reduce fishing effort can simply redistribute fishing effort and threaten stocks in other areas (Polunin & Wabnitz 2001). Models developed by Baum *et al* (2003) show that the closure of large areas to shark capture, for example, needs considerable planning.

The size of existing reserves is highly variable. Many small ones show highly significant effects at local scales but are not yet making an impact on national fisheries statistics. In their review of existing data Roberts and Hawkins (2000) suggest that protecting from 20 to 40% of the total marine area "is likely to substantially increase long-term yields of over-exploited fisheries."

Despite numerous positive examples, there remains distrust by many fishers based on fears of loss of income, belief that conservation rather than fisheries management is the driving motive or concerns that no-take zones will be unfairly enforced. There is also reluctance at a political level to expand the existing protected areas into comprehensive networks. Critical to the future development and acceptance of such networks will be processes of raising public awareness as well as the involvement of the fishers in the management processes.

*Conservation*

The wider case for marine protected areas (MPAs) being of value for species conservation is clearly supported by the same studies already discussed (showing increases in abundance, biomass and/or fecundity within MPAs) but also other studies showing changes in benthic cover and reductions in damage from anchors and trawling.

The majority of MPAs are not strict no-take zones. From a fisheries perspective they may restrict access, limit fishing gears or protect selected species. Other regulations are aimed at non-fishing uses, such as restricting anchoring, the use of certain vessels, the release of pollutants or particular recreational activities.

It is impossible to fully isolate MPAs from adjacent terrestrial, atmospheric and oceanic systems (Jameson, Tupper & Ridley 2002). Threats may be carried in from adjacent land and water areas in the form of pollutants, sediments, diseases and alien species. The influence of external factors can be reduced through both careful planning of the site boundaries and the integration of the site into adjacent planning and management efforts. A number of MPAs in the Caribbean, such as Portland Bight in Jamaica and the Virgin Islands National Park, include substantial parts of or even entire on-land watersheds, enabling greater opportunities to control the inputs of sediments and pollutants into the nearshore waters.

It is the conservation benefits of protected areas (including fisheries designated sites) that have become important drivers for tourism. Growing perceptions of the general decline in marine biodiversity have led many tourists, especially those seeking to enjoy the natural environment, to place an increased value on protected areas. Divers in particular seek out marine protected areas and many have shown a considerable willingness to pay extra to dive in such locations. Their demands have led to growing calls from the tourist industry to implement more protected areas and have even led to the establishment of private initiatives and unofficial protected areas.

Probably the most reliable figures on the number and extent of protected areas across the Caribbean are held by the UNEP World Conservation Monitoring Centre. These data, derived from a consortium of conservation organizations, show a total of 370 marine and coastal protected areas from the Caribbean island nations, with an estimated additional 125 sites on the continental mainland coasts (Figure 12). The total area covered by these sites is about 23,000 square kilometers of the Gulf of Mexico (1.5% of the total area) and 61,000 square kilometers of the Caribbean Sea (1.9% of the total). Because of their proximity to the coast, and because of a particular focus on ecosystems such as coral reefs, an estimated 3,800 square kilometers (19%) of the total area of coral reefs fall within protected areas (Spalding, Green & Paterson forthcoming).

In reality these figures give little information about the level of protection provided, and the total area of no-take zones remains unknown (Green & Conway 2001). Some clearly fail due to inadequacies in the design of the site so that, even if fully enforced, they would fail to protect marine resources. But it is also widely known that many MPAs

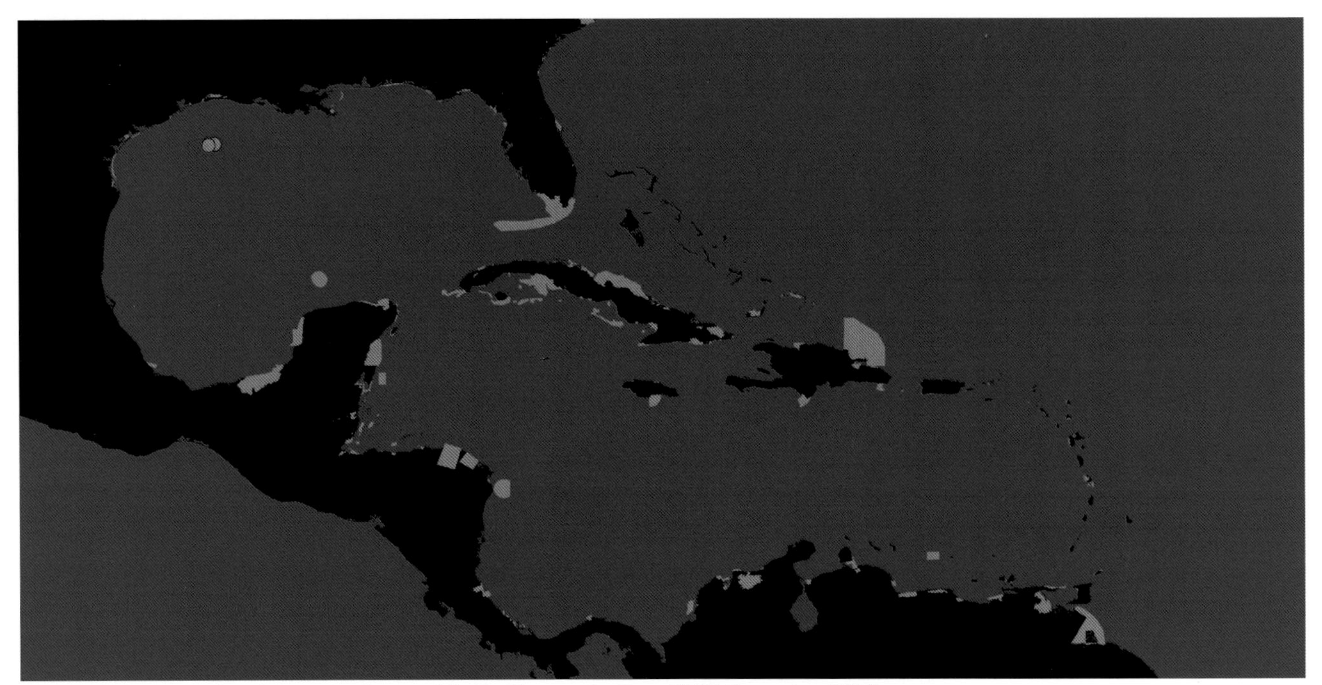

**Figure 12** Coastal and marine protected areas of the Caribbean region (from WDPA Consortium 2004).

are not accepted locally or enforced. Kelleher, Bleakley and Wells (1995) suggest that two-thirds of Caribbean MPAs are not achieving optimum management effectiveness. They attribute at least part of this problem to the fact that an estimated 80% of protected areas staff do not consider themselves adequately trained for the work.

### Other sectoral approaches

Fisheries management and the establishment of protected areas represent two direct and tangible approaches to protecting and sustaining natural resources. There are numerous other controls and approaches which operate in particular sectors, regions or societal groups. Legal and administrative regimes exist in most countries, for example, to regulate the treatment and release of solid waste and toxic products. An increasing number of Caribbean countries are setting up sewage treatment plants, and full tertiary (three-stage) treatment of sewage is often a requirement of new developments.

Legal protection for particular habitats or species can sometimes extend beyond the boundaries of protected areas. In the U.S., the Endangered Species Act provides such protection for a number of individual species. A "no-net-loss" approach is now also applied to certain wetland habitats, like seagrass beds and mangrove forests. Developments that require destruction of these habitats will only be permitted if an equal area of such habitats can be restored somewhere else. Habitat clearance may be further restricted, with regulations on the clearing of forests from the steepest slopes or from areas within a certain distance of watercourses, for example. Such measures, although far from the coast, greatly reduce levels of erosion and hence sedimentation in coastal waters.

Coastal development, particularly for the tourist industry, may be further regulated with a broad suite of controls covering building close to the coast, protecting strand-line vegetation on beaches, treating sewage, extracting freshwater from rivers or groundwater and building of coastal structures such as erosion controls, piers and seawalls.

Private initiatives are further supporting marine protection and restoration. The diving industry in many places across the Caribbean has led the way with the establishment of mooring buoys to prevent anchor damage at dive sites. Many are working alongside local communities or international agencies to promote positive conservation measures and to campaign for the establishment of protected areas.

Education is critical in winning support for environmental protection, and raising public awareness of environmental issues plays an important role in efforts to protect the environment. These include incorporation of environmental issues into national school curricula and outreach to communities.

### Integrated Coastal Management

Most conservation and sustainable-use initiatives in the marine environment are undertaken in isolation, disconnected from other initiatives and with little cross-sectoral cooperation or integration (Blake 1998, Lomelí *et al* 1999). With such a broad array of threats and an equally broad array of policy and management interventions available, unconsolidated approaches can lead to gross inefficiencies, with duplication of effort and perverse or conflicting activities undertaken by different sectors. For example, planning and development sectors may seek to promote tourism in pristine coastal areas that are actually protected, or costly efforts to clean up coastal waters may be rendered ineffective if new agricultural subsidies lead to a massive increase in agrichemicals in the watershed.

Integrated Coastal Management (ICM) is the term applied to formal efforts to develop coordinated management efforts. The ICM concepts can be applied more broadly, however, to mean a process of planning and management that takes a coastal perspective and includes the involvement of all stakeholders in decision-making, the integration of all relevant sectors (governmental, societal and industrial) and the wide use of consultation processes (Colmenares & Escobar 2002). ICM aims to harmonize cultural, economic and environmental values and to balance economic development with environmental protection. By ensuring broad stakeholder involvement in the planning stages, management can be achieved with relatively low levels of regulation (Yáñez-Arancibia 2000). The ICM approach is intimately bound up with the concept of sustainable development in its broadest sense as "development that meets the need of the current generation without compromising the needs of future generations" (WCED, cited in Yáñez-Arancibia 2000).

A few models of more integrated planning now exist in the Caribbean. These include the Environmental Action Plan of Nicaragua, the Coastal Zone Management Unit in Barbados, the Colombian National Environmental Council, and systems in Belize, the U.S. Virgin Islands and Mexico (Colmenares & Escobar 2002). In Mexico, Lomelí *et al* (1999) stress the need for integration of institutional, planning and legislative systems. They further encourage the nation's involvement in international agreements to support regional integration of coastal management. In Nicaragua, the need for integrated coastal zone management has been recognised, but there are deficiencies in the legal and institutional framework; cross-sectoral communication appears to be increasing, but, like many other countries, progress is extremely slow (Jameson, Tupper & Ridley 2000). In other places, the growth of ICM may be organic, beginning with small initial developments. In both Jamaica and Honduras, for example, integrated management is developing at sub-national levels (in Portland Bight and the Bay Islands) with government support. The success of these models may lead to the wider adoption of ICM over time (Vierros 2000, Harbourne, Afzal & Andrews 2001).

### International approaches

The rapid spread of diseases, exemplified by the *Diadema* die-off across the entire region within a year, illustrates the interconnectedness of Caribbean waters. The movement of species and the actions of oceanographic processes are blind to political boundaries, and many environmental issues—ranging from efforts to control climate change to the management of fish stocks on the high seas—need to be addressed at international scales.

Some of these needs are recognised in the international agreements that already exist (Montero 2002) and are further being addressed by a growing number of international groupings within the Caribbean working to develop multinational policy and management approaches. At the same time, the complexity of these agreements and associations can make assessment of their success difficult.

#### *Agreements and conventions*

There are many environmental agreements and protocols relevant to the Caribbean region (Table 2). Involvement in these agreements is highly variable, ranging from Panama, which is signatory to 20, to Grenada, which has signed only eight (Table 3).

Probably the most important regional environmental treaty is the Cartagena Convention (Convention for the Protection and Development of the Marine Environment of the Wider Caribbean Region). One of the more "advanced" regional seas conventions, it builds on the broad aims of the United Nations Convention on the Law of the Sea (UNCLOS) with more regionally relevant controls and agreements. It incorporates protocols dealing with oil spills (in

Table 2

## Global and regional environmental conventions

**Convention for the Protection and Development of the Marine Environment of the Wider Caribbean Region (Cartagena Convention)**
To encourage the contracting Parties in concluding bilateral or multilateral agreements, including regional or subregional agreements, for the protection of the marine environment of the Convention area.
- Protocol concerning Cooperation in Combating Oil Spills in the Wider Caribbean Region
- Protocol concerning Specially Protected Areas and Wildlife (SPAW Protocol)
- Protocol concerning Pollution from Land-Based Sources and Activities

**Convention on Biological Diversity (CBD)**
To conserve biological diversity, promote the sustainable use of its components and encourage equitable sharing of the benefits arising from the utilisation of genetic resources. Such equitable sharing includes appropriate access to genetic resources, as well as appropriate transfer of technology, taking into account existing rights over such resources and such technology.

**Convention on International Trade in Endangered Species of Wild Fauna and Flora (CITES)**
To protect certain endangered species from over-exploitation by means of a rigorous system of import/export permits.

**Convention on the Conservation of Migratory Species of Wild Animals (CMS)**
To protect those species of wild animals that migrate across or outside national boundaries.

**Inter-American Convention for the Protection and Conservation of Sea Turtles (Inter-American Convention)**
To set standards for the conservation of these endangered sea turtles and their habitats.

**International Convention for the Prevention of Pollution from Ships of 1973, as modified by the Protocol of 1978 relating thereto (MARPOL 73/78)**
To preserve the marine environment by achieving the complete elimination of international pollution by oil and other harmful substances and the minimisation of accidental discharge of such substances. MARPOL 73/78 Annexes I through V:
- Annex I. Prevention of pollution by oil
- Annex II. Control of pollution by noxious liquid substances
- Annex III. Prevention of pollution by harmful substances in packaged form
- Annex IV. Prevention of pollution by sewage from ships (entry into force September 2003)
- Annex V. Prevention of pollution by garbage from ships

**United Nations Convention on the Law of the Sea (UNCLOS)**
To set up a comprehensive new legal regime for the sea and ocean and to establish material rules concerning environmental standards and enforcement provisions dealing with pollution of the marine environment.

**The United Nations Agreement for the Implementation of the Provisions of the United Nations Convention on the Law of the Sea of 10 December 1982 relating to the Conservation and Management of Straddling Fish Stocks and Highly Migratory Fish Stocks**
To set out principles for the conservation and management of those fish stocks, and establish that such management must be based on the precautionary approach and the best available scientific information. The Agreement elaborates on the principle that States should cooperate to ensure conservation and promote the objective of the optimum utilisation of fisheries resources both within and beyond Exclusive Economic Zone.

**Convention on Nature Protection and Wild Life Preservation in the Western Hemisphere (Western Hemisphere Convention)**
To preserve all species and genera of native American fauna and flora from extinction and to preserve areas of extraordinary beauty, striking geological formations or aesthetic, historic or scientific value.

**Convention Concerning the Protection of the World Cultural and Natural Heritage (World Heritage Convention)**
To establish an effective system of collective protection of the cultural and natural heritage of outstanding universal value,

Table 2 (continued)

**Global and regional environmental conventions**

organised on a permanent basis and in accordance with modern scientific methods.

**Convention on Wetlands of International Importance Especially as Waterfowl Habitat (Ramsar Convention)**
To stem the progressive encroachment on and loss of wetlands now and in the future, recognising the fundamental ecological functions of wetlands and their economic, cultural, scientific and recreational value.

**United Nations Framework Convention on Climate Change**
To regulate levels of greenhouse gas concentration in the atmosphere, to avoid the occurrence of climate change on a level that would impede sustainable economic development or compromise initiatives in food production.

- Kyoto Protocol. To limit greenhouse gases from developed nations to the levels of 1990 (Outcome of third session of the Conference of the Parties in Kyoto, Japan, in December 1997)

**International Convention for the Conservation of Atlantic Tunas**
To maintain populations of tuna and tuna-like fish in the Atlantic Ocean at levels permitting the maximum sustainable catch for food and other purposes.

**International Convention for the Regulation of Whaling**
To protect all species of whales from overfishing and safeguard for future generations the great natural resources represented by whale stocks. To establish a system of international regulation for the whale fisheries to ensure proper conservation and development of whale stocks.

**Convention on the Prevention of Marine Pollution by Dumping of Wastes and Other Matter of 1972 (London Convention)**
To control pollution of the sea by dumping and to encourage regional agreements supplementary to the Convention.

**Stockholm Convention on Persistent Organic Pollutants (POPs)**
Adopted in 2001 in response to the urgent need for global action to protect human health and the environment from "POPs." These chemicals are highly toxic, persistent, move long distances in the environment and "bioaccumulate" in plants and animals, including those used as food by human populations. The Convention seeks the elimination or restriction in production and use of all intentionally produced POPs (industrial chemicals and pesticides). It also seeks the continuing minimising and ultimate elimination of POPs byproducts and wastes, such as dioxins and furans. Stockpiles must be managed and disposed of in a safe, efficient and environmentally sound manner. The Convention imposes certain trade restrictions. As of 15 July 2002, the Stockholm Convention had 151 signatories and twelve Parties. It will enter into force after the 50th Party ratifies.

**Basel Convention on the Control of Transboundary Movements of Hazardous Wastes and their Disposal**
To set up obligations for State Parties to: (a) reduce transboundary movements of wastes subject to the Convention to a minimum consistent with the environmentally sound and efficient management of such wastes, (b) minimise the amount and toxicity of hazardous wastes generated and ensure their environmentally sound management (including disposal and recovery operations) as close as possible to the source of generation and (c) assist developing countries in environmentally sound management of the hazardous and other wastes they generate.

force since 1986) and providing for specially protected areas and wildlife (the SPAW protocol, in force since 2000). A third protocol, addressing the prevention of marine pollution from land-based sources was adopted in 1999 but has yet to enter into force (Illueca 2001).

While the efficacy of international instruments has sometimes been called into question, there are a number of successes that provide a clear illustration of what can be achieved. These include the global ban on whaling under the International Convention for the Regulation of Whaling, as well as the reduction of industrial waste disposal at sea from seventeen million tonnes in 1979 to six million by 1987, at least partly linked to the London Convention (Thiel 2003).

Table 3

**Regional involvement in international agreements. Caribbean countries which have accepted, signed or ratified each convention are indicated.**

| | **Cartagena Convention** | SPAW Protocol | Oil Spills Protocol | LBS Protocol | **CBD** | **CITES** | **CMS** | **IAC** | **MARPOL 73/78 incl. Annex I/II** | MARPOL Annex III | MARPOL Annex IV | MARPOL Annex V | **UNCLOS** | Straddling Fish Stocks and Highly Migratory Fish Stocks | **Western Hemisphere Convention** | **World Heritage Convention** | **Ramsar Convention** | **UN Framework Convention on Climate Change** | Kyoto Protocol | **Int'l Convention for the Conservation of Atlantic Tunas** | **Int'l Convention for the Regulation of Whaling** | **London Convention** | **Stockholm Convention** | **Basel Convention** | **Total agreements** |
|---|---|---|---|---|---|---|---|---|---|---|---|---|---|---|---|---|---|---|---|---|---|---|---|---|---|
| Anguilla (UK) | X | X | X | | X | X | X | | | | | | X | X | | X | X | | | X | | | | | 11 |
| Antigua and Barbuda | X | X | X | | X | X | | | X | X | X | X | X | | | X | | X | X | | X | X | X | X | 17 |
| Aruba (NL) | X | X | X | X | X | X | X | X | X | X | | X | X | X | | X | X | | | | | | | | 15 |
| Bahamas | | | | | X | X | | | X | X | | X | X | X | | | | X | X | | | | X | X | 11 |
| Barbados | X | | X | | X | X | | | X | X | X | X | X | X | | X | | X | X | X | | X | | X | 16 |
| Belize | X | | X | | X | X | | | X | X | X | X | X | X | | X | X | X | | X | X | | X | X | 17 |
| Bermuda (UK) | X | X | X | | X | X | X | | X | X | X | X | X | X | | X | X | | | X | | | | | 15 |
| British Virgin Islands (UK) | X | X | X | | X | X | X | | | | | | X | | | X | X | | | | | | | | 9 |
| Cayman Islands (UK) | X | X | X | | X | X | X | | X | X | X | X | X | X | | X | X | | | | | | | | 14 |
| Colombia | X | X | X | X | X | X | | | X | X | X | X | | | X | X | X | X | X | | | | X | X | 17 |
| Costa Rica | X | | X | X | X | X | | X | | | | | X | X | X | X | X | X | | | X | X | X | X | 16 |
| Cuba | X | X | X | | X | X | | | X | | | | X | | X | X | X | X | X | X | | X | X | X | 16 |
| Dominica | X | | X | | X | X | | | X | X | | X | X | | | X | | X | | | X | | | X | 12 |
| Dominican Republic | X | X | X | X | X | X | | | X | X | X | X | | | X | X | X | X | X | X | | X | X | X | 19 |
| Grenada | X | | X | | X | X | | | | | | | X | | | X | | X | X | | | | | | 8 |
| Guadeloupe (FR) | X | X | X | X | X | X | X | | | | | | X | X | | X | X | | | | | | | | 11 |
| Guatemala | X | X | X | | X | X | | | | | | | X | | X | X | X | X | X | | | X | X | X | 14 |

*Organisational approaches*

Numerous organisations work internationally across the region to support conservation management, planning and education. The Caribbean Environment Programme, a UNEP Regional Seas office, includes the Secretariat for the Cartagena Convention but also supports a broad range of activities at regional and national levels (Colmenares & Escobar 2002). These include a number of training projects and the promotion of MPAs. Many of these are linked with other organisations.

Quite a number of the international non-governmental organisations (NGOs) are very active in countries across the Caribbean, and there are also a number of regional NGOs. These are often able to mobilise action over wide areas and use lessons learned from one place to develop ideas in other places. Collaboration among these groups can also be important. The International Coral Reef Action Network has had a strong regional focus, encouraging good practice and providing training in marine resource management (ICRAN 2004). One area that has received particular attention has been the Mesoamerican reef from the Yucatan Peninsula (Mexico) to Honduras, and major projects there have bought together expertise from several global, regional and local groupings of organisations (MBRS 2004)).

Research is critical in driving change, and there are a growing number of regional studies that are placing the concerns felt by individual countries into a wider framework. These include the Reefs at Risk in the Caribbean Study being

Table 3 (continued)

**Regional involvement in international agreements. Caribbean countries which have accepted, signed or ratified each convention are indicated.**

| | **Cartagena Convention** | SPAW Protocol | Oil Spills Protocol | LBS Protocol | **CBD** | **CITES** | **CMS** | **IAC** | **MARPOL 73/78 incl. Annex I/II** | MARPOL Annex III | MARPOL Annex IV | MARPOL Annex V | **UNCLOS** | Stradding Fish Stocks and Highly Migratory Fish Stocks | **Western Hemisphere Convention** | **World Heritage Convention** | **Ramsar Convention** | **UN Framework Convention on Climate Change** | Kyoto Protocol | **Int'l Convention for the Conservation of Atlantic Tunas** | **Int'l Convention for the Regulation of Whaling** | **London Convention** | **Stockholm Convention** | **Basel Convention** | **Total agreements** |
|---|---|---|---|---|---|---|---|---|---|---|---|---|---|---|---|---|---|---|---|---|---|---|---|---|---|
| Honduras | X | | X | | X | X | | X | X | | | X | X | | | X | X | X | X | X | | X | X | X | 16 |
| Jamaica | X | X | X | | X | X | X | | X | X | X | X | X | X | | X | X | X | X | | | X | X | X | 19 |
| Martinique (FR) | X | X | X | X | X | X | X | | | | | | X | X | | X | X | | | | | | | | 11 |
| Mexico | X | X | X | | X | X | | X | X | | | X | X | | X | X | X | X | X | X | X | X | X | X | 19 |
| Montserrat (UK) | X | X | X | | X | X | X | | | | | | X | | | X | X | | | | | | | | 9 |
| Netherlands Antilles (NL) | X | X | X | X | X | X | X | X | X | X | | X | X | X | | X | X | | | | X | | | | 16 |
| Nicaragua | X | | X | | X | X | | X | X | X | X | X | X | | X | X | X | X | X | | X | | X | X | 18 |
| Panama | X | X | X | | X | X | X | | X | X | X | X | X | | X | X | X | X | X | X | | X | X | X | 20 |
| Puerto Rico (US) | X | X | X | X | X | X | | X | X | X | | X | | | X | X | X | | | | | | | X | 14 |
| St. Kitts and Nevis | | | | | X | X | | | X | X | X | X | X | | | X | | X | | | X | | | X | 11 |
| St. Lucia | X | X | X | | X | X | | | X | X | X | X | X | X | | X | X | X | | | X | X | X | X | 18 |
| St. Vincent & Grenadines | X | X | X | | X | X | | | X | X | X | X | X | | | X | | X | | | X | | | X | 14 |
| Trinidad and Tobago | X | X | X | | X | X | | | X | X | X | X | X | | X | | X | X | X | X | | | X | X | 17 |
| Turks and Caicos (UK) | X | X | X | | X | | X | | | | | | X | X | | X | X | | | X | | | | | 10 |
| United States of America | X | X | X | X | X | X | | X | X | X | | X | | | X | X | X | | | | | | | X | 14 |
| U.S. Virgin Islands (US) | X | X | X | X | X | X | | X | X | X | | X | | | X | X | X | | | | | | | X | 14 |
| Venezuela | X | X | X | | X | X | | X | X | X | X | X | | | X | X | X | X | | X | X | | X | X | 18 |
| **Total** | **32** | **25** | **32** | **10** | **34** | **33** | **12** | **10** | **25** | **22** | **15** | **24** | **28** | **14** | **13** | **32** | **27** | **21** | **14** | **12** | **11** | **11** | **16** | **23** | |

undertaken by the World Resources Institute in collaboration with numerous partners from across the region (Burke & Maidens forthcoming). Field-based studies with a regional focus include the Atlantic and Gulf Rapid Reef Assessment (AGRRA) (Lang 2003), the Caribbean Coastal Marine Productivity Program of CARICOMP, a grouping of research laboratories (UNESCO 1998) and volunteer centered groups such as Reef Check and the Reef Environmental Education Foundation.

Industry is also becoming involved in promoting conservation or sustainable development. Most of the major oil transporters are members of the International Tanker Owners Pollution Federation, which exists to coordinate and help in response to oil spills. The Caribbean Association of Sustainable Tourism is one of a number of groups working toward reducing environmental damage in the tourism industry, particularly looking at hotels and the development of environmental certification. In 2001 the International Council of Cruise Lines, which represents two-thirds of the industry, adopted a set of standard waste management practices which are intended to bring them up to or beyond the standards demanded under international agreements such as the International Convention for the Prevention of Pollution from Ships (MARPOL) (Sweeting & Wayne 2002, WTTC *et al* 2002).

International collaboration is fraught with challenges and complications, however. The various international groups concerning fisheries, many with associated regulations, provide a valuable illustration of the potential, and also the complexity, of international cooperation (Figure 13). Two groups have already been actively involved in establishing collaborative fisheries management regimes: the Caribbean Community and Common Market (CARICOM) and the Organization of Eastern Caribbean States (OECS). The CARICOM Fisheries Resource Assessment and Management Program has been working toward a Caribbean Regional Fisheries Mechanism which will have legal status and be open to signature by non-CARICOM nations. Other organisations have a broader regional coverage but are less involved at the fisheries level, such as the Caribbean Forum of ACP States (CARIFORUM) and the Association of Eastern Caribbean States (ACS). The FAO West Central Atlantic Fisheries Commission (WECAFC) and its subcommittees such as WECAFC Lesser Antilles Commission (LAC) are primarily for information exchange. The International Commission for the Conservation of Atlantic Tunas (ICCAT) has an even broader geographic scope, though restricted to tuna and tuna-like species. To some degree, ICCAT illustrates the problems of broadening the scope of environmental treaties too far, in that there is a very limited Caribbean membership or influence (Chakalall, Mahon & McConney 1998). CARICOM has an observer status, but despite the growing interest of small island nations in entering this fishery on a small scale, they remain largely on the outside (Singh-Renton, Mahon & McConney 2003).

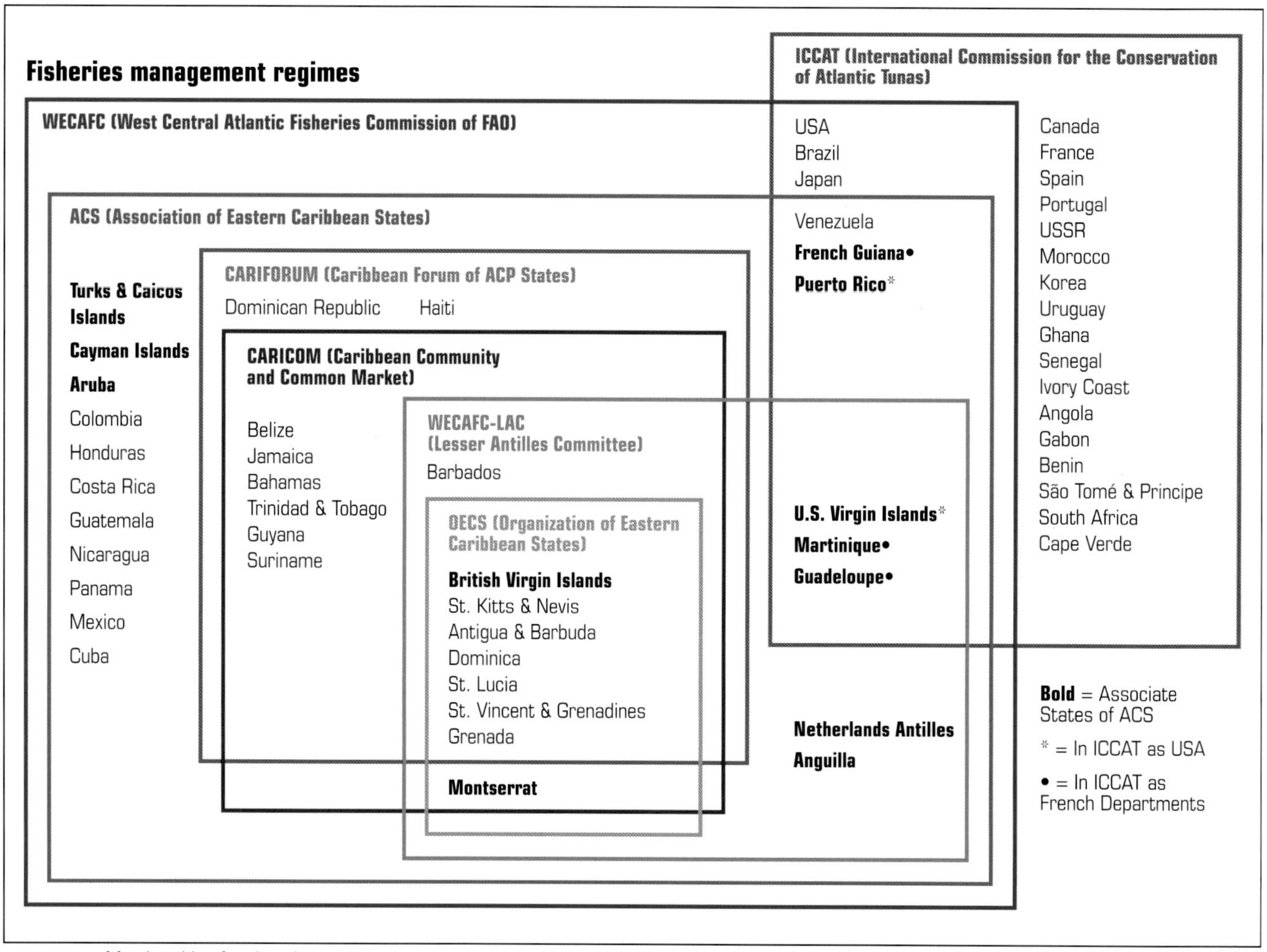

**Fisheries management regimes**

**ICCAT (International Commission for the Conservation of Atlantic Tunas)**

**WECAFC (West Central Atlantic Fisheries Commission of FAO)**

USA
Brazil
Japan

Canada
France
Spain
Portugal
USSR
Morocco
Korea
Uruguay
Ghana
Senegal
Ivory Coast
Angola
Gabon
Benin
São Tomé & Principe
South Africa
Cape Verde

**ACS (Association of Eastern Caribbean States)**

**Turks & Caicos Islands**
**Cayman Islands**
**Aruba**
Colombia
Honduras
Costa Rica
Guatemala
Nicaragua
Panama
Mexico
Cuba

Venezuela
**French Guiana•**
**Puerto Rico***

**CARIFORUM (Caribbean Forum of ACP States)**

Dominican Republic Haiti

**CARICOM (Caribbean Community and Common Market)**

Belize
Jamaica
Bahamas
Trinidad & Tobago
Guyana
Suriname

**WECAFC-LAC (Lesser Antilles Committee)**

Barbados

**OECS (Organization of Eastern Caribbean States)**

**British Virgin Islands**
St. Kitts & Nevis
Antigua & Barbuda
Dominica
St. Lucia
St. Vincent & Grenadines
Grenada
**Montserrat**

**U.S. Virgin Islands***
**Martinique•**
**Guadeloupe•**

**Netherlands Antilles**
**Anguilla**

**Bold** = Associate States of ACS
* = In ICCAT as USA
• = In ICCAT as French Departments

**Figure 13** Membership of various intergovernmental organisations with some influence over fisheries management in the Caribbean (from Chakalall, Mahon & McConney 1998).

LANCE K. B. JORDAN

**Figure 14** A foureye butterflyfish among corals off Vieques, Puerto Rico.

## Defying Ocean's End in the Caribbean

Our review demonstrates that a considerable amount is known about biodiversity, threats and responses in the Caribbean. Baseline information on the distribution of habitats, species (Figure 14) and oceanographic processes is available across the region, and the resolution of datasets is improving all the time. There is a wealth of "status" information, including direct measures of coral reef status, for example. As understanding of the processes behind environmental degradation improves, it is increasingly possible to look for indicators of threat and for patterns likely to create future problems. Perhaps most importantly, there is now a burgeoning array of examples of successful "solutions" to environmental problems. Few threats arise which have never been faced before and examples from across the Caribbean provide management responses that will not only halt environmental decline but will often provide positive socioeconomic benefits as well.

While there remain gaps in knowledge and blanks on the maps, the excuse of needing more science before taking action is clearly not valid in this region. We already hold a vast amount of knowledge and certainly enough to take actions that are likely to challenge the trends of anthropogenic impacts.

We present here a number of recommendations for approaches to defy ocean's end. We first consider underlying approaches and priorities, many of which are applicable to other regions of the world, and indeed to the world's ocean as a whole. We then recommend a number of priority actions and targets that could be assigned budgets and timelines.

### Changing mindsets

Public support is a critical factor in any long-term environmental improvement. Successful environmental management interventions are "essentially a social process" (Montero 2002) and are heavily reliant on community cooperation. Concern and interest from the general public can be a very powerful force, leading to far greater changes in activities and behavior than could possibly be achieved through policy or legislative means alone.

In many parts of the Caribbean, however, a substantial sector of the population fails to understand the importance—or the fragility—of their adjacent marine ecosystems. Few make the critical connections between their activities and the health of their coasts.

In order to enact change there is an urgent need to raise awareness, understanding and appreciation with a targeted drive toward education at all levels and in all sectors. While this includes traditional education through schools, the printed press and broadcast media, it also means more directed outreach to community leaders, politicians and heads of industry. The importance of this drive to educate is laid out clearly in the Rio Declaration, Principle 10: "Environmental issues are best handled with the participation of all concerned citizens. . . . States shall facilitate and encourage public awareness and participation." Without such involvement at the local level, most efforts are likely to be unsustainable.

Education must be targeted. Fishermen need to learn about better methods and need to be made aware of the benefits (to them) of new approaches such as no-take zones. Tourists need to be educated on how to behave responsibly and will take these lessons to heart if their enthusiasm can also be ignited. Schoolchildren need to be taught to swim and snorkel. Journalists need to be made aware of the importance of these issues and given the "stories" to aid their communications efforts. Decision-makers in the fields of governance and industry must be key targets for expanded communications.

#### *Economic arguments*

The use of economic arguments is one of the most powerful forces in driving decision-makers to adopt environmental measures and is thus a key component of the education process. It is widely accepted that "ecosystem services are not fully 'captured' in commercial markets or adequately quantified in terms comparable with economic services and manufactured capital; they are often given too little weight in policy decisions" (Costanza *et al* 1997). Natural resource accounting—taking into account the use and depletion of natural resources over time—is only slowly moving into the mainstream of economics. The concept of Total Economic Value (TEV) has been widely used to convert all values and benefits into simple economic terms (WCPA 1998, Cesar 2000). Such approaches must be encouraged in order to place the natural environment on an equal footing with other human activities in development, planning and policy processes.

*Direct knowledge transfer*

Part of the process of changing mindsets involves the transfer of knowledge, implicit in all forms of education. Aside from traditional routes (teachers to pupils, governments to communities), it is important that knowledge transfer is considered more broadly to allow information to permeate right across society. This may include the encouragement of knowledge transfer from north to south, from scientists to managers and from village to village.

There is often a distrust of outside expertise by local communities, so communication among communities or among fishers can be a more powerful tool. This has happened as a matter of course in a few places where the communities adjacent to successful management regimes are mirroring the efforts of their neighbors. Local communities are developing another marine management area in St. Lucia, for example, on the model of the Soufriere Marine Management Area; and MPAs in Belize and Mexico have been learning from the success of the Hol Chan Marine Reserve. The International Coral Reef Action Network and the Caribbean Environment Programme have been promoting workshops for fishers and protected areas personnel to share their experiences with colleagues in other countries.

To further obviate the need for external advice, it will be important not only to build capacity within countries and within communities to continue this process of education, but also to undertake scientific monitoring, planning and management at local and national levels.

*Actions and targets*

- Ensure marine education is incorporated into schools' curricula across the region
- Provide educational resources—books, broadcast media, posters—to schools, NGOs and communities
- Support exchange of knowledge among communities
- Train local scientists and managers
- Support international transfer of knowledge at practical levels through training workshops
- Incorporate natural resource accounting into national budgeting; provide studies that will feed relevant information into this process

**Integrating and improving existing information**

Information is a critical tool for natural resource management. Considerable information about biodiversity exists across the region, but greater collaboration and information exchange is required for this to be utilised more effectively. Data from studies across the region should be incorporated into region-wide datasets, but the findings of individual studies should also be more widely available. For example, national and regional groups looking at priority-setting exercises in biodiversity conservation should be able to readily access *all* existing information (not simply end-products) from other studies to support their decision-making. This would reduce the extravagant waste associated with repeated exercises in re-accessing and re-compiling information needed for individual decision-making efforts.

Free and easy access to existing information is crucial, and making such information available should become standard practice and a prerequisite for funding to conservation projects. Regional and national nodes should be encouraged to hold and disseminate information from within the region. The form of data presentation and ease of access can greatly affect its utility. While the Internet is a powerful tool, it is not universally accessible, and in some places access can be expensive. Information must be interpreted and presented in formats that are readily understood by stakeholders. For ease of comparison among countries, it is important to develop clear definitions and standards of data presentation.

In certain scientific fields and some parts of the Caribbean, we still have insufficient knowledge to determine management responses. This is particularly the case for deeper water habitats such as deep shelf bioherms (limestone formed from fossilized organisms, often the remains of former coral reefs). For these areas primary research is still a high priority.

*Actions and targets*

- Strengthen regional centres for data holding and dissemination
- Pull together and utilise existing information for priority setting and management planning on a large (ecoregion) scale
- Ensure project funding does not support "reinventing the wheel" and that it requires rapid and free dissemination of all results
- Expand the knowledge base in key areas, particularly deep-water ecology and natural resource accounting

**Improving management systems**

There is a need to strengthen policy, planning and management frameworks to ensure economically, socially and environmentally sustainable growth. Comprehensive legal systems are important, but full stakeholder participation is equally important in developing the management process.

Highly targeted management approaches, such as fisheries controls or the installation of mooring buoys, may be sufficient to control spatially limited or direct human impacts. By contrast, many indirect human impacts such as pollution and sedimentation may be linked to remote locations or to multiple sources for which control may only be possible with more integrated management. Community understanding and involvement is essential to ensure compliance. In certain locations, active intervention—such as habitat restoration, restocking, removal of alien species or even restoration engineering—may be required to restore ecosystem functioning.

*Integrated coastal management*

The principles of integrated coastal management (ICM) are simple and revolve around the critical concept of integration at all levels:

- Integration of sectors—agriculture, fisheries, development, environment
- Integration of programmes and planning—food production, waste disposal, tourism, transport
- Integration of users—fishers, tourists, farmers, foresters and the general public
- Integration of management tasks—planning, implementation, staffing, budgeting, monitoring, enforcement
- Integration across scales—local community, local government, national, multinational, international

There are few examples of fully implemented ICM efforts in the Caribbean at present, and one of the primary obstacles appears to be inadequate institutional frameworks. There is an urgent need to establish a process and system of inter-institutional communication and collaboration. This is not a process of centralization, for while there may be some need for a clearer designation of roles and even a central authority to coordinate such activities, there must also be distribution of decision-making to relevant authorities and communities.

*Overfishing and new approaches to fisheries management*

While there is a growing number of successful examples of co-management (TURFs) and other forms of resource ownership, much of this remains at the level of case examples. In some countries it would appear there is a reluctance of governments to relinquish their rights to control and manage common property resources.

Fisheries departments are often subsumed in larger departments and fisheries issues are considered a low priority for funding and staff attention. There is also an ongoing distrust of governments by the fishers. There is thus a clear need for education and awareness building, as well as for the empowerment of fishers. In some cases, international cooperation may help in this regard. The Organization of Eastern Caribbean States has developed Harmonised Fisheries Legislation, which includes giving Ministers the responsibility of designating "Local Fisheries Management Areas" and "Local Fisheries Management Authorities," while CARICOM's Fisheries Resource Assessment and Management Programme (CFRAMP) has recognised the importance of awareness raising and empowerment (Brown & Pomeroy 1999).

Protected areas, and especially no-take zones, are proving extremely effective at local scales (Figure 15). There is now a need to develop this into larger-scale national and Caribbean-wide efforts—and to focus on building effective networks of protected areas—so the benefits to communities as well as to the ecosystems can be more widely felt.

*Other industries*

Aside from fishing, a broad array of industries has an influence on the environment in the Caribbean, including oil and gas, agriculture and tourism. In many cases, the impacts of industry may be best controlled by legal requirements. Controls on waste treatment and limits on smokestack emissions have gone a long way to reducing pollution in some countries. Accepting that some levels of environmental

DAVID BRYAN

**Figure 15** A large school of permit fish off the coast of Florida, where "no-take" zones have created dramatic increases in fish stocks.

damage may be inevitable, some innovative approaches have been developed to mitigate the losses—including pollution taxes that are invested directly in environmental improvements elsewhere and no-net-loss schemes, whereby any damage must be offset by habitat restoration elsewhere.

In a growing number of cases, industry has moved ahead of regulation in reducing environmental impacts. Consumer demand and the threat of litigation have both helped to drive such change, but in many cases there is a perception of internal benefits that may accrue from adopting change. Many oil companies have now taken on the burden of oil-spill prevention measures and are funding the establishment of rapid oil-spill response systems. Tourism in many areas is using environment-friendly credentials as a key selling point for a perceived advantage over competitors. Certification schemes are being developed for the hotel industry and the aquarium trade to arm consumers with the mechanisms for choice.

Consumer choice can drive change through direct intervention efforts such as letters, protests or through the systems of governance. A further growing driver for change is that of market forces. A scheme to identify sustainable fish-products has been established but does not yet cover Caribbean fisheries. The labeling of "organic" food products is growing in Western Hemisphere markets enabling even very remote purchasers of Caribbean products (bananas, coffee, cotton) to identify products that are less polluting.

*Actions and targets*

- Support integrated management projects
- Support co-management approaches to fisheries, such as TURFS
- Involve recreational fishers and associations in new and existing reserves
- Greatly increase the area of no-take marine reserves—target 30% of nearshore demersal fishery areas for no-take zones
- Regulate and restrict the total area open to trawling
- Raise levels of sewage treatment to 90% by 2030
- Reduce solid waste production, and reduce damaging ocean dumping of wastes
- Develop restoration projects in damaged areas, particularly nearshore nursery areas
- Encourage implementation of park user fees to support management
- Encourage and develop industry involvement
- Adopt a no-net-loss approach to seagrass and mangrove ecosystems
- Support regional certification schemes for fisheries, tourism and the marine aquarium trade (Figure 16)

**Working internationally**

In the planning of new marine protected areas, decision-makers should consider multinational and region-wide networks of protected areas and existing management strategies to build a more cohesive and functional whole. This and many environmental issues require collaboration among countries, which may range from agreed regulations to more practical activities at the level of planning and management. This collaboration should not be restricted to government agencies but should include NGOs, stakeholder organisations, scientists and funding agencies.

There is also a growing need to harmonise and streamline international agreements and the range of international and intergovernmental organisations dealing with environmental issues. There is a danger that the act of signing on to international agreements will be seen as a solution in itself. Clearly, implementation of these agreements is critical, and such implementation typically requires national capacity building and awareness raising. A related problem has been a lack of "ownership"; when problems and solutions are discussed at international levels, national bodies can lose interest or become disenfranchised. While this may be a natural consequence of the lack of national capacity, it rarely leads to sustained action by the "target" country (Montero 2002).

**Figure 16** Collecting wild "giant anemones" for the aquarium trade has caused local extinctions of this species (Gasparini *et al* forthcoming).

In the Caribbean region, poverty and the small size of many nations undermine the ability of some countries to sustain full scientific and administrative capacities to address regional issues. In this case, the need to develop a "regional capacity" becomes even more imperative, with shared funding of administrative, research and enforcement work. Such regional capacity building can already be seen in some of the regional groupings described earlier.

Montero (2002) outlines some of the key principles that must underlie every international effort or programme:

- There must be a common purpose among participants
- The programme or project must have a high national priority to ensure support and follow-up
- Human and material resources must be available, or made available, at the national level
- There must be clear leadership, with strong scientific and moral authority
- National and international bodies must exist to promote, implement and support the programme or project
- There must be sustained financial and in-kind contributions from international and national sources

*Actions and targets*

- Protect regionally important resources through regional regulation
- Protect all fish and marine invertebrate spawning sites across the region with no-take zones or closed seasons during spawning aggregations

- Increase the number of Caribbean nations ratifying and implementing relevant international agreements
- Improve international funding mechanisms

**General priorities**

In setting priorities for *Defying Ocean's End,* it will be critical to think in terms of spatial and temporal scales. We are faced with a multiplicity of problems, highly interconnected, and all requiring differing levels of action and reaction. It will be impossible to address climate change or the demise of straddling fish stocks at the level of local communities, or even national governments, without working within international agreements. Similarly, high-level agreements on local fisheries controls are worthless without ground-level support, education and co-management at the level of local communities.

Education, in its broadest sense, is a critical tool. However, other methods, including financial aid or incentives, coercive measures and legislative frameworks all have a part to play. In some deep-water areas, scientific research on which to base decisions is clearly required, and science and monitoring have a critical role to play in the ongoing management of protected areas and their expansion into regional networks. One particular failure of science to date has been at the level of communication and outreach. In this regard, the scientific community must do a much better job in converting science to education, public awareness and environmental action.

## Acknowledgements

*We gratefully acknowledge support from the Gordon and Betty Moore Foundation for preparation of this work and for supporting our attendance at the* Defying Ocean's End *Conference. Thanks especially to Linda Glover, Jennifer Jeffers, Tim Werner and Mike Smith of Conservation International for their support, as well as to Sonya Krough and Debra Fischman of CI's GIS Lab. Thanks to Lauretta Burke and the World Resources Institute and to the UNEP World Conservation Monitoring Centre for the provision of data. Thanks also to Patricia Kramer, Tries Razak and Stephen Grady for preparation of datasets and checking of the manuscript.*

## Literature Cited & Consulted

The American Heritage Dictionary. 2000. *The American Heritage Dictionary of the English Language,* Fourth Edition, Boston: Houghton Mifflin Company.

Baum J., R. A. Myers, D. G. Kehler, B. Worm, S. J. Harley and P. A. Doherty. 2003. Collapse and conservation of shark populations in the northwest Atlantic. *Science* 299:389–392.

Blake, B. 1998. A strategy for cooperation in sustainable ocean management and development, Commonwealth Caribbean. *Marine Policy* 22:505–513.

Bohnsack, J. A. 2000. A comparison of the short-term impacts of no-take marine reserves and minimum size limits. *Bulletin of Marine Science* 66 (3): 635–650.

Bright, C. 1999. *Life Out of Bounds: Bioinvasion in a Borderless World.* New York: WW Norton and Company.

Brown, D. N. and R. S. Pomeroy. 1999. Co-management of Caribbean community (CARICOM) fisheries. *Marine Policy* 23:549–570.

Bryant, D., L. Burke, J. McManus and M. Spalding. 1998. *Reefs at Risk: A map-based indicator of threats to the world's coral reefs.* Washington, D.C.: World Resources Institute, International Center for Living Aquatic Resources Management, World Conservation Monitoring Centre and United Nations Environment Programme.

Burke, L., L. Selig and M. Spalding. 2002. *Reefs at Risk in Southeast Asia.* Washington, D.C.: World Resources Institute.

Burke, L. and J. Maidens. Forthcoming. *Reefs at Risk in the Caribbean.* Washington, D.C.: World Resources Institute.

Caddy J. F. and L. Garibaldi. 2000. Apparent changes in the trophic composition of world marine harvests: The perspective from the FAO capture database. *Ocean & Coastal Management* 43:615–655.

Carpenter, R. C. 1997. Invertebrate predators and grazers. In *Life and Death of Coral Reefs,* ed. C. Birkeland, 198–229. New York: Chapman-Hall.

Cesar, H. S. J., ed. 2000. *Collected Essays on the Economics of Coral Reefs.* Kalmar, Sweden: CORDIO, Kalmar University.

Chakalall, B., R. Mahon and P. McConney. 1998. Current issues in fisheries governance in the Caribbean community (CARICOM). *Marine Policy* 22:29–44.

Colmenares, N. A. and J. J. Escobar. 2002. Ocean and coastal issues and policy responses in the Caribbean. *Ocean and Coastal Management* 45:905–924.

Cortés, J., ed. 2003. *Latin American Coral Reefs.* Amsterdam: Elsevier Science Publishing Co.

Costanza, R., R. d'Arge, R. de Groot, S. Farber, M. Grasso, B. Hannon, S. Naeem, K. Limburg, J. Paruelo, R. V. O'Neill, R. Raskin, P. Sutton and M. van den Belt. 1997. The value of the world's ecosystem services and natural capital. *Nature* 387:253–260.

Craig, A. K. 1966. Geography of fishing in British Honduras and adjacent coast waters. *Coastal Studies,* Series no. 14. Baton Rouge, Louisiana: Louisiana State University Press.

Creed, J. C., R. C. Phillips and B. I. Van Tussenbroek. 2003. The seagrasses of the Caribbean. In *World Atlas of Seagrasses,* ed. E. P. Green, F. T. Short, 234–244. Berkeley: University of California Press.

Denevan, W. 1992. *The Native Population of the Americas in 1492.* Madison: University of Wisconsin Press.
Diaz, R. J. 2001. Overview of hypoxia around the world. *Journal of Environmental Quality* 30:275–281.
Environmental Justice Foundation (EJF). 2003. *Squandering the Seas: How shrimp trawling is threatening ecological integrity and food security around the world.* London: Environmental Justice Foundation.
——. 2004. *Farming the Sea, Costing the Earth: Why We Must Green the Blue Revolution.* London: Environmental Justice Foundation.
Food and Agriculture Organization (FAO). 2002. *The State of the World's Fisheries and Aquaculture.* Rome: Food and Agriculture Organization of the United Nations.
Gasparini, J. L., C. E. L. Ferreira, S. R. Floeter and I. Sazima. Forthcoming. Marine ornamental trade in Brazil. *Biodiversity and Conservation* 1–17. *http://paginas.terra.com.br/educacao/otima/brazilianreeffish/library/Gasparinietal_Ornamental_proof.pdf* (accessed September 22, 2004).
Gell, F. R. and C. M. Roberts. 2002. *The Fishery Effects of Marine Reserves and Fishery Closures.* Washington, D.C.: WWF – US.
GESAMP. 2001. GESAMP Reports and Studies, 71: *Protecting the Oceans from Land-Based Activities. Land-based sources and activities affecting the quality and uses of the marine, coastal and associated freshwater environment.* The Hague, Netherlands: IMO/FAO/UNESCO-IOC/WMO/WHO/IAEA/UN/UNEP Joint Group of Experts on the Scientific Aspects of the Marine Environmental Protection.
Gibson, J. and J. Carter. 2003. The reefs of Belize. In *Latin American Coral Reefs,* ed. J. Cortés, 171–202, Amsterdam: Elsevier Science Publishing Co.
Gitay, H., A. Suárez, R. T. Watson and D. J. Dokken, eds. 2002. *Climate Change and Biodiversity.* Geneva: Intergovernmental Panel on Climate Change (IPCC).
Goldenberg, S. B., C. W. Landsea, A. M. Mestas-Nuñez and W. M. Gray. 2001. The recent increase in Atlantic hurricane activity: Causes and implications. *Science* 293:474–479.
Green, E. P. and A. W. Bruckner. 2000. The significance of epizootiology for coral reef conservation. *Biological Conservation* 96:347–361.
Green, E. P. and L. Conway. 2001. *Restricted Fishing Zones in the Wider Caribbean: Results of a Literature and Database Survey.* Unpublished report.
Griffin, D. W., V. H. Garrison, J. R. Herman and E. A. Shinn. 2001. African desert dust in the Caribbean atmosphere: Microbiology and public health. *Aerobiologia* 17 (3): 203–213.
Gustavson, K. 2002. Economic production from the artisanal fisheries of Jamaica. *Fisheries Research* 57 (2): 103–115.
Guzmán, H. M. 2003. Caribbean coral reefs of Panama: Present status and future perspectives. In *Latin American Coral Reefs,* ed. J. Cortés, 241–274. Amsterdam: Elsevier Science Publishing Co.
Harbourne, A. R., D. C. Afzal and M. J. Andrews. 2001. Honduras: Caribbean Coast. *Marine Pollution Bulletin* 42:1221–1235.
Health Ecological and Economic Dimensions of Global Change Program (HEED). 1998. Marine ecosystems: Emerging diseases as indicators of change. In *Health of the Oceans from Labrador to Venezuela,* ed. P. R. Epstein, B. H. Sherman, E. S. Siegfried, A. Langston, S. Prasad, B. Mckay. Year of the Ocean Special Report. NOAA-OGP and NASA Grant number NA56GP 0623. Boston, MA: The Center for Conservation Medicine and CHGE Harvard Medical School.
Hughes, T. P. 1994. Catastrophes, phase-shifts, and large-scale degradation of a Caribbean coral reef. *Science* 265:1547–1551.
Hughes, T. P. and J. H. Connell. 1999. Multiple stressors on coral reefs: A long-term perspective. *Limnology and Oceanography* 44 (3): 932–940.
Illueca, J. 2001. Regional Seas Conventions and Action Plans: An evolving framework for the protection and sustainable development of the marine and coastal environment. Paper presented at the Global Conference on Oceans and Coasts at Rio + 10: Assessing Progress, Addressing Continuing and New Challenges. Dec 3–7, 2001.
Intergovernmental Panel on Climate Change (IPCC). 2001. *Climate Change 2001.* New York: Cambridge University Press.
International Coral Reef Action Network (ICRAN). 2004. International Coral Reef Action Network. *http://www.icran.org/* (accessed September 22, 2004).
Jackson, J. B. C. 1997. Reefs since Columbus. *Coral Reefs* 16 (suppl.): S23–S32.
Jackson, J. B. C., M. X. Kirby, W. H. Berger, K. A. Bjorndal, L. W. Botsford, B. J. Bourque, R. H. Bradbury, R. Cooke, J. Erlandson, J. A. Estes, T. P. Hughes, S. Kidwell, C. B. Lange, H. S. Lenihan, J. M. Pandolfi, C. H. Peterson, R. S. Steneck, M. J. Tegner and R. R. Warner. 2001. Historical overfishing and the recent collapse of coastal ecosystems. *Science* 293:629–638.
Jameson, S. C., L. B. Trott, M. J. Marshall and M. J. Childress. 2000. Nicaragua: Caribbean Coast. In *Seas at the Millennium: An Environmental Evaluation,* Volume 1, ed. C. Sheppard, 517–530. Amsterdam: Elsevier Science Publishing Co.
Jameson, S. C., M. H. Tupper and J. M. Ridley. 2002. The three screen doors: Can marine "protected" areas be effective? *Marine Pollution Bulletin* 44 (11): 1177–1183.
Kelleher, G., C. Bleakley and S. Wells, eds. 1995. *A Global Representative System of Marine Protected Areas. Volume 2: Wider Caribbean, West Africa and South Atlantic.* Washington, D.C.: The Great Barrier Reef Marine Park Authority, The World Bank and IUCN–The World Conservation Union.
Lang, J. C. 2003. Status of coral reefs in the western Atlantic: Results of initial surveys, Atlantic and Gulf Rapid Reef Assessment (AGRRA) Program. *Atoll Research Bulletin* 496:630.
Lomelí, D. Z., T. S. Vázquez, J. L. R. Galavéz, A. Yáñez-Arancibia and E. R. Arriaga. 1999. Terms of reference towards an integrated management policy in the coastal zone of the Gulf of Mexico and the Caribbean. *Ocean and Coastal Management* 42:345–368.

Longhurst, A. 1998. *Ecological Geography of the Sea,* 2nd edition. San Diego: Academic Press.

Luckhurst, B. E. 2002. *Recommendations for a Caribbean Regional Conservation Strategy for Reef Fish Spawning Aggregations.* Unpublished paper prepared for the Nature Conservancy.

Lutchman, I. and D. Hoggarth. 1999. *Net Losses: Untying the Gordian Knot of Fishing Overcapacity.* Gland, Switzerland and Cambridge, UK: IUCN–The World Conservation Union.

McAllister, D. E., J. Baquero, G. Spiller and R. Campbell. 1999. A global trawling ground survey. A study carried out for the World Resources Institute, Marine Biological Conservation Institute and Ocean Voice International.

Merino, M. 1997. Upwelling on the Yucatan Shelf: hydrographic evidence. *Journal of Marine Systems* 13:101–121.

Meschede, M. and W. Frisch. 1998. A plate-tectonic model for the Mesozoic and early Cenozoic history of the Caribbean plate. *Tectonophysics* 296:269–291.

MesoAmerican Barrier Reef Systems Project (MBRS). 2004. MesoAmerican Barrier Reef Systems Project. *http://www.mbrs.org.bz/* (accessed September 22, 2004).

Montero, G. G. 2002. The Caribbean: Main experiences and regularities in capacity building for the management of coastal areas. *Ocean and Coastal Management* 45:677–693.

Morgan, L. E. and R. Chuenpagdee. 2003. *Shifting Gears: Addressing the collateral impacts of fishing methods in U.S. waters.* Washington D.C.: Island Press.

Mosquera, I., I. M. Côté, S. Jennings and J. D. Reynolds. 2000. Conservation benefits of marine reserves for fish populations. *Animal Conservation* 4:321–332.

Myers, R. A. and B. Worm. 2003. Rapid worldwide depletion of predatory fish communities. *Nature* 423:280–283.

Onuf, C. P., R. C. Phillips, C. A. Moncreiff, A. Raz-Guzman and J. A. Herrera-Silveira. 2003. The seagrasses of the Gulf of Mexico. In *World Atlas of Seagrasses,* ed. E. P. Green and F. T. Short, 224–234. Berkeley: University of California Press.

Peters, E. C. 1997. Diseases of coral-reef organisms. In *Life and Death of Coral Reefs,* ed. C. Birkeland, 114–139. New York: Chapman-Hall.

Pindell, J. L. 1994. Evolution of the Gulf of Mexico and the Caribbean. In *Caribbean Geology: An Introduction,* ed. S. K. Donovan and T. A. Jackson. Kingston, Jamaica: University of the West Indies Press.

Polunin, N. V. C. and C. C. C. Wabnitz. 2001. Marine protected areas in the North Sea. *Fisheries Society of the British Isles Briefing Paper* 1:1–19.

Rawlins, B. G., A. J. Ferguson, P. J. Chilton, R. S. Arthurton, J. G. Rees and J. W. Baldock. 1998. Review of agricultural pollution in the Caribbean with particular emphasis on small island developing states. *Marine Pollution Bulletin* 36:658–668.

Rebel, T. P. 1974. *Sea Turtles and the Turtle Industry of the West Indies, Florida, and the Gulf of Mexico.* Coral Gables: University of Miami.

Roberts, C. M. and J. P. Hawkins. 2000. *Fully-protected marine reserves: A guide.* WWF Endangered Seas Campaign and Environment Department, University of York, Washington, D.C. and York, U.K.

Roberts, C. M., J. A. Bohnsack, F. Gell, J. P. Hawkins and R. Goodridge. 2001. Effects of marine reserves on adjacent fisheries. *Science* 294:1920–1923.

Roberts, C. M., C. J. Mclean, G. R. Allen, J. P. Hawkins, D. E. McAllister, C. Mittermeier, F. Schueler, M. Spalding, J. E. N. Veron, F. Wells, C. Vynne and T. Werner. 2002. Marine biodiversity hotspots and conservation priorities for tropical reefs. *Science* 295:1280–1284.

Rogozinski, J. 1999. *A Brief History of the Caribbean from the Arawak and Carib to the Present.* New York: Penguin Putnam Inc.

Rosen, B. R. 1984. Reef coral biogeography and climate through the late Cainozoic: Just islands in the sun or a critical pattern of islands? In Fossils and Climate, ed. P. J. Brenchley. *Geological Journal Special Issue* 11: 201–62.

Saenger, P. and G. Luker. 1997. Some phytogeographical considerations. In *World Mangrove Atlas,* ed. M. D. Spalding, F. Blasco and C. D. Field. Okinawa, Japan: International Society for Mangrove Ecosystems.

Sapp, J. 1999. *What is Natural? Coral Reef Crisis.* New York: Oxford University Press.

Schumacher, M., P. Hoagland and A. Gaines. 1996. Land-based marine pollution in the Caribbean. *Marine Policy* 20:99–121.

Sealy, N. E. 1994. *Caribbean World: A Complete Geography.* Cambridge: Cambridge University Press.

Sen, S. and J. R. Nielsen. 1996. Fisheries co-management: A comparative analysis. *Marine Policy* 20:405–418.

Short, F. T., R. G. Coles and C. Pergent-Martini. 2001. Global seagrass distribution. In *Global Seagrass Research Methods,* ed. F. T. Short and R. G. Coles, 5–30. Amsterdam: Elsevier Science Publishing Co.

Singh-Renton, S., R. Mahon and P. McConney. 2003. Small Caribbean (CARICOM) states get involved in management of shared large pelagic species. *Marine Pollution Bulletin* 27:39–46.

Smith, M. L., K. E. Carpenter and R. W. Waller. 2002. An introduction to the oceanography, geology, biogeography, and fisheries of the tropical and subtropical western central Atlantic. In *The Living Resources of the Western Central Atlantic, Volume 1: Introduction, molluscs, crustaceans, hagfishes, sharks, batoid fishes and chimaeras,* ed. K. E. Carpenter. Rome: Food and Agriculture Organization of the United Nations.

Smith, T. J. III, M. B. Robblee, H. R. Wanless and T. W. Doyle. 1994. Mangroves, hurricanes and lightning strikes. *BioScience* 44 (4): 256–262.

Spalding, M. D. 2003. Document 1: Role, status and trends and protected areas. Working document prepared for the Convention on Biological Diversity Secretariat.

——. 2004. *A Guide to the Coral Reefs of the Caribbean.* Berkeley: University of California Press.

Spalding, M. D., F. Blasco and C. D. Field, eds. 1997. *World Mangrove Atlas.* Okinawa, Japan: International Society for Mangrove Ecosystems.
Spalding, M. D., C. Ravilious and E. P. Green. 2001. *World Atlas of Coral Reefs.* Berkeley: University of California Press.
Spalding, M. D, M. L. Taylor, C. Ravilious, F. T. Short and E. P. Green. 2003. The global distribution and status of seagrass ecosystems. In *World Atlas of Seagrasses,* ed. E. P. Green and F. T. Short. Berkeley: University of California Press.
Spalding, M. D., E. P. Green and A. Paterson. Forthcoming. Marine and coastal ecosystems. In *State of the World's Protected Areas,* ed. M. D. Spalding, M. Jenkins and S. Chape.
Stoddart, D. R. 1971. Coral reefs and islands and catastrophic storms. In *Applied Coastal Geomorphology,* ed. J. A. Steers, 155–197. London: Macmillan.
Sweeting, J. and S. Wayne. 2003. *A Shifting Tide: Environmental Challenges & Cruise Industry Responses.* Washington, D.C.: Center for Environmental Leadership in Business at Conservation International. *http://www.conservation.org/ImageCache/news/content/press_5freleases/2004/march/octa_5fkit/cruise_5f5finterim_5f5fsummary_2epdf/v1/cruise_5f5finterim_5f5fsummary.pdf* (accessed October 7, 2004).
Thiel, H. 2003. Anthropogenic impacts on the deep sea. In *Ecosystems of the World, 28: Ecosystems of the Deep Oceans,* ed. P. A. Tyler, 427–465, Amsterdam: Elsevier Science Publishing Co.
Tomczak, M. and J. S. Godfrey. 1994. *Regional Oceanography: An Introduction.* Oxford: Permagon Press.
Tomlinson, P. B. 1986. *The Botany of Mangroves.* Cambridge, U.K.: Cambridge University Press.
Tyler, P. A., ed. 2003. *Ecosystems of the World, 28: Ecosystems of the Deep Oceans.* Amsterdam: Elsevier Science Publishing Co.
United Nations Environment Programme (UNEP). 1999. *Regional Seas Reports and Studies, 172: Assessment of Land-Based Sources and Activities Affecting the Marine, Coastal and Associated Freshwater Environment in the Wider Caribbean Region.* The Hague, Netherlands: UNEP/GPA Coordination Office and Caribbean Environment Programme.
——.2004. *Global Environmental Outlook, GEO Year Book 2003.* Nairobi: United Nations Environment Programme.
UNESCO. 1998. *Coastal Region and Small Island Papers, 3: CARICOMP–Caribbean Coral Reef, Seagrass and Mangrove Sites.* Paris: United Nations Educational, Scientific and Cultural Organization.
Veron, J. E. N. 1995. *Corals in Space and Time: The Biogeography and Evolution of the Scleractinia.* Sydney, Australia: UNSW Press.
Vierros, M. 2000. Jamaica. In *Seas at the Millennium: An Environmental Evaluation,* vol. 1, ed. C. Sheppard, 599–574. Amsterdam: Elsevier Science Publishing Co.
Viles, J. 1996. *The Florida Keys: A History of the Pioneers; Florida's History through Its Places.* Saratoga, Florida: Pineapple Press.
Vinogradova, N. G. 1997. Zoogeography of the abyssal and hadal zones. *Advances in Marine Biology* 32:325–387.
Ward, T. J., D. Heinemann and N. Evans. 2001. *The Role of Marine Reserves as Fisheries Management Tools: A Review of Concepts, Evidence and International Experience.* Canberra, Australia: Bureau of Rural Sciences.
WDPA Consortium. 2004. *World Database on Protected Areas 2004.* Copyright World Conservation Union (IUCN) and UNEP–World Conservation Monitoring Centre (UNEP-WCMC), 2004. *http://sea.unep-wcmc.org/wdbpa/download/wdpa2004/index.html* (accessed October 25, 2004).
Wikipedia. 2004. Tethys Sea. *Wikipedia: The Free Encyclopedia. http://en.wikipedia.org/wiki/Tethys_Sea* (accessed September 20, 2004).
Wilkinson, C., ed. 2002. *Status of Coral Reefs of the World: 2002.* Townsville, Australia: Australian Institute of Marine Science.
Wilson, S. M. 1997. *The Indigenous People of the Caribbean.* Gainsville, Florida: University Press of Florida.
Wing, S. R. and E. S. Wing. 2001. Prehistoric fisheries in the Caribbean. *Coral Reefs* 20:1–8.
World Commission on Protected Areas (WCPA). 1998. *Best Practice Protected Area Guidelines, 2: Economic Values of Protected Areas: Guidelines for Protected Area Managers.* Prepared by the Task Force on Economic Benefits of Protected Areas of the World Commission on Protected Areas (WCPA) of IUCN, in collaboration with the Economics Services Unit of IUC Gland, Switzerland and Cambridge, U.K.: IUCN–The World Conservation Union.
WTTC, IFTO, IHRA and ICCL. 2002. *Industry as a partner for Sustainable Development: Tourism.* World Travel and Tourism Council (WTTC), International Hotel and Restaurant Association (IHRA), International Federation of Tour Operators (IFTO), International Council of Cruise Lines (ICCL) and the United Nations Environment Programme, U.K.
Yáñez-Arancibia, A. 2000. Coastal management in Latin America. In *Seas at the Millennium: An Environmental Evaluation,* vol. 1, ed. C. Sheppard, 457–466. Amsterdam: Elsevier Science Publishing Co.
Zezina, O. N. 1997. Biogeography of the bathyal zone. *Advances in Marine Biology* 32:389–426.

CHAPTER 2

# *Seamount Biodiversity, Exploitation and Conservation*

Gregory S. Stone, *New England Aquarium*
Laurence P. Madin, *Woods Hole Oceanographic Institution*
Karen Stocks, *University of California, San Diego*
Glenn Hovermale, *New England Aquarium*
Porter Hoagland, *Woods Hole Oceanographic Institution*
Mary Schumacher, *Woods Hole Oceanographic Institution*
Peter Etnoyer, *Aquanautix Consulting*
Carolyn Sotka, *Communication Partnership for Science and the Sea (COMPASS) & Standford University*
Heather Tausig, *New England Aquarium*

IMAGE COURTESY OF NOAA, WHOI, THE ALVIN GROUP AND THE 2004 GOA EXPEDITION SCIENCE PARTY

## Introduction

Rising up from the deep seafloor, seamounts are steep-sided underwater mountains occurring throughout the world's ocean that provide unique marine habitats. By official geologic definition, seamounts are discrete areas of underwater elevation that rise higher than 1,000 meters above the surrounding seafloor (International Hydrographic Organization 2001). Their tops can range from near the sea surface to several thousand meters below it. Terms such as "knoll" and "hill" are applied to features of less than 1,000-meter elevation. These shorter features often have biological communities similar to those on true seamounts (e.g. Koslow *et al* 2001, Probert, McKnight & Grove 1997), so many biologists are not making a distinction between the size classes and refer to all of them as seamounts. This chapter will differentiate where possible and use the geologically correct terminology.

The number of seamounts in the ocean is unknown because the seafloor has not been fully mapped at high enough resolution to find all of these features. It is known, however, that seamounts exist in every ocean basin and that they are common features on the seafloor. Epp and Smoot (1989) found about 800 seamounts and hills more than 100 meters high in the Atlantic. Smith and Jordan (1988) extrapolated from the mapped areas of the Pacific to estimate there are more than 30,000 true seamounts over 1,000 meters high in the Pacific, and that as much as 6% of the Pacific seafloor is covered by seamounts or smaller underwater hills.

Seamounts are most commonly of volcanic origin and are distributed most densely along areas where continental drift causes sections of the Earth's crust (tectonic plates) to converge, and in areas where oceanic crust is raised up, such as along seafloor spreading ridges (WWF, IUCN & WCPA 2001). Many seamounts are formed when plumes of hot magma from the Earth's mantle push up through the Earth's crust at locations called "hotspots," and the slow movement of a oceanic plate across a stationary hotspot beneath it will cause a linear chain of volcanic islands or seamounts to be formed (Morgan 1971). This is not to be confused with the recent use of the term "hotspots" by conservation biologists to describe areas of enhanced productivity, unique species richness and high degree of threat from human activities (Roberts 2002). Most seamounts, however, appear to meet both the geologic and biologic definitions of the term. Seamounts created over hotspots are composed of hard volcanic rock, often with polymetallic deposits formed by precipitation from hydrothermal vents or seeps of material from below the Earth's crust. Some younger seamounts—located near seafloor spreading ridges or still forming over volcanic hotspots—have active hydrothermal venting and support unique chemosynthetic communities that do not require sunlight or photosynthesis for life but depend on minerals from below the seafloor as the basis of their food chain. (Chapter 12 has a fuller discussion of these biological communities.)

### Estimating the number of seamounts worldwide

Because of limitations in all available methods of mapping and measuring underwater features, the total number of seamounts in the world's ocean can only be estimated, and these estimates vary greatly depending upon the data, methods and approaches used. Seamounts were originally described and first sampled around the time of the *HMS Challenger* expedition in the 1870s, but significant mapping efforts did not begin until the 1950s. Two United Nations organizations—the International Hydrographic Organization and the Intergovernmental Oceanographic Commission—maintain the General Bathymetric Chart of the Oceans (GEBCO) and record the official names of undersea features in their Gazetteer. Based on records dating from 1950 to the present, GEBCO (2004) documents 1,013 named seamounts worldwide, including guyots and tablemounts, both of which are flat-topped seamounts (Figure 1).

Methods that estimate the global number of unsurveyed and undiscovered seamounts are based on:

- Extrapolations from known seamount abundance in a given region (Batiza 1982, Fornari, Batiza & Luckman 1987, Smith & Jordan 1988)
- High readings in the Earth's gravity-gradient field, derived from the Seasat satellite's gravity sensor, that indicate local areas of high masses of rock rising above the seafloor (Craig & Sandwell 1988)
- Altimetry (sea surface height) measurements, from the Geosat/ERS-1 satellite's altimetry sensor, that indicate where underwater masses affect ocean circulation and sealevel (Smith & Sandwell 1997, Wessel 1997, 2001)
- Direct interpretation of contrast-enhanced altimetric marine gravity and predicted bathymetry with image processing software (Chapel & Small 1996)

The Earth's gravitational field is locally affected by the large underwater masses of rock in seamounts, deflecting the normal straight-down pull of gravity toward the center of the seamount. The deflected gravitational pull also acts on the water around the seamount pulling it toward the center of mass. This causes "sealevel" not to be level over seamounts but pitched upward over the centers of the seamounts (Smith & Sandwell 2001). Sensors on satellites can measure these minor deflections in both the gravity field and the sea surface height. The size of the deflections appear to correlate with the size of the underwater feature causing them, so from this information the depths of water (bathymetry) and shape of the seafloor can be crudely mapped from space.

Smith and Jordan (1988) collected shipboard bathymetry in the Pacific, identified the distribution of seamounts, and then used a mathematical model to extrapolate from the mapped areas to the whole Pacific Ocean. They predicted more than 30,000 seamounts in the Pacific alone. Craig and Sandwell (1988) used Seasat altimeter data to identify 8,556 seamount locations. Smith and Sandwell (1997) produced a gridded vertical gravity-gradient dataset and an estimated seafloor bathymetry dataset—both at four-kilometer resolution worldwide between 72° N and 72° S latitudes—using gravity and altimetry data collected from the Geosat and ERS-1 satellite missions. Their datasets have been used since to estimate the number of seamounts in the world ocean. Wessel (2001) predicted the locations of 14,675 large seamounts globally by locating near-circular, anomalously high values in the Sandwell and Smith (1997) gravity data (Figure 2). Wessel pointed out, however, that limitations in both his method and the resolution of the gravity/altimetry datasets made finding any seamounts less than 2,000 meters in height unlikely. He further stated that an "approximate extrapolation of the empirical trend would suggest there may be on the order of 100,000 seamounts globally that exceed the height of one kilometer" (Wessel 2001).

Chapel and Small (1996) used image processing software to pick potential seamounts from the gridded gravity data using interactive contrast enhancement techniques to discriminate seamounts and smaller features. They predicted locations for a total of 15,329 potential seamount-like features in the Pacific Ocean basin alone, with more than 5,000 estimated to be true seamounts higher than 1,000 meters. A recent global analysis using this technique (Small & Chapel forthcoming) estimates a total of just under 5,300 true seamounts worldwide (Figure 3). More recent analyses (Figure 4) based on ETOPO2, a dataset the U.S. National Oceanic and Atmospheric Administration derived from the Smith and Sandwell data of 1997, predict locations for 15,962 seamounts greater than 1,000 meters worldwide (Kitchingman & Lai 2004). But this altimetry data and approach have limitations similar to Wessel's and are likely to underestimate the number of seamounts considerably.

Very different seamount abundance estimates can be made depending on the data and techniques used. To assess the various methods, a study area was selected in the Northeast Pacific and three abundance estimate techniques were compared. GEBCO identified 93 seamounts surveyed and named within the area. Contour methods based on Smith and Sandwell (1997) predicted 168 features and Wessel's gravity-gradient method predicted 211 seamounts (Etnoyer forthcoming).

It is defensible to say that about 15,000 seamounts can currently be located with existing technologies, regardless of the approach used. The true number of seamounts is probably several times this value—and may well be as high as Wessel's estimate of 100,000—but until better seamount measurement or prediction techniques are available the exact number will remain uncertain. But even assuming only 15,000 seamounts in the world ocean, our biological knowledge of them is small: less than 7% of that number are surveyed and named, and less than 2% have been biologically sampled. Based on these numbers, biologists can safely assume that a large number of endemic species, and an important reservoir of marine biodiversity, lie undiscovered on seamounts not yet found.

### The state of knowledge of seamounts

While the existence of seamounts has been known for over a hundred years, seamount biology received very little attention or research effort until the late 1950s (Hubbs 1959). Even today, only a very small proportion of the estimated tens of thousands of existing seamounts have been sampled for biological information (Figure 5). In a 1987 review of seamount biology, Wilson and Kaufmann (1987) reported that about 96 seamounts had been biologically sampled. Since then, the number has increased to about 300. But even these small numbers overestimate what is known about seamount ecology. Most seamounts have only been sampled opportunistically—with a few biological samples taken during geologic or water-column studies, for example—or sampled for only a few types of marine plants or animals. Comprehensive ecological studies are rare. Many of these sampling events have been reported in the literature only as

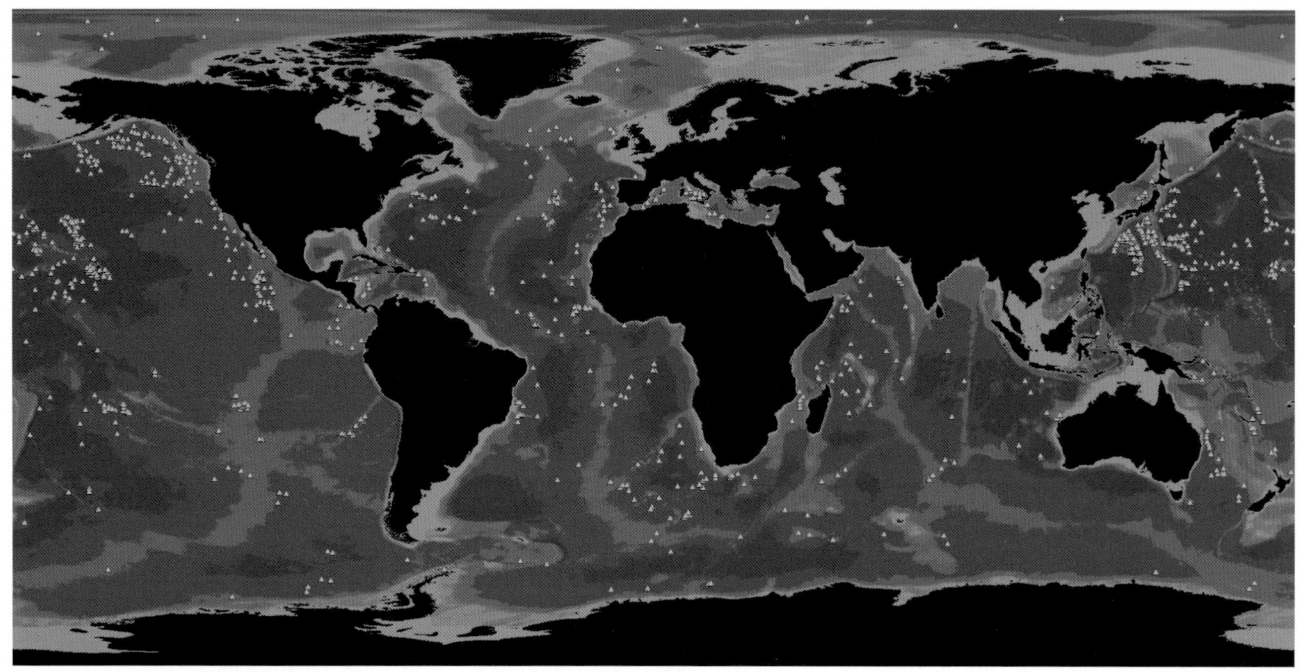

**Figure 1** Global distribution of known seamounts from ship surveys (GEBCO 2004).

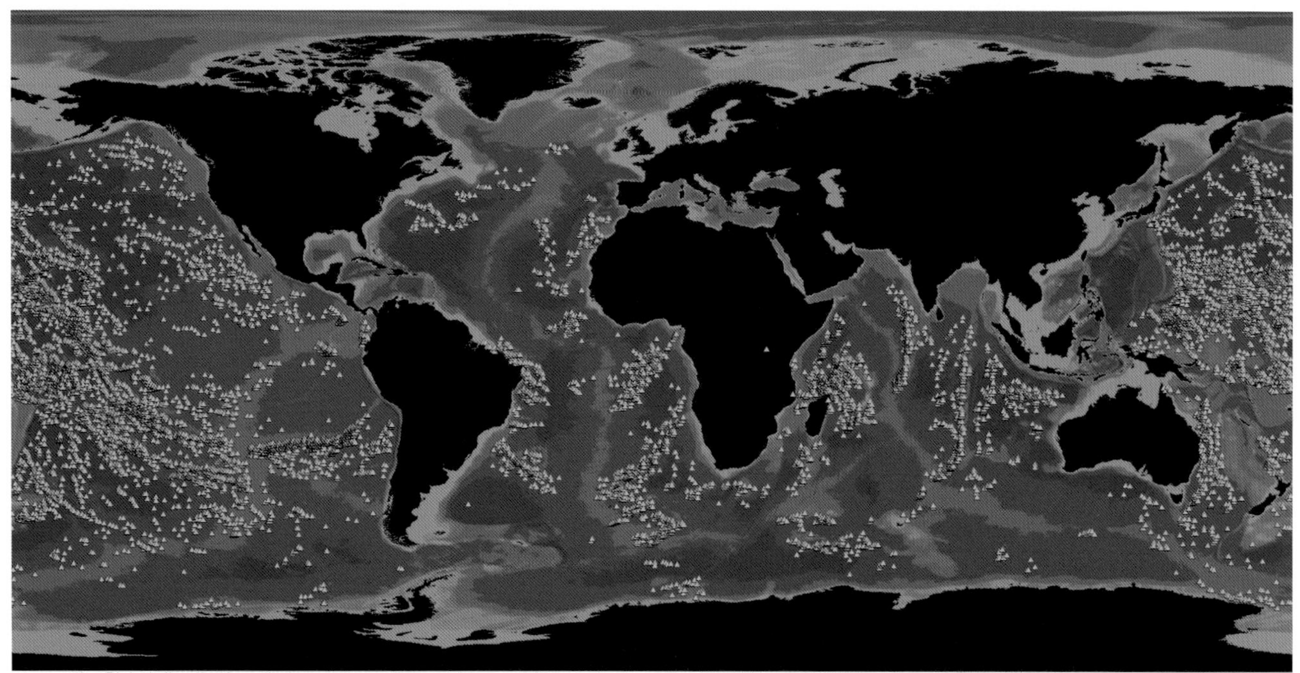

**Figure 2** Global distribution of seamounts estimated from satellite-derived gravity data (Wessel 2001).

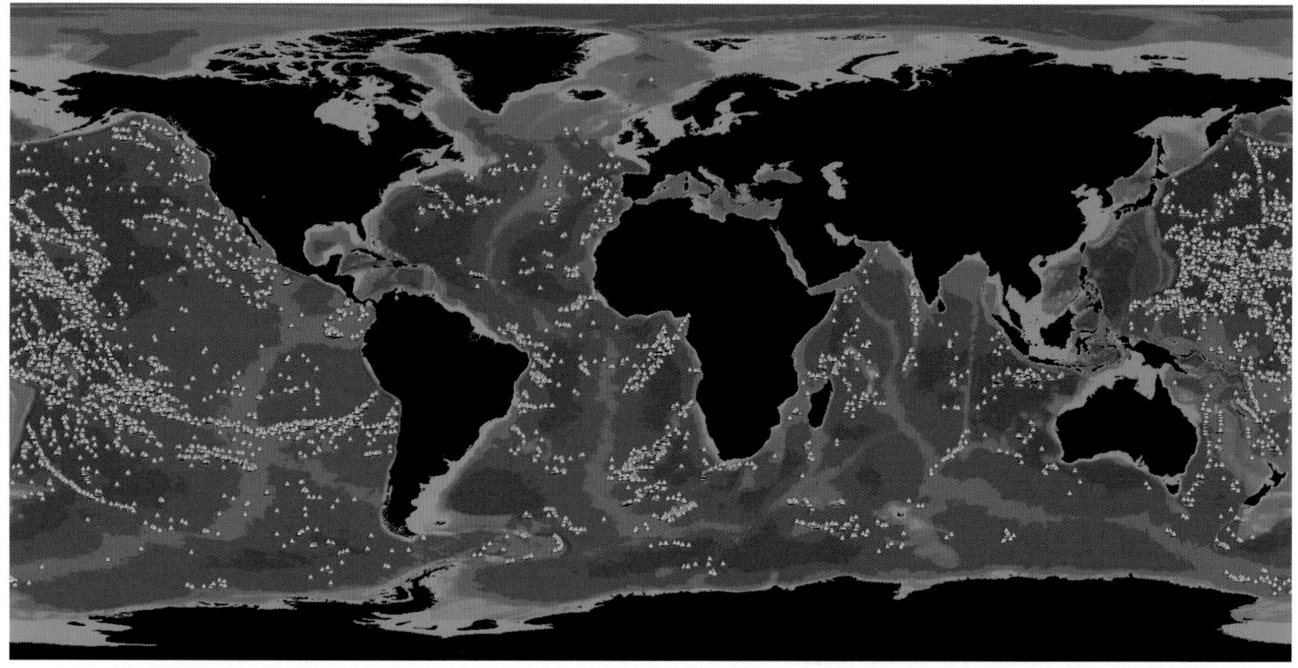

**Figure 3** Global distribution of seamounts estimated from image processing of gravity and predicted bathymetry data (Small & Chapel forthcoming).

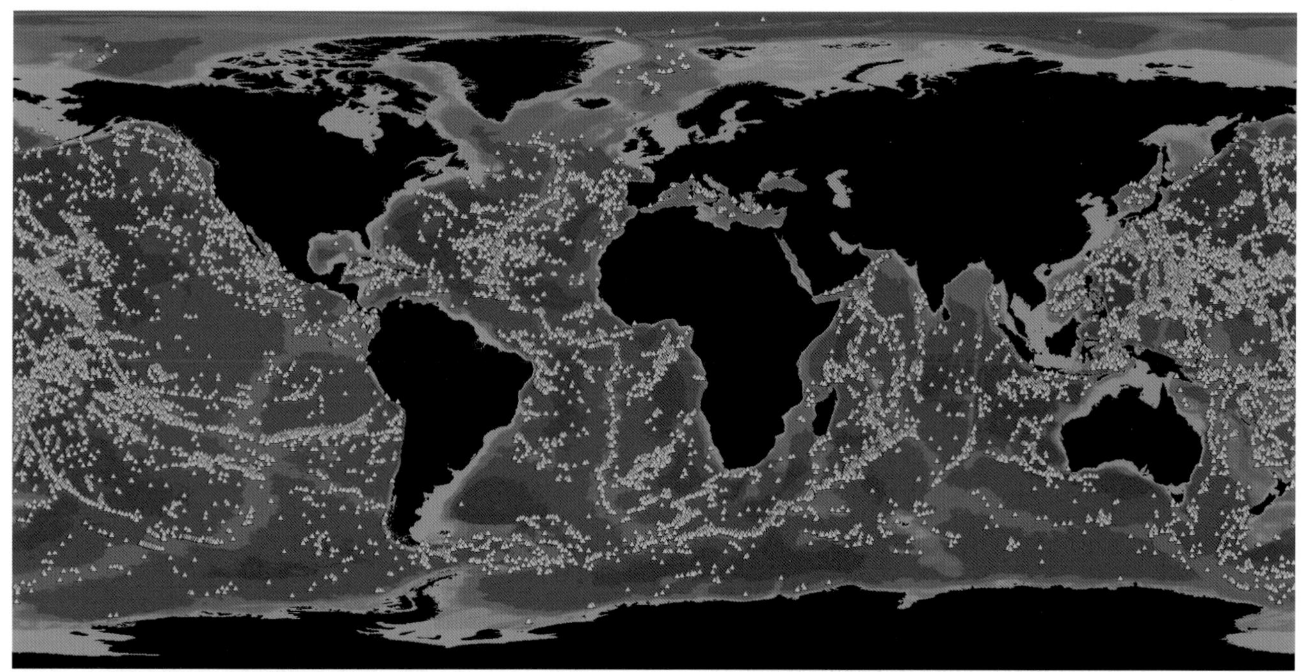

**Figure 4** Global distribution of seamounts estimated based on ETOPO2 satellite data (Kitchingman & Lai 2004).

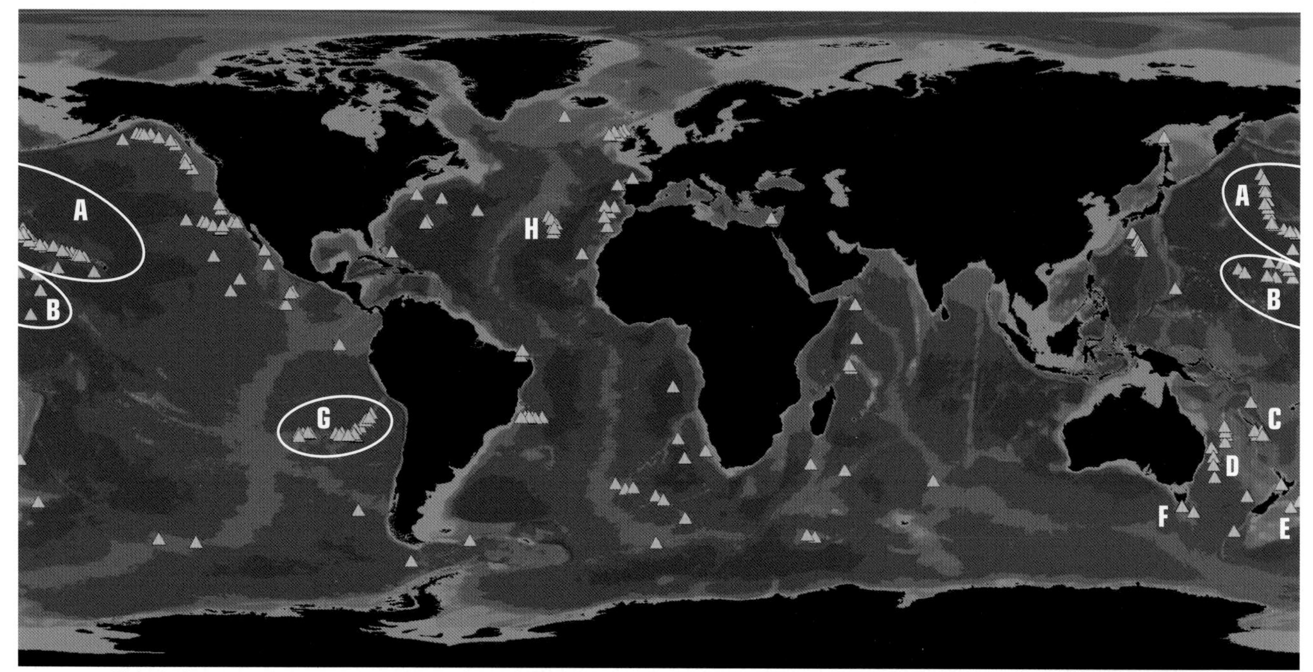


**Figure 5** Biologically sampled seamounts worldwide (Stocks 2004).

A. Hawai'ian & Emperor Chain
B. Mid-Pacific Mountains
C. Norfolk Ridge
D. Lord Howe Ridge
E. Chatham Rise Hills
F. Tasmanian Hills
G. Nasca & Sala-y-Gomez Chain
H. Great Meteor Seamount

brief data summaries, without full species lists or abundance information. And much of the work that has been published is difficult to access and collate in a central repository because it has been published in various languages, often in obscure sources like hard-to-find government reports. There is a pressing need for well-planned, comprehensive biological research into more seamount areas to understand how patterns and features seen in one area may extrapolate to others.

The seamount areas with the most comprehensive biological sampling—and for which results are published in widely available, English-language literature—include those south of New Caledonia on the Norfolk and Lord Howe Ridges; seamounts or hills south of Tasmania; the Nasca and Sala-y-Gomez seamount chain in the southeast Pacific; the mid-Pacific, Hawai'ian and Emperor chains; the Great Meteor seamount in the north Central Atlantic; and several isolated seamounts off the U.S. Pacific coast. In addition, a large body of Soviet data from seamounts worldwide will soon be made available from the P. P. Shirshov Institute of Oceanology through a website (SeamountsOnline 2004). Seamount biology is currently an active area of research, and additional sampling has recently been conducted or is planned for the Northeast and Northwest Atlantic, the Coral and Tasman Seas, the northeast Pacific and the Indian Ocean.

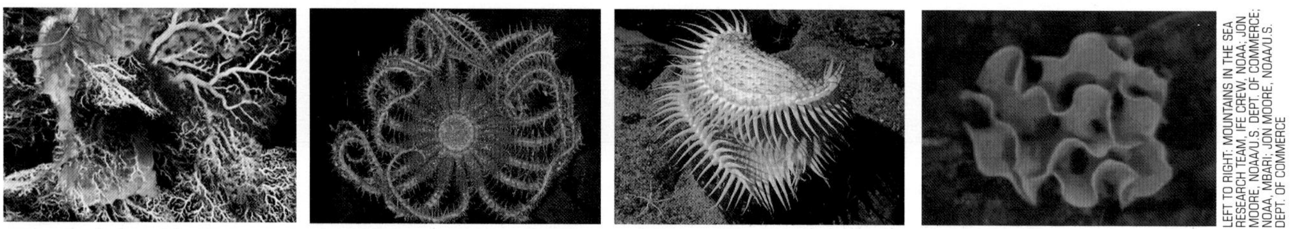

LEFT TO RIGHT: MOUNTAINS IN THE SEA RESEARCH TEAM, IFE CREW, NOAA; JON MOORE, NOAA/U.S. DEPT. OF COMMERCE; NOAA, MBARI; JON MOORE, NOAA/U.S. DEPT. OF COMMERCE

**Figure 6** *Left to right:* Deepsea corals, brittlestars, crinoids and sponges on Balanus Seamount • Brisingid sea star from Manning Seamount • A flytrap anemone on Davidson Seamount • Xenophyophore—a large one-celled organism—from Galápagos Rift.

One of the major gaps in existing research is that, as far as the authors know, no true "before and after" sampling exists to study the effects of commercial fishing, particularly bottom trawling on the seamount habitats. Some seamounts have been sampled that are too deep for commercial fishing (more than 1,400 meters below the sea surface), but many of the shallower seamounts studied may already have been subject to some fishing. The authors know of no large, shallow seamounts that are in pristine condition. This creates a challenge for those who wish to demonstrate the impacts of fishing. The existing study closest to a "before and after" comparison is one by Koslow *et al* (2001) off New Caledonia, where several seamounts with protected areas are compared to those where fishing is allowed. Although the study is confounded by depth variations (most of the reserves are established on deeper seamounts), the effect of trawling is clearly demonstrated by the greatly reduced coverage of seafloor lifeforms, species diversity and overall abundance.

One of the truths of seamount research is that it will never be possible to sample all of them. Efforts must be concentrated on building predictive seamount biodiversity models of diversity and endemism and targeting future sampling to best address data gaps and test the models. To support this work, the SeamountsOnline project is currently creating a database of all known biological sampling on seamounts and making it freely available online to researchers and conservation managers (SeamountsOnline 2004, Stocks 2004). While our knowledge of seamounts is far from complete, enough sampling has been done to show that seamounts support biologically unique and valuable habitats.

## Seamount Biology and Ecology

Seamounts are biologically unique habitats for two main reasons. First, they appear to support very high abundances of some species, particularly fishes, drawing the attention of significant commercial fishing effort. Second, they support high levels of undiscovered and endemic species (species found on one seamount or seamount chain and nowhere else in the ocean). Because of their unique characteristics, it has been hypothesized that seamounts may play important roles in ocean biodiversity, including acting as centers of speciation (evolution of new species), refugia for relict populations (safe havens where ancient species continue to exist despite changes in the environment elsewhere) and stepping-stones for the dispersal of species across open-ocean areas (Hubbs 1959).

### Seamount communities and abundance

The communities of species found on seamounts are rich and varied (Figures 6 and 7). On those seamounts where the underwater currents are slow and there are sandy or muddy sediments, we find mainly deposit-feeding macrofauna communities (animals that eat matter falling down from the water column above) consisting of polychaetes, echinoderms (ophiuroids, holothuroids, echinoids), crustaceans (tanaids, isopods), sipunculids, nemerteans, molluscs (bivalves, aplacophorans, gastropods) and/or sponges; also found are nematodes and harpacticoid copepods (Levin & Thomas 1988, 1989, Levin, McCann & Thomas 1991, 1994). On deeper seamounts, xenophyophores—large single-celled organisms found exclusively on the seafloor in deepwater areas—often dominate (Levin & Thomas 1988, Kaufmann, Wakefield & Genin 1989).

One unusual aspect of seamounts is that underwater currents often interact with the topography, or shape, of seamounts to create faster waterflows. This can remove sediments and provide hard-bottom habitats that are rare in the deepsea. Many of these rocky-bottom seamount areas support rich, dense assemblages of "suspension feeding" species that eat suspended matter in the water, including various corals (gorgonians, antipatharians, scleractinians, sylasterids), crinoids, hydroids, ophiuroids and sponges (Rogers 1994). One of the primary concerns with seamount fisheries is that these deepwater species are fragile and very slow

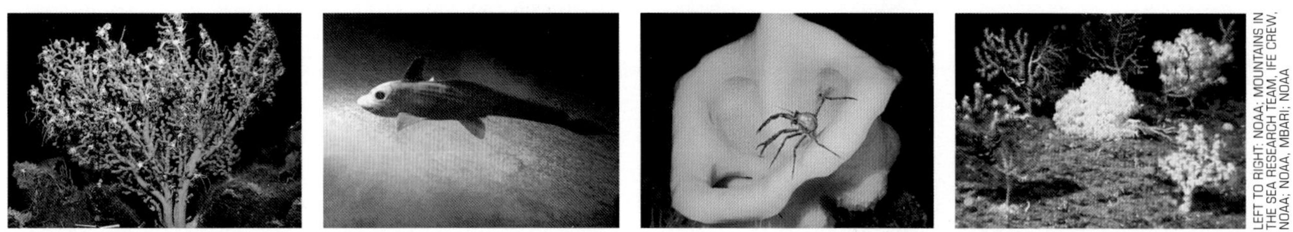

LEFT TO RIGHT: NOAA; MOUNTAINS IN THE SEA RESEARCH TEAM, IFE CREW, NOAA; NOAA, MBARI; NOAA

**Figure 7** *Left to right:* Seafan deepsea coral • Chimaera on Manning Seamount • Crab on a sponge at Davidson Seamount • Deepsea corals.

growing, and when damaged by bottom-trawl fishing, will have extremely long recovery times, potentially thousands of years.

A second characteristic of seamounts is enhanced biological productivity and biomass compared to surrounding open-ocean waters. This may be due to increased nutrient levels caused by upwelling waters around seamounts or it may be due to the lateral currents that move migrating zooplankton (small drifting animals) by the seamounts (Rogers 1994). Regardless of the cause, some seamounts have exhibited:

- Elevated levels of primary productivity—the transferring of light or chemicals into food through photosynthesis or chemosynthesis
- High concentrations of zooplankton
- High abundances of benthic fauna—animals living on the seafloor (Boehlert 2001 and references cited therein)
- Consequent large populations at higher trophic levels—animals higher on the food chain (Figure 8).

These increases are not universal, however. Some seamounts have lower levels of productivity than surrounding areas. Aggregations of spawning and feeding fish such as orange roughy *Hoplosthesus atlanticus,* hoki *Macruronus novaezelandiae,* oreos *(Oreosomatidae),* slender armorhead *Pseudopentaceros wheeleri* and rockfish *Sebastes* spp. are commonly found over seamounts (Probert *et al* 1997, WWF, IUCN & WCPA 2001, Dower & Perry 2001). These aggregations are at times extremely dense. Catch rates of just one species, for example, the pelagic (open-ocean) armorhead over Colohan seamount were 96 metric tons per trawling hour in the early years of this fishery (Humphreys *et al* 1984). As discussed in later sections, these high abundances have led to many commercial fisheries focused on seamounts. In addition to fishes, seabirds (Blaber 1986, Haney *et al* 1995), whales and dolphins (Hui 1985) have also been found to aggregate over seamounts, likely in response to elevated levels of available food.

NOAA/U.S. DEPT. OF COMMERCE

**Figure 8** An octopus on Explorer Ridge in the northeastern Pacific.

**Endemism and species diversity**

One of the most exciting discoveries in deepsea ecology in the last decade or so has been the documentation of highly endemic faunas and many new species on seamounts (Richer de Forges, Koslow & Poore 2000). Endemics are species found at only one restricted location in the ocean, in this case on a single seamount or several seamounts within a single seamount chain. While the presence of endemics on seamounts has been documented for a long time, our understanding of the magnitude is changing rapidly. In 1987, Wilson and Kaufmann estimated that 12–15% of all species found on seamounts were endemic, and considered this a high value. Since then, several major sampling programs have found much higher levels. Koslow *et al* (2001) found rates of about 35% for benthic invertebrates on underwater hills off Tasmania. Richer de Forges, Koslow and Poore (2000) worked south of New Caledonia and recorded rates of 36% on the Norfolk Ridge seamounts and 31% on Lord Howe Island seamounts. And Parin, Mironov and Nesis (1997) recorded rates of 44% for fishes and 52% for invertebrates (animals without backbones) on the Nasca and Sala-y-Gomez seamount chains off Chile. These high rates are not universal: a recent compilation of data from the Hawai'ian and Emperor seamounts found rates of fish endemism of only 5% (Stocks 2003), and a study of the Great Meteor seamount found fish endemism of 9%. However, taken together, the recent findings suggest that Wilson and Kaufmann's estimate of 15% endemism is likely to be significantly low.

In addition, it appears that each individual seamount may support a unique community. Richer de Forges, Koslow and Poore (2000) found that adjacent seamounts in the New Caledonia area shared an average of just 21% of their species, and for seamounts on separate ridges roughly 1,000 kilometers apart this decreased to about 4%. So each of the thousands of unsampled seamounts is a potential source for the discovery of many species new to science.

Recent evidence also suggests that seamount-like features do not have to be large or isolated to support highly endemic faunas. The Tasmanian hills which host 35% endemic species (Koslow *et al* 2001) average only 400 meters in height above the seafloor. And recent sampling on underwater hills around New Zealand (100 to 350 meters in elevation) just off of the large Chatham Rise have found rates of endemism over 15% (Rowden, O'Shea & Clark 2002, Rowden personal communication) and elevated species diversity compared to nearby non-seamount areas (Probert, McKnight & Grove 1997). What is noteworthy about this finding is that some of these smaller seamount-like features had not been found, either by scientists or by fishermen, until very recently when high-resolution bathymetric mapping surveys were completed. Unlike most large offshore seamounts, there is the potential to protect these areas before significant fishing impacts occur.

At present, it is not possible to estimate the total number of new species on seamounts worldwide. We cannot simply multiply the average number of new species per seamount by the estimated number of seamounts (potentially as many as 100,000 worldwide), as we do not understand how overall biodiversity and rates of endemism may vary with depth, surface productivity, latitude or other factors. It has been hypothesized that deeper seamounts may have lower rates of endemism (Wilson & Kaufmann 1987) but they are underrepresented in current seamount sampling. As more deepsea sampling occurs, we may find that some of the newly discovered species thought to be seamount endemics actually have larger ranges. Despite all of these considerations, it is still clear that seamounts represent large pools of undiscovered biodiversity in the ocean.

**Seamount isolation**

An important question for conservation is whether seamounts harbor isolated populations depending only on local reproduction or host parts of larger populations that receive "recruits" from a larger area. This distinction is important for determining whether protected areas on individual seamounts can succeed in ensuring survival of their populations regardless of the levels of commercial fishing in surrounding areas and whether they can provide source populations for fisheries based elsewhere. It is also of scientific interest in understanding the high levels of speciation and endemism found on seamounts.

Evidence that suggests seamounts *are not* isolated includes:

- Genetic studies finding no differences between fish populations on adjacent seamounts or between seamount and nearby non-seamount sites—such as the blue-eye trevalla *Hyperoglyphe antarctica* around Australia (Bolsch, Elliot & Ward 1993) and a boarfish species (Pentacerotidae) on the Hawai'ian chain (Borets 1980)—suggesting there is interbreeding across sites
- Tagging studies finding individual bigeye and yellowfin tuna moving between seamount and non-seamount areas (Itano & Holland 2000, Sibert, Holland & Itano 2000)
- A population model by Boehlert, Wilson and Mizuno (1994) indicating that adult numbers of the pearlsides fish *Maurolicus muelleri* over Southeast Hancock Seamount may not be large enough for a self-sustaining population
- Plankton studies finding no increase in concentration of larval fish concentrations of seamount-associated species over seamounts compared to areas some distance from them (Boehlert 1988)

The evidence indicating a lack of isolation of seamount populations comes primarily from large fishes, however, which are probably the most mobile component of the seamount community. Evidence suggesting seamounts *are* isolated includes:

- A study finding that invertebrates and fish on Cobb seamounts have a higher proportion of species with short or no larval durations and a lower proportion of species with medium to long larval duration (Parker & Tunnicliffe 1994) than non-seamount areas nearby. Because it is thought that species with shorter larval times have shorter dispersal distances, this indicates that seamount species may be adapted for local lifecycle and reproduction.
- A genetic study of the widespread amphipod *Eurythenes gryllus* finding that individuals from the base of Horizon Guyot were genetically similar to deepsea individuals from widely separated sites, but that individuals from the top of Horizon were genetically distinct (Bucklin, Wilson & Smith 1987)
- Studies finding morphological differences between seamount and non-seamount species of fishes (Ehrich 1977, Wilson & Kaufmann 1987, Boehlert, Wilson & Mizuno 1994) and gorgonians (Grasshoff 1972)
- Some studies showing a low number of species shared by adjacent seamounts, such as Richer de Forges, Koslow and Poore (2000) who found that adjacent seamounts in the New Caledonia area shared only an average of 21%

of species in common. It is hard to imagine that there are large enough environmental differences among these seamounts of similar depth, bottom type and other physical characteristics to account for the large differences in species.

- And perhaps the most compelling argument for seamount isolation is the high level of endemism found. It is very difficult to explain how this level of local speciation could occur without genetic isolation.

It appears seamounts must have some degree of isolation, except perhaps in the larger, more mobile fishes. Genetic studies—which provide the most compelling evidence of isolation or lack thereof—have been very limited on seamounts to date. More such studies on a variety of taxonomic groups would be very informative to seamount ecology.

There are several hypotheses to explain why this assumed genetic isolation occurs. The first is simply distance. For species requiring shallow or hard-bottom areas, seamounts represent small, isolated islands of potential habitat surrounded by large areas of sediment-covered deepsea. To colonize a seamount, or to move between seamounts, a species must be able to travel across these deepsea areas. Second, seamounts may support species that have reproductive adaptations to ensure that offspring stay local (Parker & Tunicliffe 1994). Finally, the biogeophysical environment of seamounts creates complex water-movement patterns that influence the retention and dispersal of species in the surrounding waters. The massive structure of seamounts (Figure 9) creates downwelling zones with jets and eddies which may form closed circulation patterns known as stratified Taylor cones (reviewed by Rogers 1994). These features may play a critical role in retaining larvae and promoting self-recruitment—reproduction within a closed community—in seamount communities (Richardson 1980, Dower & Perry 2001), though how widespread and persistent this phenomenon is has been debated (Boehlert & Mundy 1993).

## History of Exploitation

Photosynthesis occurs in the world's ocean to an approximate depth of 100 meters, thus there has been little investigation into deepsea fisheries until recent decades (Ross & Smith 1997). The collapse of many traditional continental shelf fisheries and the advancement of fishing gear technology have combined to encourage fleets to travel to remote parts of the ocean in search of virgin fish

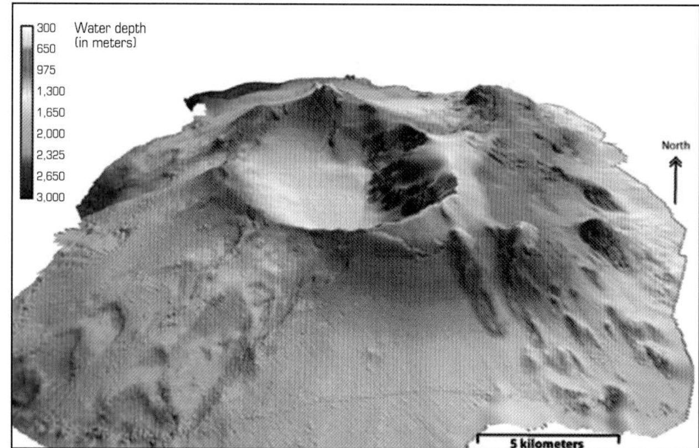


**Figure 9** Bathymetry (water depth) data from *R/V Thompson* shows the shape of West Rota seamount on the Mariana Arc in the tropical western Pacific.

stocks. With this global expansion of fishing effort, distant water fleets discovered productive deep-water habitats including seamounts (Koslow *et al* 2000). Forty percent of the world's trawling grounds are in waters deeper than the continental shelf (Roberts 2002). Dense schools of deep-water species, including orange roughy *H. atlanticus,* hoki *M. novaezelandiae,* oreos (Oreosomatidae), Patagonian toothfish *Dissostichus eleginoides* and armorheads (Pentacerotidae), were found around seamounts. These enormous aggregations of fish were easy to target with acoustic sounders and trawls, and their large size and desirable texture made them highly marketable (Koslow *et al* 2000). The deep-water trawl fleets began exploiting these habitats as early as the 1960s and have since contributed 800,000 to one million tons annually to global marine fish landings. Fishing effort on seamounts increased dramatically through the 1970s and 1980s and currently many populations of orange roughy and hoki are overfished, some to the point of being commercially extinct (Koslow *et al* 2000).

### New Zealand orange roughy fishery

At the time of discovery, orange roughy *H. atlanticus* was a bycatch species—caught unintentionally while seeking a different species—on the Chatham Rise of New Zealand. Little suggested the species could develop into a marketable commercial fishery. Inhabiting depths between 700 and 1,500 meters, orange roughy was found to be a slow-growing species with a maximum age limit that exceeds 100 years and an average reproductive age of 24 (Clark & Tracey 1994). Of specific interest to commercial fishermen was the discovery of large feeding and spawning aggregations in

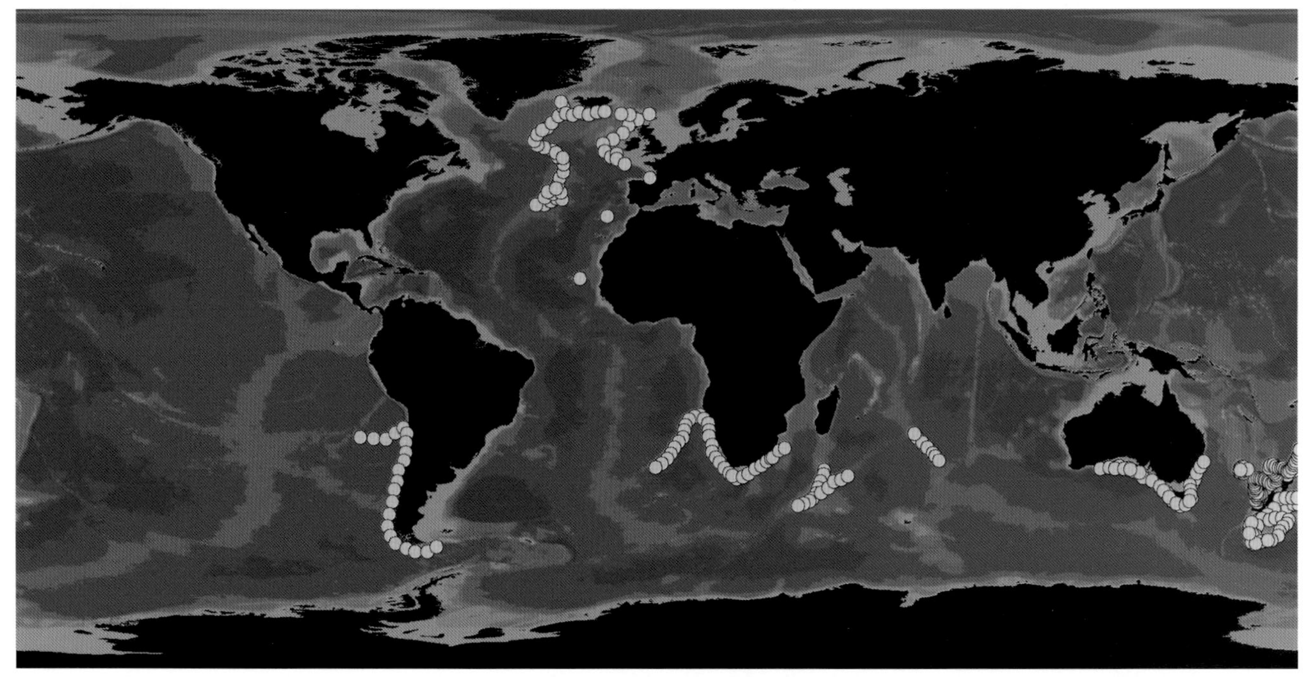

**Figure 10** Global distribution of orange roughy catches (from Kotlyar 1996, Branch 1996, Magnússon & Magnússon 1998, Clark 1999, 2001, Gordon 1999, Tracey & Horn 1999, Bax 2000, Strutt 2000, Anderson 2001, Hareide & Garnes 2001, Hope 2001—as compiled in Branch 2001).

several areas of the world ocean (Figure 10) around deepsea physical structures such as seamounts (Koslow 1997). While feeding aggregations tend to occur, it is during the New Zealand summer months of June through August that the easily predictable and significantly larger spawning aggregations form (Clark & Tracey 1994). Being such a newly exposed species, much of the biology of orange roughy that is necessary for proper management is not fully understood. Lacking historical distribution and abundance information makes understanding the resiliency and stock trends for rebuilding the stocks difficult (Batson 2003).

For a species to thrive in a deepsea environment, it is typically observed to have adaptations such as high percentages of water in the body and the development of condensed tissue and bones. Orange roughy differs from this standard by possessing firm musculature with high protein and lipid fat levels (Koslow 1997). An exploratory Russian trawl fishery originally recognized these features in what were then unrestricted fishing grounds around New Zealand in 1972 (Batson 2003). Orange roughy markets were developed at an opportune time when traditional species were in decline and a white fleshy fish substitute was being sought (Batson 2003). Marketing tactics such as changing the promotional name of the fish from "slimehead" to "orange roughy" assisted in driving the demand for the species (Robertson 1990). Initially fished from mudflats throughout the Chatham Rise, orange roughy was later identified in large aggregations around deepsea seamounts (Probert, McKnight & Grove 1997). Clark and Tracey (1994) found that a nearly complete shift in fishing approach for this species—from the mudflats to seamounts—occurred between 1982 and 1990. Through the continued improvement of fish-stock detection techniques, major aggregations were identified on the Challenger Plateau, Cooke Canyon, Pursegur Bank, the East Cape and around New Zealand (Clark 2001). As an exploratory fishery, catch limits were unmonitored until 1982 when a total allowable catch (TAC) quota management system was established by the New Zealand Ministry of Fisheries (Clark & Tracey 1994). Prior to 1982, the effort existed as a "pulse-fishery," in which trawls would repeatedly sweep an area until it was emptied of the aggregations of spawning or feeding fish (Robertson 1990). In 1983, each of the distinct fishing grounds were given specific TACs for this species to be managed separately (Clark & Tracey 1994). The fishery peaked in 1989 with a level of catch at nearly 60,000 tons for all grounds (Figure 11). But without proper consideration of over-exploitation, catch levels had fallen below 20,000 tons by 1997 (Clark 2001). During the

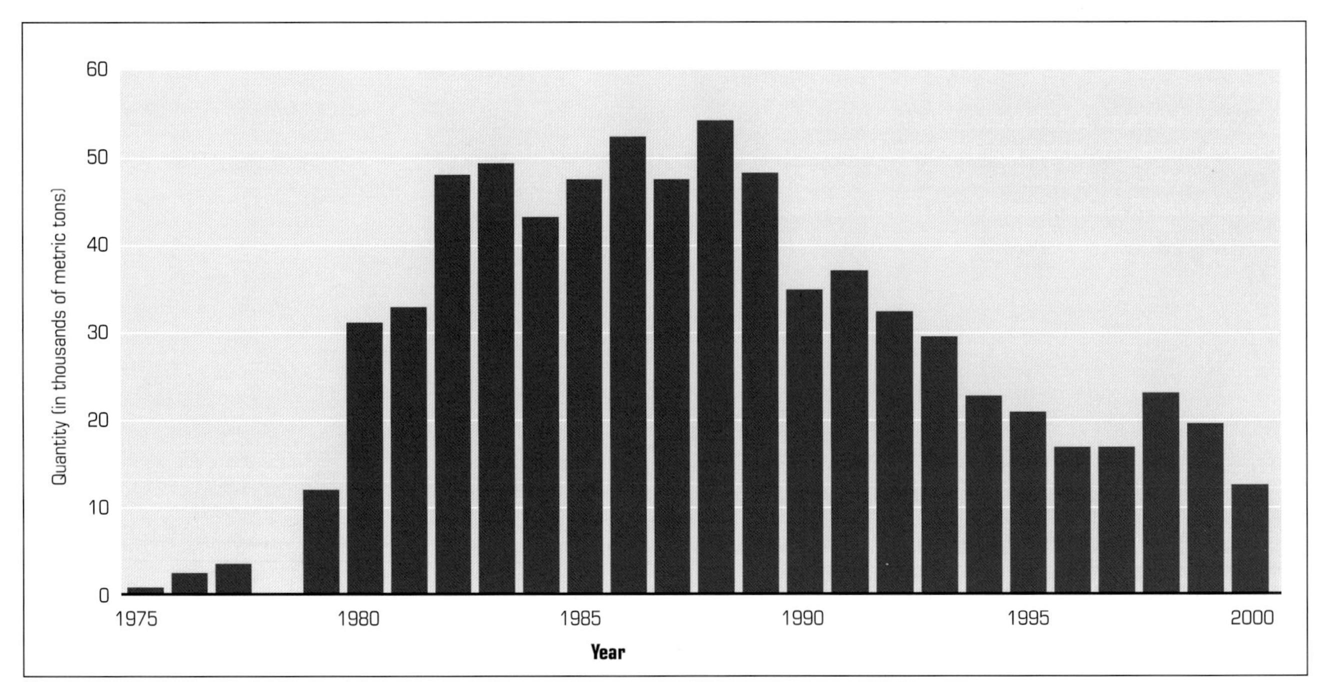


**Figure 11** Catch history of orange roughy in New Zealand waters (New Zealand Ministry of Fisheries 2000).

prosperous years, the orange roughy fishery around New Zealand had an export value of NZ $125 million (Clark 1999). The original virgin (non-fished) biomass was estimated at 300,000 tons and it is believed to have been fished down to one-sixth of that level through the early 1990s (Clark 1999). New Zealand fishery managers now believe that the TAC levels for the orange roughy fishery should not exceed 1–2% of the original virgin biomass.

The method of fishing seamounts for large aggregations of orange roughy is very distinct and specialized. Due to the difficult fishing conditions caused by invertebrate growth like corals and the rocky terrain on the sides and tops of seamounts, specialized equipment, called "otter trawls" designed for bottom-trawl fishing is designed with heavy, durable materials. The trawls (Figure 12) average 60 meters in length and have a footrope arrangement including heavy steel chains to ensure contact with the seafloor and destroy potential seabed snags (Probert, McKnight & Grove 1997). The complex assemblages of benthic (bottom-dwelling) invertebrates—including delicate coral species—are plowed flat by the trawls and many of the fished seamounts have been stripped bare by bottom-trawl fishing.

In a recent study, there was 95% bare rock on seamounts that had been fished compared with only 10% on unfished ones; the unfished seamounts also had double the benthic biomass and 46% more species than the fished areas (Koslow *et al* 2001). A comparison of seamount benthic macrofauna was also conducted between lightly trawled seamounts and heavily trawled seamounts. The impact was found to be significant, with coral communities being almost completely removed from the heavily trawled seamounts. The substrate that did remain was comprised of bare rock, coral rubble and invasive species. A marked reduction of species richness and abundance was also noted on the heavily fished seamounts. These results suggest that during the fishery's infancy, invertebrate bycatch was abundant, but due to the fragile nature of the invertebrates and the high

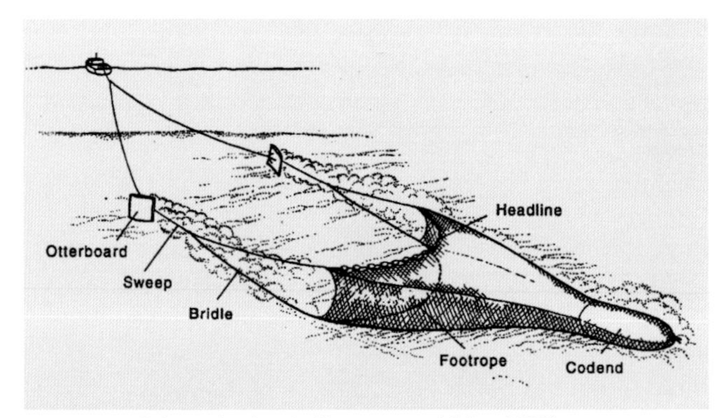


**Figure 12** Schematic of typical bottom trawl (King 1995).

density of trawling, the ecosystems were quickly demolished (Koslow *et al* 2001). Another significant consequence of this removal of habitat-forming organisms like corals is the establishment of scavenging and opportunistic species in the environment (Batson 2003).

**Australian and Namibian orange roughy fisheries**

Orange roughy was also discovered in Australian waters in 1972, but commercial harvesting did not occur until 1982 off Tasmania (Lyle 1994). In its infancy this fishery was harvesting less than 200 tons per year, peaking in 1990 with 41,000 tons and plummeting in 1995 to less than 7,000 harvested tons (Ross & Smith 1997).

A commercial orange roughy fishery was explored off Namibia in 1994, resulting in the discovery of four distinct spawning aggregations, but the commercial fishery did not develop until 1997, with no more than five vessels involved. TAC levels were established with the intention of fishing the orange roughy spawning biomass down to 50% of its virgin level. A theoretically sufficient management plan was developed and implemented by the Namibian government, but without proper understanding of the species population dynamics, the stock was in decline and over-exploited within six years. It is believed the existing stock of Namibian orange roughy constitutes only about 25% of the original biomass. The Namibian orange roughy fishery TAC was highest in 1998 at 12,000 tons but was set at only 1,875 tons in 2000 (Branch 2001).

**Current management**

The importance and vulnerability of seamounts has been recognized only gradually over the past 40 years, and little research has been done to assess the effects of trawling and other extractive activities on these delicate habitats. But we now know enough to realize the importance of protecting these special habitats (Richer de Forges, Koslow & Poore 2000). The United States has designated small marine protected areas (MPAs) for two seamounts off Alaska (Roberts 2002). New Zealand and Australia are on the forefront of seamount ecosystem conservation, but only a fraction of seamounts in their Exclusive Economic Zones (EEZs) are fully protected. In New Zealand, nineteen seamounts are closed to all fishing activities, but this only represents 2% of New Zealand's seamount habitats (New Zealand Ministry of Fisheries 2002). Australia established the Tasmanian Seamounts Reserve that protects 20% of the total (fifteen of the 70), excluding bottom-trawling although still allowing tuna longlining in the upper 500 meters of the water column (CSIRO 2002b). Countries that have seamounts with deep-sea fisheries within their EEZs have attempted to assess the total species biomass to set an appropriate total allowable catch (TAC) for their waters. But due to both the recent discovery of these fish populations and inadequate assessment techniques for deep-water fish stocks, accurate biomass estimates have been difficult to obtain (Branch 2001). Because of the uncertainties associated with the ability of slow-maturing deepsea species to recover from fishing pressure, even the most accurate stock assessment may not be adequate to prevent over-exploitation of the stocks, as has occurred with orange roughy and hoki fisheries over the last twenty years.

**Conservation issues**

The development and over-exploitation of the orange roughy fishery should serve as a forewarning for the commercial exploitation of other deepsea species. Other existing deepsea fisheries on seamounts include oreos, slender armorheads *P. wheeleri* and roundnose grenadier *Coryphaenoides rupestris* (Figure 13). Over 70 species of marketable fish, shellfish and corals are found in seamount habitats (WWF/IUCN/WCPA 2001) and more than a dozen (Table 1) are currently targeted by commercial fishing efforts. Major bottom-trawl fisheries exist throughout the world's ocean ranging from alfonsino fish (Berycidae) and orange roughy *H. atlanticas* in Australia and New Zealand to king and snow crab and rock lobster in the Gulf of Alaska. Also, the shallow seamounts in the Hawai'ian Islands and Emperor Seamount chain are targeted for their highly valuable corals (DFO 2002).

In some parts of the world the seamount trawl fisheries also have negative impacts on marine mammals. In the Australian fishery there is an "incidental" catch (bycatch) of Australian and New Zealand fur seals. In the areas of high fishing effort, the Australian fur seal suffered a 1.9% increase in mortality in comparison to on-land populations. According to population models for similar species, a 1% increase in mortality will cause a population to decline (Shaughnessy 1999).

There is also a potential for extraction of minerals found on seamounts—ferromanganese oxide encrustations and polymetallic sulphide deposits on seamounts contain many highly prized metals (principally cobalt and manganese) now

**Figure 13** *Top to bottom:* Common seamount fishery species include orange roughy, hoki (blue grenadier), deepsea trevally (Antarctic butterfish) and Redstripe rockfish.

mined on land but expected to be mined from the seafloor in the future. Ferromanganese crusts are highly valuable and found in Pacific and Mediterranean seamount groups (WWF/IUCN/WCPA 2001) with cobalt being the most valuable metal found in these deposits. With anticipated advances in technology, the ability to mine seamounts for valuable materials may become a major threat to the health of these unique ecosystems in the future (WWF/IUCN/WCPA 2001).

## International Law Relevant to Seamount Conservation

Some seamounts have become the target of commercial fishing, and a number of other human activities may be directed at seamounts in the future, including recreational fishing, increased scientific research, minerals exploitation and bio-prospecting. In recent years, commercial fishing at selected seamounts has attracted the most attention—from proponents and opponents alike—and this unregulated fishing is believed to create adverse impacts on targeted fish stocks and on the ecosystems of seamounts.

Table 1
**Primary species targeted for commercial exploitation on and over seamounts**

| Seamount locations | Species targeted |
|---|---|
| New Zealand | Orange Roughy, Hoki, Gemfish, Paua, Shellfish |
| North Atlantic (Georges Bank, Iceland Basin, Norwegian Sea and Greenland Sea) | Shellfish, Darwin's Roughy, Roundnose Grenadier, Roughhead Grenadier |
| Australia/Tasmania (Tasman Sea) | Orange Roughy, Alfonsino, Deep Sea Oreo |
| Gulf of Alaska | King Crab, Snow Crab, Rock Lobster |
| Hawai'ian Islands | Corals, Slender Armorhead, Alfonsino, Onagu, Ehu (Squirrelfish Snapper) |
| Azores | Orange Roughy |

**Source:** DFO 2002, Hébert 2002, CSIRO Marine Research 2002a, Moore & Vechione 2002, WWF 2002.

Environmental activists have begun calling for the conservation of seamount biological communities, in particular through the establishment of marine protected areas both within and beyond national jurisdictions (Gianni 2002, de Fontaubert 2001) and a moratorium on bottom-trawl fishing on the high seas (Gianni 2000, DSCC 2004). A few coastal nations have developed conservation and management measures for the fishery resources on seamounts located within their ocean jurisdictions, generally out to 200 nautical miles.

There is no international law pertaining specifically to the conservation of seamount ecosystems. The international law of the sea comprises a broad legal framework governing a multitude of uses of the ocean (Churchill & Lowe 1999), many of which may occur on or over seamounts. The United Nations Convention on the Law of the Sea (UNCLOS adopted in 1982) establishes jurisdictions that define the current legal status of seamounts and it must serve as the context for any discussion of prospective seamount conservation (Gjerde 2001). We focus mainly on the activity of commercial fishing, as that use is seen as the most threatening to seamount ecosystems. A related body of legal and policy analysis is emerging with respect to the conservation of hydrothermal vent systems (Allen 2001, Glowka 2000). Vent systems are subject to different threats, primarily intrusive scientific sampling and bioprospecting, but these activities are clearly high seas freedoms in ocean areas beyond national jurisdiction given current international law. Consequently, we can look to this activity for lessons in understanding the international legal framework as it relates to uses of and protections for seamounts.

The relevant international law varies by geographic area. We can distinguish broadly between the high seas where freedoms to use the ocean and its resources for various purposes persist and areas where the access to ocean resources is subject to the sovereign jurisdiction of coastal nations. Further, international fisheries regulation varies according to the characteristics of the fish, including whether they are sedentary or migratory. For example, international law draws a distinction between those living resources of the legal continental shelf that are sedentary—in constant contact with the seabed at their harvestable stage—and those that are not.

We organize our discussion of relevant international law with respect to seamount protection according to international legal zonation. All relevant ocean zones are delimited in reference to a "baseline," which is either the coastal low-water line marked on official charts, a line enclosing deep embayments less than 24 miles across, or a line enclosing the area among islands in an archipelago. Other ocean law zones that apply are the territorial seas, the Exclusive Economic Zones and the continental shelf (Figure 14).

A number of international agreements apply, including the U.N. Convention on the Law of the Sea (UNCLOS), the Straddling and Highly Migratory Fish Stocks Agreement (FSA), the Convention on Biological Diversity (CBD) and other sources of international law. The ocean law zones are discussed below.

### Territorial sea

Nations may establish a territorial sea (Zone A in Figure 14) out to twelve nautical miles (nm) from their baselines. A coastal nation's sovereignty—a nation's right to control and regulate its own affairs, those of its nationals and those of any foreign visitors, independent of any foreign influence—extends to its territorial sea, and all of its relevant laws and policies apply in the territorial sea to the same extent as if it were a part of its land territory.

Any fishing activity in a coastal nation's territorial sea, whether conducted by domestic or foreign vessels, is subject to the laws of that nation. Foreign-flagged ships are allowed "innocent passage" through a nation's territorial sea, but fishing is not considered to be innocent passage. A coastal nation has no obligation to share the fishery resources of its territorial sea with any other nation, although it may adopt fishing rules for foreign-flag vessels otherwise engaged in innocent passage. It is not known whether seamounts are situated within the territorial sea of any nation, but this appears to be a distinct possibility for volcanic island systems, especially those in the Pacific Ocean. Thus, the regulation of commercial fishing on any seamount located in a coastal nation's territorial sea, except possibly for migratory species discussed below, is wholly a matter of national law and policy.

### Exclusive Economic Zone

Under UNCLOS, a nation may establish an Exclusive Economic Zone (EEZ) in the ocean waters, seabed and subsoil off its shores (Zones B & C in Figure 14) out to a distance of 200 nautical miles from its baseline. An EEZ is

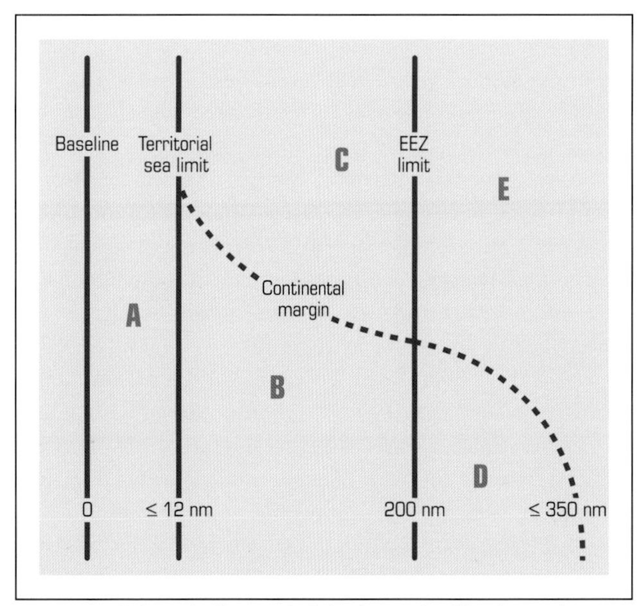


**Figure 14** Schematic diagram depicting international legal zones in mapview (looking down from above); not to scale.

adjacent to and extends beyond a nation's territorial sea, typically from twelve to 200nm offshore. Seamounts are known to exist within the EEZs of several nations, including the U.S., New Zealand and Australia, and commercial fisheries have been developed on some of these seamounts. In its EEZ, a coastal nation has sovereign rights to explore, exploit, conserve and manage all natural resources, including living marine resources.

These sovereign rights come with a set of important responsibilities, however. Among other things, the coastal nation must promote the optimal utilization of its EEZ fisheries by deciding upon levels of allowable catch and by implementing management measures to maintain stocks at levels that can produce a maximum sustainable yield (MSY). In particular, these measures must ensure that stocks are not endangered by over-exploitation, and they must lead to the restoration of stocks that are at biomass levels below MSY. These measures also must consider the effects on other species that associate with or depend upon the "target stocks." The biomass of the stocks must be maintained at levels sufficiently high that their reproduction rates are not seriously threatened.

There are several qualifications to a nation's responsibility for optimal utilization within its EEZ. First, a coastal nation may consider any mitigating environmental and economic factors when establishing management measures. Such factors might relate to the economic needs of coastal fishing communities, for example, and they could lead to exploiting biomass below MSY levels. Further, where a coastal nation does not have the capacity to take all of its allowable catch, it has an obligation to enter into agreements to share any surplus with other nations. In practice, insufficient fishing capacity has rarely been an issue for most developed nations, and most nations are understandably reluctant to share surpluses, but some developing nations have entered into arrangements with foreign fishing fleets to allow access to fish stocks in their EEZs.

About three-quarters of the world's coastal nations have formally declared their EEZs, representing about 84% of the 107 million square kilometers of ocean space that can potentially be zoned in this way. Where coastal states fail to make such declarations (often because of geographic crowding and the need to negotiate boundaries with neighbors), the areas in question remain subject to the largely unregulated fisheries regime of the high seas.

**Continental shelf**

A coastal nation has a continental shelf extending beyond its territorial sea to the outer edge of its continental margin (Zones B, C and potentially the seabed and subsoil of D, in Figure 14). The continental shelf comprises the seabed and subsoil of the continental margin, and a coastal nation has sovereign rights to explore and exploit the natural resources on the seabed and in the subsoil found there. Although these rights do not depend upon any proclamation of sovereignty (as they do in the case of an EEZ), a coastal nation must survey and legally establish the outer limit of its continental shelf.

The legal extent of the continental margin must be determined according to rules articulated in UNCLOS. Under these rules, the outer limit of the legal continental shelf is at least 200 nautical miles and may be as much as, but not more than, 350nm from the baseline. Thus the continental shelf is at least co-extensive with the EEZ from the outer limit of the territorial sea to the 200nm limit, but where the continental margin exceeds 200nm, the continental shelf's outer limit may extend up to 350nm from the baseline, based on seafloor topography and sediment geology.

Natural resources of a continental shelf may include "sedentary species"—organisms that are either immobile or move only in constant contact with the seabed or subsoil (Figure 15). There are no provisions comparable to those that apply in the EEZ for optimal utilization, conservation or management of living resources of the continental shelf. Although it has not been established by custom or articu-

OAR/NATIONAL UNDERSEA RESEARCH PROGRAM (NURP)

**Figure 15** Small sea anemone on seamount off Hawai'i.

lated in a treaty, an implied duty to conserve the sedentary species of the continental shelf may exist.

Seamounts are known to exist as natural geologic features of the continental margin. A coastal State's fishery management laws and policies apply to the living marine resources on seamounts both within 200nm and beyond to the outer limit of its legal continental shelf. If a coastal nation does not fully exploit the sedentary living resources of its continental shelf, they may not be exploited by anyone else without the nation's express consent. For seamounts on the continental shelf beyond 200nm, the conservation and management measures of coastal nations apply only to sedentary resources, not to living marine resources in the water column over the seamounts.

### The Area

The seabed and subsoil located seaward of claimed continental shelves is referred to in international law as "the Area" (the seabed and subsoil of Zone E in Figure 14). By definition, it is beyond the limits of national jurisdiction, and the Area and its mineral resources are considered to be the "common heritage of mankind." An International Seabed Authority (ISA), established under UNCLOS and headquartered in Kingston, Jamaica, acts on behalf of mankind to manage the exploitation of the mineral resources of the Area.

A majority of the world's seamounts are located in the Area and many are potential locations of mineral resources. For example, ferromanganese crusts exist on the tops and flanks of some seamounts, and these mineral occurrences have been investigated as a potential source of both nickel and cobalt. Marine polymetallic sulfide deposits, with comparatively high concentrations of zinc and copper, occur as a result of hydrothermal vents or seeps on other seamounts. But the exploitation of minerals from seamounts is extremely speculative, and commercial mineral development is not expected in the foreseeable future. Despite the distant prospect of mineral development in the Area, the ISA has already promulgated a mining code that governs access to and recovery of polymetallic nodules from the deep seabed. It is now developing additional regulations for minerals that occur on seamounts. Importantly, the ISA has no authority for conserving or managing the living resources of the Area or its seamounts, except by promulgating regulations to protect the environment in the course of mineral development.

### High seas

The high seas are that part of the ocean waters beyond any national jurisdiction (Zone E in Figure 14). Most of the Earth's seamounts are thought to be distributed in areas beneath the high seas. The high seas are open for use by all nations, meaning that no nation may claim sovereignty or jurisdiction over resources there. Nations are free to engage in a wide number of uses on the high seas. These high seas freedoms are not wholly unrestricted however; some are subject to international legal obligations.

There is no comprehensive list of high seas freedoms, although international law specifically recognizes fishing, navigation, scientific research, the laying of submarine cables and pipelines and the construction of artificial islands and other installations as being among these freedoms. All nations have the right to exercise high seas freedoms. Nevertheless, due regard must be given to the interests of other nations wherever the latter also wish to exercise high seas freedoms.

Under provisions of UNCLOS, the citizens of all nations have the right to engage in fishing on the high seas. This right is conditioned in several ways, including that it is subject to the rights, duties and interests of coastal nations in conserving migratory fish and marine mammals. Further, all nations have a duty to regulate their own nationals and to work with other States to conserve high seas living resources. A number of regional fishing organizations have been established to study, conserve and manage migratory and high seas fish stocks, and cooperation through such regional institutions is encouraged by UNCLOS.

The fishing practices of nations on the high seas are also addressed in nonbinding international "soft law," including three U.N. General Assembly resolutions calling for a moratorium on large-scale driftnet fishing on the high seas. This moratorium was driven by concerns about the wasteful bycatch of non-target species, and it is reflective of a "precautionary approach" to fishery management, as the evidence demonstrating the wasteful effects of driftnetting is incomplete. The driftnet moratorium is not legally binding on any nation and does not have universal compliance, but the major high seas driftnet nations have curtailed the practice in the last decade.

Participants at the Earth Summit in Rio de Janeiro in 1992 identified a wide range of problems associated with high seas fishing, including inadequate international cooperation in conservation and insufficient protection of habitats and ecologically sensitive areas. The Summit's Agenda 21 (1992) articulates a nonbinding statement of goals for conserving living resources on both the high seas and within EEZs. In particular, it calls for integrated management across all levels of government and the sustainable development of resources using precautionary and anticipatory approaches. Agenda 21 laid the groundwork for international negotiations resulting in an agreement on migratory fish stocks as well as a set of nonbinding norms for responsible fishing.

The United Nations Food and Agriculture Organization produced a nonbinding Code of Conduct for Responsible Fishing (FAO Code) in 1995, which details principles and standards of behavior for the conservation, management and development of fishery resources. The FAO Code provides a blueprint for the sustainable use of fishery resources, but it has been criticized for having no prioritization among its numerous agenda items (Hey 1996). One binding element of the FAO Code is an agreement to prevent vessel "reflagging"—a practice of registering vessels in a country that is not party to relevant international agreements, such as those that establish high seas fishing rules. This agreement has only 29 signatories to date, however, and it entered into force only in April 2003. Among its objectives that focus mainly on sustainable use, the FAO Code calls for the protection of living aquatic resources and their environments. The FAO Code is an example of soft law, but its terms were reinforced in 1999 through a declaration of implementation signed by the Fishery Ministers of 126 nations.

### Migratory fish stocks

For stocks of fish that migrate between the territorial seas or EEZs of adjacent nations or from these zones to the high seas, the question of sovereign rights and responsibilities for use, conservation and management becomes less clear. International law has developed to deal with several types of migratory fish. This issue is likely to be irrelevant for the living resources of the continental shelves, as sedentary stocks—those spending most of their lives in contact with the seafloor—do not migrate long distances at their harvestable stage.

Coastal nations have the lead responsibility for management of both anadromous and catadromous fish—those migrating upriver to spawn or out to sea for spawning, respectively—including the setting of allowable catch levels and the responsibility for management. Exploitation of either type should take place in the coastal nation's EEZ (although special circumstances may operate in the case of anadromous fish such as salmon that have traditionally been exploited by distant water States). It is unknown whether anadromous or catadromous fish are directly associated with seamount ecosystems, but it seems unlikely.

UNCLOS and a related implementing agreement adopted in 1995 clarify the law relating to fish stocks that "straddle" a boundary between two national jurisdictions or between a coastal nation's EEZ and the high seas, and those that are highly migratory—moving through multiple EEZs and the high seas. Some migratory stocks, such as some species of tuna, are known to be transient visitors to seamounts. Thus the relevant UNCLOS provisions and the Straddling and Highly Migratory Fish Stocks Agreement (FSA) of 1995 may

Table 2
**International conventions and regional fishery bodies that directly establish fishery management measures**

| | |
|---|---|
| CCAMLR | Convention on the Conservation of Antarctic Marine Living Resources |
| CCSBT | Commission for the Conservation of Southern Bluefin Tuna |
| GFCM | General Fisheries Commission for the Mediterranean |
| IATTC | Inter-American Tropical Tuna Commission |
| IBSFC | International Baltic Sea Fishery Commission |
| ICCAT | International Commission for the Conservation of Atlantic Tunas |
| IOTC | Indian Ocean Tuna Commission |
| IPHC | International Pacific Halibut Commission |
| IWC | International Whaling Commission |
| NAFO | Northwest Atlantic Fisheries Organization |
| NASCO | North Atlantic Salmon Conservation Organization |
| NEAFC | North-East Atlantic Fisheries Commission |
| NPAFC | North Pacific Anadromous Fish Commission |
| PSC | Pacific Salmon Commission |
| SEAFO | Southeast Atlantic Fisheries Organization |
| SWIOFC | Southwest Indian Ocean Fisheries Commission |
| WCPFC | Convention for the Conservation and Management of Highly Migratory Fish Stocks in the Western and Central Pacific |

Table 3

**Multispecies regional fisheries organizations with potential relevance to seamounts**

| Establishment, scope and objectives | Membership | Species covered and management measures | General comments/ relevance to seamounts |
|---|---|---|---|
| **CCAMLR Convention on the Conservation of Antarctic Marine Living Resources** | | | |
| Part of the Antarctic Treaty System, CCAMLR was adopted in 1980 in response to concerns over pulse fishing of krill and various species of finfish. Convention applies to all Antarctic marine living resources other than (1) seals south of 60°S and (2) all whales (where other regional organizations and agreements share competence). Northern boundary of geographic area ranges between 45°S and 60°S. | The EU and 30 nations: Argentina, Australia, Belgium, Brazil, Bulgaria, Canada, Chile, Finland, France, Germany, Greece, India, Italy, Japan, Namibia, Netherlands, New Zealand, Norway, Peru, Poland, Russia, S. Africa, S. Korea, Spain, Sweden, Ukraine, U.K., U.S., Uruguay, Vanuatu. | Management measures are applied in the fisheries for Patagonian toothfish, icefishes, electrona, krill, crab and squid. The most widely applied measures are those concerning minimization of bycatch; reporting of catch, effort and biological data; and measures to promote compliance. In addition, various fisheries (including toothfishes) are subject to seasonal closures, catch and effort limits, and special limits and reporting requirements for exploratory fisheries. | Relevant in that the Patagonian toothfish is found on seamounts and continental shelves around most sub-Antarctic islands within the Convention Area. CCAMLR has been largely ineffective in curbing IUU fishing for Patagonian toothfish by non-parties but the recently established catch documentation scheme is expected to improve information and enforcement. |
| **GFCM General Fisheries Commission for the Mediterranean** | | | |
| Established as a regional arm of the FAO under a 1949 agreement that entered into force 1952. Area of application includes Mediterranean Sea and "contiguous waters" (i.e., the Black Sea). Objectives are to promote the development, conservation and management of living marine resources; to formulate and recommend conservation measures; and to encourage training and cooperative projects. | The EU and 23 nations: Albania, Algeria, Bulgaria, Croatia, Cyprus, Egypt, France, Greece, Israel, Italy, Japan, Lebanon, Libya, Malta, Monaco, Morocco, Romania, Slovenia, Serbia and Montenegro, Spain, Syria, Tunisia, Turkey. | Commercial catches consist of about 115 species—mostly demersals, shellfish and small pelagics, also bluefin tuna and swordfish. Roughly 90% of fishing effort is coastal, but as few coastal States have declared fishery zones or EEZs, most of the Med remains high seas and is fished by coastal States and distant water fleets alike. GFCM measures generally follow EU policy, except that catch limits and quotas are not applicable for species other than recent ones set for tuna. | Relevant in that western Med has several seamounts in international waters, is a major spawning area for bluefin tuna and is known for especially heavy overfishing of tuna. Despite a moratorium on driftnet fishing for tuna, several countries continue the practice in the area. More generally, GFCM has been criticized as following a "non-adaptive management style": Measures in place are essentially permanent and affirm the longstanding practice of the main fishing States, rather than being based on scientific assessment of stock/ecosystem conditions (Lleonart, Sanat & Franquesa 2003). The recent establishment of a Scientific Advisory Committee suggests that GFCM may be responding to such criticisms. |

govern. FSA entered into force in December 2001, but only 52 nations have ratified the agreement to date.

FSA calls for cooperation among coastal nations and with international organizations to manage straddling or highly migratory fish stocks in the relevant "region," which may include one or more EEZs as well as the high seas. Importantly, FSA states that nations can only gain access to a fishery either by joining a relevant regional fishery organization or by adhering to its management measures. The agreement further states that even a nation that is not a member of a regional body is obliged to cooperate in conserving and managing fish stocks. One interpretation of this obligation is that the regime of high seas freedoms is being replaced—at least where fisheries and fish conservation are concerned—by a regime of international collective regulation (Bratspies 2001).

Table 3 (continued)

**Multispecies regional fisheries organizations with potential relevance to seamounts**

| Establishment, scope and objectives | Membership | Species covered and management measures | General comments/ relevance to seamounts |
|---|---|---|---|
| **NAFO Northwest Atlantic Fisheries Organization** | | | |
| In 1979, succeeded the International Commission of the Northwest Atlantic Fisheries (founded in 1949). NAFO's primary objective is to contribute through consultation and cooperation to the optimal utilization, rational management and conservation of the fishery resources of the Convention Area, which is bordered by U.S., Canada, St. Pierre and Miquelon, and Greenland. | The EU and 16 nations: Bulgaria, Canada, Cuba, Denmark, Estonia, France, Iceland, Japan, Korea, Latvia, Lithuania, Norway, Poland, Russia, Ukraine, U.S. | Covers most fishery resources of the NW Atlantic, which include over 40 species of groundfish, over 20 pelagic and over 50 other finfish species. Fishery resources not covered are (1) resources managed by other regional bodies (salmon, tunas and marlins, whales) and (2) sedentary shelf species (which are subject to national jurisdiction). Measures include catch limits for different fish species and NAFO areas, regulations on fishing gear and vessel equipment, a notification scheme, a vessel monitoring system, inspection at sea, inspection of landings in all NAFO member ports, an observer scheme and a non-Contracting Parties compliance scheme. | Clear relevance to several seamount species— roundnose and roughhead grenadiers—that aggregate at seamounts beyond national jurisdiction. |
| **NEAFC North-East Atlantic Fisheries Commission** | | | |
| Established under a 1959 Convention that entered into force in 1964. (Convention accorded stronger regulatory powers to an existing commission to address problems of overfishing.) NEAFC regulatory decisions cover various species in three large high-seas areas within a Convention Area bounded by Greenland, southwestern Spain and the western tip of Novaya Zemlya. | The EU and 5 nations: Denmark, Iceland, Norway, Poland, Russia. | Catch and effort limits apply to redfish, mackerel and 22 deepsea species that were subject for the first time in 2003 to a ceiling at previous years' levels, including alfonsinos, orange roughy and roundnose grenadier. (For a number of species, relevant scientific committees of the International Council for the Exploration of the Sea [ICES] have strongly urged much more aggressive measures than the ceiling recently established.) Gear restrictions apply to capelin and blue whiting; rockall haddock is subject to area closures. | Clear relevance to several species that aggregate at seamounts beyond national jurisdiction. |

There are 44 regional fishery organizations, but 27 provide only scientific information or management advice and do not exercise regulatory authority. There are 17 regional organizations (Table 2) with regulatory authority to establish management measures for straddling or highly migratory species that move partly within their member nations' waters, and for other high seas stocks that are fished by their nationals.

Most regional fishery organizations deal with single species or narrow species groups (mainly tunas), but six existing and one proposed organization have broad substantive authority that could deal with ecosystem-wide issues concerning seamounts in areas beyond national jurisdiction (Table 3).

We are unaware of measures instituted by any of these organizations that focus on the conservation of seamount communities *per se*, although such measures might be covered under a broader management plan. A more serious issue is that most of these regional bodies operate by consensus and have little means for compelling a member to adhere to proposed management measures. Consequently, in their existing state, Regional Fishery Management

Table 3 (continued)

**Multispecies regional fisheries organizations with potential relevance to seamounts**

| Establishment, scope and objectives | Membership | Species covered and management measures | General comments/ relevance to seamounts |
|---|---|---|---|
| **SEAFO Southeast Atlantic Fisheries Organization** | | | |
| Organization established by a treaty adopted in 2001. Not yet in force. Convention Area is defined as waters of the SE Atlantic (roughly, from 6°S and 10°W to 50°S and 30°E) that are beyond national jurisdiction. Objective is to ensure long-term conservation and sustainable use of fishery resources, including straddling fish stocks, in Convention Area. | Signed by the EU and 8 nations, but no ratifications to date. Signatories are the EC, Angola, Iceland, Namibia, Norway, S. Africa, S. Korea, U.K., U.S. | Convention applies to *all fishery resources,* including fish, molluscs, crustaceans and other sedentary species, *except for* (1) sedentary species subject to the fishery jurisdiction of coastal States (LOS Article 77) and (2) highly migratory species (LOS Annex I) already subject to regulation by other regional fisheries bodies. Convention contains measures pertaining to other living marine resources, defined to mean all living components of marine ecosystems, including seabirds.<br>Regulatory measures may include restrictions on: the quantity of any species caught, areas and periods in which fishing may occur, size and sex of any species taken, fishing gears and technologies used, level of fishing effort, including vessel numbers, types and sizes. May also include the designation of regions and sub-regions; other measures regulating fisheries with the objective of protecting any species; other measures the Commission considers necessary to meet Convention objective. | Clear potential relevance to seamount-aggregating and other deepwater species, including alfonsino, orange roughy, armorhead, wreckfish, deep-water hake and red crab. |
| **SWIOFC Southwest Indian Ocean Fisheries Commission** | | | |
| Not yet established. A draft agreement indicates the purpose shall be to promote sustainable development, conservation, rational management and best utilization of fisheries resources in all waters of the Southwest Indian Ocean, with special emphasis on non-tuna fisheries. Three substantial issues need to be resolved: accommodating all interests in an agreement on high seas fisheries and straddling stocks; a framework for cooperation in sustainable development of fisheries under the jurisdiction of coastal developing states; and the role of the Fisheries & Agriculture Organization. | Participants in preparatory consultations include the EC and 8 nations: Comoros, France, Kenya, Madagascar, Mauritius, Mozambique, Seychelles, Tanzania. | Not yet formally specified, but most important commercial fisheries are those targeting tunas and shrimp. Both fisheries have developed rapidly in recent years. | Most obvious relevance would be for tunas, which are already under the jurisdiction of IOTC. Potentially relevant for shrimp. |

Table 3 (continued)

**Multispecies regional fisheries organizations with potential relevance to seamounts**

| Establishment, scope and objectives | Membership | Species covered and management measures | General comments/ relevance to seamounts |
|---|---|---|---|
| **WCPFC Convention for the Conservation and Management of Highly Migratory Fish Stocks in the Western and Central Pacific** | | | |
| Convention opened for signature in September 2000, entered into force June 2004 and is presently administered by an Interim Commission. Open to all coastal States in the Convention Area (defined in terms of geographic coordinates) and to other States and economic integration organizations whose nationals fish therein. Objective is to ensure, through effective management, the long-term conservation and sustainable use of highly migratory fish stocks in the western and central Pacific Ocean in accordance with LOS and the 1995 U.N. Fish Stocks Agreement. | 14 nations: Australia, Cook Islands, Fiji, Kiribati, Marshall Islands, Micronesia, Nauru, New Zealand, Niue, Papua New Guinea, Samoa, Solomon Islands, Tonga, Tuvalu. | Covers all species of highly migratory fish stocks (defined as all those listed in Annex I of LOS and such others as the Commission may determine), *except sauries.* (The U.S. and Japan are the only states in whose waters Pacific sauries are found.) Conservation and management measures are to be applied throughout the range of the stocks, or to specific areas within the Convention Area, as determined by the Commission. | Potentially of great relevance, as it applies to a large number of tunas and tuna-like species, at least some of which aggregate at the numerous seamounts beyond national jurisdiction in the Convention Area. |

Organizations (RFMOs) often fall short of the goals set in the FSA.

### Biological diversity

One outcome of the 1992 Earth Summit in Rio was the signing of an international agreement to conserve biological diversity and promote the sustainable use of its components. In the Convention on Biological Diversity (CBD), biodiversity is recognized as a "common concern of humankind," reflecting the interest of developing countries in maintaining their rights to biological resources that occur within their jurisdictions, for the development of pharmaceuticals for example. It is thus a more dilute concept than the "common heritage of mankind" found in UNCLOS. The CBD emphasizes the notions of sustainable development, integrated management and the precautionary approach, and it attempts to promote international scientific and technical cooperation, fair and equitable access to genetic resources and biotechnology transfers.

Sustainable development of a nation's biological resources—where sustainability is interpreted as promoting use across current and future generations—is the main emphasis of the CBD. Where biodiversity is threatened or endangered, the parties have a duty to conserve it as much as possible and appropriate. In the absence of such threats, however, the CBD recognizes the sovereign rights of nations to develop and exploit the biological resources within their own jurisdiction (territorial sea, EEZ and continental shelf) subject to their own conservation laws and policies. The CBD also applies to the activities of nationals of a signatory party on the high seas. Thus, to the extent that contracting parties to the CBD also participate in regional fishery bodies, these organizations arguably must conserve and promote the sustainable use of biological resources in their implementation of fishery management measures.

According to language in the CBD, the responsibilities incorporated in it must be consistent with the rights and obligations of nations under UNCLOS, suggesting that the provisions of the latter may preempt the CBD where they may conflict. The Jakarta Mandate on Marine and Coastal Diversity, adopted by the CBD Conference of Parties in 1995, provides guidance to contracting parties on ways to conserve biological diversity in coastal and marine areas, including the establishment of marine protected areas. There has been little practical progress toward achieving the goal of the Jakarta Mandate.

**Legal discussion**

The international law pertaining to human uses of seamounts and their resources, especially commercial fishing activities, depends on the geographic location of seamounts relative to international legal zones and the behavior of the species targeted by fishing.

Where seamounts are located in areas subject to the jurisdiction of coastal nations, the domestic law and policy concerning the conservation and management of living resources applies. Some nations—most notably New Zealand, Australia, Canada and the United States—have begun to implement measures such as quotas and marine reserves to manage fish stocks or protect ecosystems on seamounts in their EEZs.

The majority of seamounts are located in the Area beneath the high seas. Commercial fishing on these seamounts is still a high seas freedom, subject to the due regard that must be accorded other nations in their exercise of high seas freedoms and a duty to cooperate with other nations to conserve high seas resources. Although many regional fishery organizations have recognized the need to manage fishery resources on an ecosystem basis rather than just with species-specific approaches, there is little evidence that they recognize a need to manage the fishery resources of seamounts *per se*.

International law relating to the living resources of seamounts is use-oriented, implying that the focus is on the conservation and management of commercially exploitable species. Some commentators have invoked UNCLOS, Article 194(5) concerning protection or preservation of rare or fragile ecosystems and habitats as a justification in international law for the establishment of marine protected areas at seamounts. However, this provision of UNCLOS relates specifically to the control of pollution, defined as the introduction of substances or energy into the environment, and it is unclear whether it could provide a legal basis for the conservation of seamounts where the main ecological threat is from commercial fishing.

International "soft law"—such as encouragements found in the Rio Summit's Agenda 21 (1992) or other similar pronouncements—calls for sustainable development, integrated management and the application of precautionary and anticipatory approaches in the face of uncertainty. Broad planning, impact assessment, accounting and management tools are suggested by soft law, but specific conservation and management measures are not mandated. For the most part, these concepts are still latent with respect to the conservation of seamount species or ecosystems (Hey 1996).

Because of their remoteness, there has been little need until recently to think seriously about the conservation of living resources on high seas seamounts. As more is learned about the location and value of seamount resources—and the threats to them from human activities—their conservation undoubtedly will come to rely on the effectiveness of measures implemented by regional institutions and the most likely candidates appear to be the existing regional fishing organizations.

It is clear that nations are free to agree among themselves to limit their high seas freedoms in a way that conserves seamounts (Gjerde 2001). The unanswered questions are how widespread such an agreement could realistically be and whether it could be meaningfully enforced. Given the historical ineffectiveness of many regional organizations in fishery management (Meltzer 1994) it is hard to be sanguine about the conservation of seamounts on the high seas through such measures. It is likely that the geographical remoteness of most seamounts will contribute more than international law to their conservation.

## Summary and Recommended Measures for Seamount Conservation

Seamounts—underwater mountains rising to an elevation of 1,000 meters or more above the seafloor—are distributed throughout the world's ocean. There may be as many as 30,000 to 100,000 seamounts worldwide. They are rich and unusual deepsea biological communities, yet little is known about their geology, biology and ecology. Although relatively few seamounts have been sampled carefully, this research shows they support highly unusual (Figure 16) and endemic faunas and harbor large pools of undiscovered biodiversity in the ocean. They may be sites for the evolution of new species, refuges for rare species and stepping-stones for distribution of species across the open ocean. Many characteristic species on seamounts are very long-lived, and their biological communities can take hundreds and even thousands of years to develop.

Newly developed fisheries for deepsea species are now altering these habitats faster than they can be studied. It is likely that many shallow seamounts have experienced some degree of fishing pressure already. Trawling on deep seamounts became possible with the development of new

**Figure 16** A blob sculpin on Davidson Seamount.

fishing gear technology. Large-scale commercial fishing began in the 1960s for orange roughy and related species, leading to the collapse of many of these populations by the 1990s. Current management efforts in Australia and New Zealand include closing parts of their seamount areas to fishing, but newly developing fisheries in other parts of the world are largely unregulated. Collateral damage from deep-sea bottom-trawling is also a significant effect, destroying complex, long-lived and poorly known seafloor communities in the process of catching fish.

The legal framework for management and protection of seamount resources varies depending on whether they are within territorial waters, Exclusive Economic Zones (EEZs) or international waters. Most are in international waters, where a complex array of treaties may affect their regulation, but the high seas freedoms of the United Nations Convention on the Law of the Sea are the predominant principles that apply. With the accelerating collapse of more conventional fisheries and the development of sophisticated deep-water fishing technology, seamount populations are expected to come under increased exploitation pressure in the near future. Experience has shown that unregulated bottom-trawl fishing poses a significant threat to both the commercial fish stocks and their supporting benthic communities. The experience of New Zealand points to practical measures that can be taken to protect these environments and manage their resources. These approaches need to be extended to other locations, in both national and international waters. There is now an urgent need to discover, explore and protect seamount environments before they are destroyed.

There is a clear need to prevent the further decline of seamount habitats and their associated biota through better management practices, modifications of fishing techniques and other measures, such as establishing deep-water reserves. There is also a need to increase the level of research on seamounts to discover and document these unique habitats and their species and to consolidate new and existing data in publications, maps and online data sources. Since the *DOE* Conference a coalition of conservation groups has urged a moratorium on bottom-trawl fishing on the high seas—through a United Nations General Assembly resolution—to allow adequate scientific research and establishment of an international regime for fisheries management and biodiversity conservation. There are many legal and practical considerations, the analysis of which is beyond the scope of this chapter, but such an action would be compatible with the research and conservation goals above.

We recommend the following specific steps:

- Fishing gear modifications to decrease negative impacts on habitat and marine mammals
- A comprehensive set of reserves to protect representative seamount habitats throughout the world ocean
- Fish stock assessments based on techniques that are appropriate for slow-maturing deepsea species
- Public information programs to inform consumers about the ecological danger of deepsea fisheries. The public can influence demand for these vulnerable species by electing not to purchase them. Previous consumer awareness campaigns, such as dolphin-safe tuna, have been successful in bringing about changes in fishing regulations and practices
- A comprehensive effort to bring together existing international research on seamounts to guide future sampling of underrepresented seamount types and regions and to support optimal siting of high seas seamount reserves

## Literature Cited & Consulted

Agenda 21. 1992. Agenda for the 21st Century. United Nations Conference on Environment and Development (UNCED) held in Rio de Janerio, Brazil, 3–14 June 1992. *http://www.un.org/esa/sustdev/documents/agenda21/index.htm* (accessed October 8, 2004).

Alexandria Digital Library Gazetteer. 2002. *http://fat-albert.alexandria.ucsb.edu:8827/gazetteer/* (accessed May 12, 2003).

Allen, C. H. 2001. Protecting the oceanic gardens of Eden: international law issues in deep-sea vent resource conservation

and management. *Georgetown International Environmental Law Review* 13:563–660.
Anderson, O. F. 2001. A summary of biological information on the New Zealand fisheries for orange roughy (Hoplostethus atlanticus) for the 1999–2000 fishing year. *New Zealand Fishery Assessment Report 2001* (45).
Batiza, R. 1982. Abundance distribution and sizes of volcanoes in the Pacific Ocean and implications for the origin of non-hotspot volcanoes, *Earth and Planetary Science Letters* 60:196–206.
Batson, P. 2003. *Deep New Zealand: blue water, black abyss.* Christchurch: The Canterbury University Press.
Bax, N. J., compiler. 2000. Stock assessment report 2000: Orange roughy *(Hoplosthethus atlanticus)*. Unpublished report. Hobart: CSIRO Marine Research.
Blaber, S. J. M. 1986. The distribution and abundance of seabirds south-east of Tasmania and over the Soela Seamount during April 1985. *Emu* 86:239–244.
Boehlert, G. W. 1988. Current-topography interactions at mid-ocean seamounts and the impact on pelagic ecosystems. *Geojournal* 16:45–52.
——. 2001. *Productivity and Population Maintenance of Seamounts Resources and Future Research Directions.* NOAA Technical Report NMFS 43:95–101.
Boehlert, G. W. and B. C. Mundy. 1993. Ichthyoplankton assemblages at seamounts and oceanic islands. *Bulletin of Marine Science* 53:336–361.
Boehlert, G. W., C. D. Wilson and K. Mizuno. 1994. Populations of the sternoptychid fish *Maurolicus muelleri* on seamounts in the central North Pacific. *Pacific Science* 48:57–69.
Bolsch, C. J. S., N. G. Elliot and R. D. Ward. 1993. Enzyme variation in south-eastern Australian samples of the blue-eye or deepsea Trevalla, *Hyperoglyphe antarctica* Carmichael 1818 (Teleostei: Stromateoidei). *Australian Journal of Marine and Freshwater Research* 44:687–697.
Borets, L. A. 1980. The population structure of the boarfish Pentaceros richardsoni (Smith) from the Emperor Seamounts and the Hawaiian Ridge. *Journal of Ichthyology* 19:15–20.
Branch, T. A. 1998. *Assessment and adaptive management of orange roughy off southern Africa.* M.Sc. thesis, University of Cape Town.
Branch, T. 2001. A review of orange roughy *Hoplostethus atlanticus* fisheries, estimation methods, biology and stock structure. *South Africa Journal of Marine Science* 23:181–203.
Bratspies, R. 2001. Finessing King Neptune: Fisheries management and the limits of international law. *Harvard Environmental Law Review* 25:213–58.
Bucklin, A., R. R. J. Wilson and K. L. J. Smith. 1987. Genetic differentiation of seamount and basin populations of the deep-sea amphipod *Eurythenes gryllus. Deep-Sea Research Part A: Oceanographic Research Papers* 34:1795–1810.
Chapel, D. and C. Small. 1996. *The Distribution of Large Seamounts in the Pacific Ocean.* Abstract, American Geophysical Union Meeting.
Churchill, R. R. and A. V. Lowe. 1999. *The Law of the Sea.* Dover, NH: Manchester University Press.
Clark, M. R. 1999. Fisheries for orange roughy *(Hoplostethus atlanticus)* on seamounts in New Zealand. *Oceanologica Acta* 22:593–602.
——. 2001. Are deepwater fisheries sustainable? The example of orange roughy *(Hoplostethus atlanticus)* in New Zealand. *Fisheries Research* 51:123–135.
Clark, M. R. and D. M. Tracey. 1994. Changes in a population of orange roughy, *Hoplostethus atlanticus,* with commercial exploitation on the Challenger Plateau, New Zealand. *Fishery Bulletin* 92:236–253.
Craig, C. H. and D. T. Sandwell. 1988. Global Distribution of Seamounts from Seasat Profiles, *Journal of Geophysical Research* 93:10408–10420.
CSIRO Marine Research. 2002a. *Lost worlds under Pacific depths:* Media release highlighting Dr. Tony Koslow's research. *http://www.marine.csiro.au/PressReleasesfolder/00releases/seamnts2206/* (accessed May 28, 2002).
——. 2002b. *Discovering seamounts:* Highlighting Dr. Tony Koslow's research. *http://www.marine.csiro.au/leafletsfolder/42seamounts/42.html* (accessed May 28, 2002).
de Fontaubert, A. C. 2001. Legal and political considerations. In *The Status of Natural Resources on the High Seas,* ed. WWF/IUCN, 69–93. Gland, Switzerland: WWF/IUCN.
DFO, Canada. 2002. *Bowie Seamount Ecosystems Overview Part B: About Seamounts.* Department of Fisheries and Oceans. *http://www.pac.dfo-mpo.gc.ca/oceans/Bowie/bowie_part%20b.htm* (accessed May 29, 2002).
Dower, J. F. and R. I. Perry. 2001. High abundance of larval rockfish over the Cobb Seamount, an isolated seamount in the Northeast Pacific. *Fisheries Oceanography* 10 (3): 268–374.
Deep Sea Conservation Coalition (DSCC). 2004. *Urgent Action Needed to Protect Seamounts, Cold Water Corals and Other Vulnerable Deep Sea Ecosystems.* Deep Sea Conservation Coalition (DSCC) position paper. *http://www.highseasconservation.org/publicdocs/DSCC_Position_US.pdf* (accessed October 3, 2004).
Ehrich, S. 1977. Die Fischfauna der Grossen Meteorbank [Atlantic]. Meteor Forschungsergebnisse: Reihe D 25:543–5935.
Epp, D. and N. C. Smoot. 1989. Distribution of seamounts in the North Atlantic. *Nature* 337: 257.
Etnoyer, P. Forthcoming. Seamount Detection in the Northeast Pacific.
Fornari, D. J., R. Batiza and M. A. Luckman. 1987. Seamount abundances and distribution near the East Pacific Rise 0–24 deg N based on Seabeam data. In *Seamounts, Islands and Atolls,* ed. B. H. Keating, P. Fryer, R. Batiza and G. Boehlert. *AGU Geophysical Monograph* 43:13–21.
Generalized Bathymetric Chart of the Ocean (GEBCO). 2004. *Names of Undersea Features.* GEBCO Gazetteer. *http://www.*

*ngdc.noaa.gov/mgg/gebco/underseafeatures.html#gazetteer* (updated May 2004).

Gianni, M. 2002. Protecting the biodiversity of seamount ecosystems in the deep sea—the case for a global agreement for marine reserves on the high seas. Discussion Paper IUCN/WWF High Seas Marine Protected Areas Workshop, Malaga, Spain (15–17 January).

——. 2004. *High Seas Bottom Fisheries and their Impact on the Diversity of Vulnerable Deep-Sea Ecosystems.* Report prepared for The World Conservation–IUCN, Natural Resources Defense Council, WWF International and Conservation International.

Gjerde, K. M. 2001. High seas marine protected areas. Participant report of the "Expert Workshop on Managing Risks to Biodiversity and the Environment on the High Seas, Including Tools such as Marine Protected Areas: Scientific Requirements and Legal Aspects." *International Journal of Marine and Coastal Law* 16 (3): 515–28.

Glowka, L. 2000. Bioprospecting, alien invasive species, and hydrothermal vents: Three emerging legal issues in the conservation and sustainable use of biodiversity. *Tulane Environmental Law Journal* 13:329–60.

Gordon, J. D. M., coordinator. 1999. *Developing deep-water fisheries: data for the assessment of their interaction with impact on a fragile environment.* Final Consolidated Report of the Commission of the European Communities. Agriculture and Fisheries, 01.12.95–31.05.99. EC-FAIR CT 95 0655.

Grasshoff, M. 1972. Infraspezifische Variabilitat und isolierte Populationen der Hornkoralle Ellisella flagellum (Cnidaria: Anthozoa: Gorgonaria)—Auswerte der "Atlantische Kuppenfahrten 1967" von F. S. "Meteor." *Meteor Forschungsergebnisse: Reihe* D 10: 65–72.

Haney, J. C., L. R. Haury, L. S. Mullineaux and C. L. Fey. 1995. Sea-bird aggregation at a deep North Pacific seamount. *Marine Biology* 123:1–9.

Hareide, N. R. and G. Garnes. 2001. The distribution and catch rates of deep water fish along the Mid-Atlantic Ridge from 43 to 61°N. *Fisheries Research* 51: 297–310.

Hébert, A. 2002. *Trawl survey: Life onboard the Den C. Martin. http://www.glf.dfo-mpo.gc.ca/sci-sci/crab-crab/crabwise-encrab_fall99-e.html* (accessed May 29, 2002).

Hey, E. 1996. Global fisheries regulations in the first half of the 1990s. *International Journal of Marine and Coastal Law* 11 (4): 459–90.

Hope, A. 2001. Chilean owners look to deep water trawling. *Fishing News International* 40 (2): 8.

Hubbs, C. L. 1959. Initial discoveries of fish faunas on seamounts and offshore banks in the eastern Pacific. *Pacific Science* 13:311–316.

Hui, C. A. 1985. Undersea topography and the comparative distributions of two pelagic cetaceans. *Fisheries Bulletin U.S.* 83:472–475.

Humphreys, R. L., Jr., D. T. Tagami and M. P. Seki. 1984. Seamount fishery resources within the southern Emperor–northern Hawaiian Ridge. In *Second Symposium on Resources Investigations in the Northwestern Hawaiian Islands (Honolulu, Hawaii, 25–27 May 1983),* ed. R. W. Grigg and K. Y. Tanoue, 283–327.

International Hydrographic Organization. 2001. *Standardization of undersea feature names. Bathymetric Publication No. 6, 3rd Edition.* Monaco: International Hydrographic Bureau. *http://www.iho.shom.fr/publicat/free/files/B6efEd3.pdf* (accessed October 8, 2004).

Itano, D. G. and K. N. Holland. 2000. Movements and vulnerability of bigeye *(Thunnus obesus)* and yellowfin tuna *(Thunnus albacares)* in relation to FADs and natural aggregation points. *Aquatic Living Resources* 13 (4): 213–223.

Kaufmann, R. S., W. W. Wakefield and A. Genin. 1989. Distribution of epibenthic megafauna and lebensspuren on two central North Pacific seamounts. *Deep-Sea Research Part A: Oceanographic Research Papers* 36:1863–1896.

King, M. G. 1995. *Fisheries biology, assessment and management.* Oxford, England: Fishing News Books, Blackwell Science.

Kitchingman, A. and S. Lai. 2004. Inferences on Potential Seamount Locations from Mid-resolution bathymetric data. In *Seamounts: Biodiversity and Fisheries,* ed. T. Morato and D. Pauly. 2004 *Fisheries Centre Research Reports* 12 (5): 7–12.

Koslow, J. A. 1997. Seamounts and the ecology of deep-sea fisheries. *American Scientist* 85: 168–176.

Koslow, J. A., G. W. Boehlert, J. D. M. Gordon, R. L. Haedrich, P. Lorance and N. Parin. 2000. Continental slope and deep-sea fisheries: implications for a fragile ecosystem. *ICES Journal of Marine Science* 57:548–557.

Koslow, J. A., K. Gowlett-Holmes, J. K. Lowrly, G. C. B. Poore and A. Williams. 2001. Seamount benthic macrofauna off southern Tasmania: community structure and impacts of trawling. *Marine Ecology Progress Series* 213:111–125.

Kotlyar, A. N. 1996. *Beryciform Fishes of the World Ocean.* Moscow: VNIRO Publishing.

Levin, L. A. and C. L. Thomas. 1988. The ecology of xenophyophores (Protista) on eastern Pacific seamounts. *Deep-Sea Research Part A: Oceanographic Research Papers* 35:2003–2028.

——. 1989. The influence of hydrodynamic regime on infaunal assemblages inhabiting carbonate sediments on central Pacific seamounts. *Deep-Sea Research Part A: Oceanographic Research Papers* 36:1897–1916.

Levin, L. A., L. D. McCann and C. L. Thomas. 1991. The ecology of polychaetes on deep seamounts in the Eastern Pacific Ocean. *Ophelia Supplement* 5: 467–476.

Levin, L. A., E. L. Leithold, T. F. Gross, C. L. Huggett and C. Dibacco. 1994. Contrasting effects of substrate mobility on infaunal assemblages inhabiting two high-energy settings on Fieberling Guyot. *Journal of Marine Research* 52:489–522.

Lleonart, J., J. Sanat and R. Franquesa. *The problems of fisheries management in the Mediterranean: Catalonia as a case study.* Paper presented at the 1st International Congress on Maritime

Technological Innovations and Research, Barcelona, 21–23 April, 1999. *http://www.gemub.com/theproblems.pdf* (accessed May 28, 2003).

Long, R. and A. Grehan. 2002. Marine habitat protection in sea areas under the jurisdiction of a coastal member state of the European Union: The case of deep water coral conservation in Ireland. *International Journal of Marine and Coastal Law* 17 (2): 235–62.

Lyle, J. M. 1994. Orange roughy *(Hoplostethus atlanticus).* In *A Scientific Review of the South East Fishery with Particular Reference to Quota Management,* ed. R. D. J. Tilzey, 98–114. Bureau of Resource Sciences Bulletin.

Magnússon, J. V. and J. Magnússon. 1995. The distribution, relative abundance, and biology of deep-sea fishes of the Icelandic Slope and Reykjanes Ridge. In *Deep-Water Fisheries of the North Atlantic Oceanic Slope,* ed. A. G. Hopper, 161–199. Dordrecht, The Netherlands: Klüwer Academic.

Meltzer, E. 1994. Global overview of straddling and highly migratory fish stocks: The nonsustainable nature of high seas fisheries. *Ocean Development and International Law* 25:255–344.

Moore, J. A. and M. Vechione. 2002. Biodiversity of Bear Seamount. Deep-Sea Fisheries Symposium. *http://www.nafo.ca/meetings/SciCoun/200`/resdocs/scr01-155.pdf* (accessed May 29, 2002).

Morgan, W.J. 1971. Convection plumes in the lower mantle. *Nature* 230:42–43.

New Zealand Ministry of Fisheries (MFish). 2000. *Safeguarding undersea mountains. http://www.fish.govt.nz/current/press/pr070900_2.htm* (accessed April 12, 2002).

Parin, N. V., A. N. Mironov and K. N. Nesis. 1997. Biology of the Nazca and Sala y Gomez submarine ridges, an outpost of the Indo-West Pacific fauna in the Eastern Pacific Ocean: Composition and distribution of the fauna, its communities and history. *Advances in Marine Biology* 32:145–242.

Parker, T. and V. Tunnicliffe. 1994. Dispersal strategies of the biota on an oceanic seamount: Implications for ecology and biogeography. *Biological Bulletin* 187:336–345.

Probert, P. K., D. G. McKnight and S. L. Grove. 1997. Benthic invertebrate bycatch from a deep water trawl fishery, Chatham Rise, New Zealand. *Aquatic Conservation: Marine and Freshwater Ecosystems* 7:27–40.

Richardson, P. L. 1980. Anticyclonic eddies generated near the corner rise seamounts. *Journal of Marine Research* 38.

Richer de Forges, B., J. A. Koslow and G. C. B. Poore. 2000. Diversity and endemism of the benthic seamount macrofauna in the southwest Pacific. *Nature* 405:944–947.

Roberts, C. 2002. Deep impact: the rising toll of fishing in the deep sea. *Trends in Ecology and Evolution:* Trends early edition: *http://tree.trends.com* (accessed May 28, 2002).

Robertson, D. 1990. The New Zealand orange roughy fishery: An overview. *Bureau of Rural Resources Proceedings* 10:38–48.

Rogers, A. D. 1994. The biology of seamounts. *Advances in Marine Biology* 30:305–354.

Ross, D. M. and D. C. Smith. 1997. The Australian orange roughy fishery: A process for better resource management. *Developing and Sustaining World Fisheries Resources: The State Of Science and Management* 256–260.

Rowden, A. A., S. O'Shea and M. R. Clark. 2002. Benthic biodiversity of seamounts on the northwest Chatham Rise. *Marine Biodiversity Biosecurity Report* 2: 1–2. Wellington, NZ: Ministry of Fisheries.

Sandwell, D. T. and W. H. F. Smith. 1997. Marine gravity anomaly from Geosat and ERS-1 satellite altimetry. *Journal of Geophysical Research* 102: B5, 10039–10054.

SeamountsOnline. 2004. *SeamountsOnline: An online information system for seamount biology. http://seamounts.sdsc.edu* (accessed October 1, 2004).

Shaughnessy, P. 1999. *The action plan for Australian seals.* Environment Australia. CSIRO Wildlife and Ecology, Canberra, ACT.

Sibert, J., K. Holland and D. Itano. 2000. Exchange rates of yellowfin and bigeye tunas and fishery interaction between Cross seamount and near-shore FADs in Hawaii. *Aquatic Living Resources* 13 (4): 225–232.

Smith, D. K. and T. H. Jordan 1988. Seamount statistics in the Pacific Ocean. *Journal of Geophysical Research* 93:2899–2919.

Smith, W. H. F. and D. T. Sandwell. 1997. Global sea floor topography from satellite altimetry and ship depth soundings. *Science Magazine* 277 (5334): 1956–1962.

——. 2004 (sic). Conventional bathymetry, bathymetry from space, and geodetic altimetry. *Oceanography* 17: 8–23.

Stocks, K. 2003. SeamountsOnline, a Biogeographic Information System for Seamounts. Poster, *Frontiers in Biogeography, the Inaugural Meeting of the International Biogeography Society, Mesquite, Nevada, Jan 4–8. http://seamounts.sdsc.edu/Biogeo_2002.ppt* (accessed May 28, 2004).

——. 2004. *SeamountsOnline: An online information system for seamount biology.* Version 3. World Wide Web electronic publication. *http://seamounts.sdsc.edu* (accessed October 8, 2004).

Stone, G., L. Madin, K. Stocks, G. Hovermale, P. Hoagland, M. Schumacher, C. Steve-Sotka and H. Tausig. 2003. Seamount biodiversity, exploitation and conservation. In *Defying Ocean's End: An Agenda for Action Executive Summary,* ed. L. K. Glover, 16. Washington, D.C.: Conservation International.

Strutt, I. 2000. Roughy bonanza in Indian Ocean. *Fishing News International* 39 (5): 1–3.

Tooby, P. 2001. Environmental informatics to the biogeography of seamounts. *Online: News about the NPACI and SDSC Community 5 (15). http://www.npaci.edu/online/v5.15/seamounts.html* (accessed May 29, 2002).

Tracey, D. M. and P. L. Horn. 1999. Background and review of aging orange roughy (Hoplostethus atlanticus, Trachichthyidae) from New Zealand and elsewhere. *New Zealand Journal of Marine and Freshwater Research* 33:67–86.

UNCLOS. 1982. United Nations Convention on the Law of the Sea. Done at Montego Bay, Jamaica, December 10, 1982, U.N. Doc.

A/Conf.62/122, reproduced in 21 I.L.M. 1261 (1982). In force November 16, 1994, with 143 parties at August 2003.
Wessel, P. 1997. Sizes and ages of seamounts using remote sensing: Implications for intraplate volcanism. *Science* 277:802–805.
——. 2001. Global distribution of seamounts inferred from gridded Geosat/ERS-1 altimetry. *Journal of Geophysical Research* 106 (B9): 19431–19441.
Wilson, R. R. and R. S. Kaufmann. 1987. Seamount biota and biogeography. *American Geophysical Union Geophysical Monographs* 43:355–377.
WWF. 2002. *Banco Gorringe-A Potential MPA. http://www.ngo.grida.no/wwfneap/Publication/briefings.BancoGorringe.pdf* (accessed May 29, 2002).
WWF/IUCN/WCPA. 2001. *The Status of Natural Resources on the High-Seas.* Gland, Switzerland: WWF/IUCN.

CHAPTER 3

# The Southern Ocean: A Model System for Conserving Resources?

John P. Croxall, *British Antarctic Survey*
Phil N. Trathan, *British Antarctic Survey*

© PHOTODISC

## Introduction

The Southern Ocean—the seas surrounding the Antarctic Continent—generates some of the richest nutrient conditions in the world and sustains an extraordinary aggregation of marine living resources. It is also renowned for having relatively simple food chains, particularly that from phytoplankton through Antarctic krill *Euphausia superba* to a variety of marine top predator species. Although Antarctica is often referred to as "the last great wilderness," the marine environment there has been greatly influenced by human activities, with commercial harvesting in the Southern Ocean over the past two centuries. Exploitation began in the late 18th and 19th centuries when Antarctic and Sub-Antarctic fur seals *Arctocephalus gazella* and *A. tropicalis* were taken for their pelts. This ceased only as the species neared extinction. Subsequently in the middle- to late-19th century, penguins and elephant seals *Mirounga leonina* were exploited for their oil, as were baleen whales in the early 20th century.

The serial over-exploitation of the great whale species culminated in whaling and sealing becoming commercially

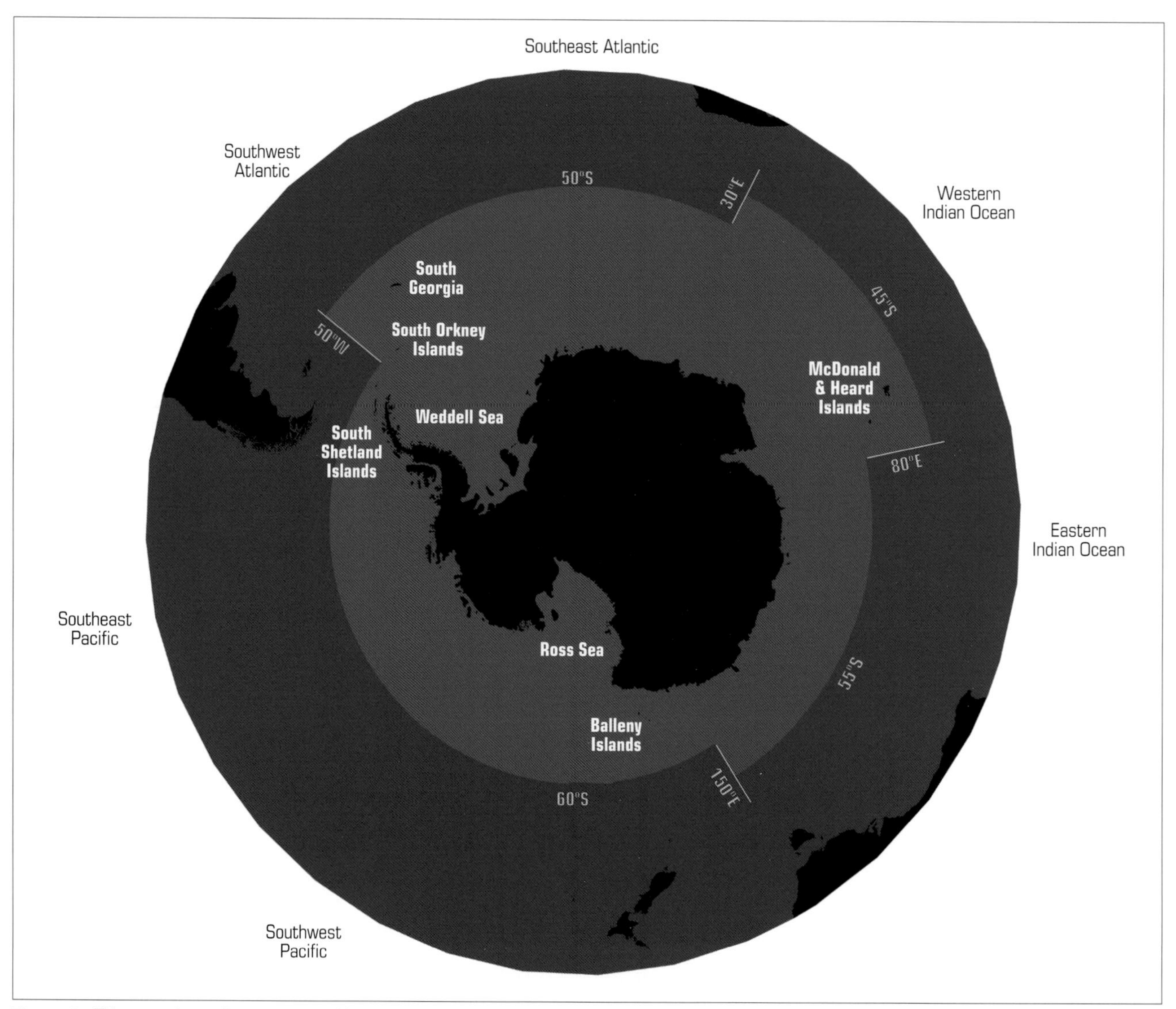


**Figure 1** This map shows the area covered by the Convention on Conservation of Antarctic Living Resources (CCAMLR). Specific locations discussed in the chapter are highlighted.

unprofitable by the 1960s. The take was then transferred to finfish and krill. With the collapse of overfished icefish *Champsocephalus gunnari* stocks in the 1970s, it was recognized that uncontrolled exploitation of species such as Antarctic krill would have an unprecedented effect on the many species dependant upon it as prey, including the recovering stocks of previously over-exploited species such as whales, Antarctic fur seals and finfish.

Thus, serial exploitation—meaning over-exploitation of one species after another until they near extinction or become commercially unviable—has long been a feature of commercial harvesting in the Southern Ocean. This led scientists operating under the Antarctic Treaty System—which had developed an excellent basis for scientific cooperation on land—to collaborate in developing a framework for regulating activities in the surrounding ocean. This eventually resulted, in 1982, in the negotiation of an international agreement named the Convention for the Conservation of Antarctic Marine Living Resources (CCAMLR). The jurisdiction of this treaty covers a huge oceanic area—approximately 32 million square kilometers, the size of the entire North Atlantic Ocean—south of the approximate boundary of the Antarctic Polar Front. The CCAMLR treaty area (Figure 1) is subdivided into various management regions that roughly delimit stocks of some of the managed species. The Commission of the Convention currently has 24 full-member nations with a subscription (or dues) structure that reflects the amount of resources taken by each nation. Parties to the Convention include fishing States and non-fishing States with a wide variety of backgrounds in terms of conservation and management philosophies and practices.

When CCAMLR was negotiated, it had three ideas that were visionary for the 1980s and which would nowadays be called "precautionary principles." CCAMLR members are required to:

- Ensure sustainable use of harvestable resources
- Take into account the needs of dependent species (i.e., those species which naturally require access to a harvested resource)
- Prevent changes in the ecosystem that would likely be irreversible within two or three decades

Actual treaty language is shown in Table 1. These were very progressive principles to include in an international treaty involving fisheries. This ecosystem-based approach to management foreshadows current attempts of this type by some 20 years.

To balance this, however, there are several aspects that combine to make CCAMLR's work particularly challenging. First, its legally binding agreements (Conservation Measures) require consensus amongst all 24 members. Once agreed upon, these Conservation Measures have generally been respected. Second, the European Community regards CCAMLR as an external fishery and exercises sole competence over fisheries matters on behalf of its members; unsurprisingly this has sometimes limited the speedy application and implementation of environmental considerations. Third, many (if not most) members primarily send fishing representatives to CCAMLR meetings to ensure that their commercial interests are well represented; fewer

Table 1

**Convention for the Conservation of Antarctic Marine Living Resources: Article II**

1. The objective of this Convention is the conservation of Antarctic marine living resources.
2. For the purposes of this Convention, the term "conservation" includes rational use.
3. Any harvesting and associated activities in the area to which this Convention applies shall be conducted in accordance with the provisions of this Convention and with the following principles of conservation:
   (a) prevention of decrease in the size of any harvested population to levels below those which ensure its stable recruitment. For this purpose its size should not be allowed to fall below a level close to that which ensures the greatest net annual increment;
   (b) maintenance of the ecological relationships between harvested, dependent and related populations of Antarctic marine living resources and the restoration of depleted populations to the levels defined in sub-paragraph (a) above; and
   (c) prevention of changes or minimisation of the risk of changes in the marine ecosystem which are not potentially reversible over two or three decades, taking into account the state of available knowledge of the direct and indirect impact of harvesting, the effect of introduction of alien species, the effects of associated activities on the marine ecosystem and the effects of environmental changes, with the aim of making possible the sustained conservation of Antarctic marine resources.

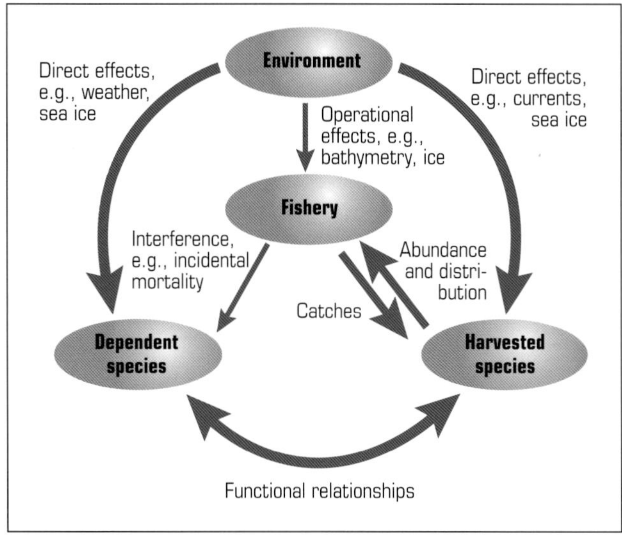


**Figure 2** A conceptual framework of CCAMLR processes and interactions. Breadth of the arrows indicates perceived importance of the interactions (CCAMLR 1995).

countries send representatives well versed in modern conservation management.

CCAMLR governance is based on annual meetings of a Commission and a Scientific Committee. The Scientific Committee rapidly developed a Working Group to deal with Fish, and later established others to address Ecosystem Management and Monitoring, and Krill; these latter two were combined to address improved management of krill on an ecosystem basis, particularly the needs of krill-dependent species. In the 1990s, a Working Group on Incidental Mortality Associated with Fishing was set up to address "bycatch" issues—the unintended catch and kill of species other than those sought in a specific fishing operation—working in association with the Fish group. In developing its ecosystem-based approach to management, CCAMLR's Scientific Committee recognised that while the main interactions of interest are those among fisheries, harvested species and dependent species, significant environmental effects operate at all levels of the system (Figure 2). Much discussion has ensued about how much data are required on each of these linkages to manage the system effectively through delivering the "best scientific advice." Inevitably, for legally binding management measures, fishers usually want to have more data, whereas more conservation-conscious interests often feel there are sufficient data to make pragmatic precautionary decisions today and that waiting until tomorrow may result in having no resource left to protect.

Three important principles currently being addressed are:

- Setting precautionary "escapement rates," that is, ensuring the amount of resource left after commercial harvesting is sufficient for the needs of all dependent species
- Adjusting the catch rate for predator support to take account of the disproportionate effects of years of naturally low resource availability
- Avoiding excessive competition (i.e., overlap) between fisheries and dependent predators. This is important because one could fulfil the first two principles and still have a fishery focussed in small areas of critical importance for dependent species.

A discussion of the three main fisheries currently managed by CCAMLR, presented below, examines how well those principles are working in practice.

## Antarctic Krill *Euphausia superba* Fishery

Antarctic krill *Euphausia superba* is a five- to six-centimeter-long shrimp-like crustacean that is at the hub of much of the Antarctic food web. The krill fishery has been the largest in the CCAMLR area since the late 1970s (Miller & Agnew 2000). CCAMLR's approach to its management developed slowly through the 1980s (Hewitt & Low 2000). An initial focus was on the use of potential measures of relative abundance, but the standard fisheries measure—catch per unit effort—was deemed to be of limited utility for determining relative abundance, or changes in abundance, of krill (CCAMLR 1989). The preferred approach to managing the krill fishery uses a simulation model—originally the Krill Yield Model, or KYM (Butterworth, Punt & Basson 1991)—which used acoustic estimates of krill abundance from the Biological Investigations of Marine Antarctic Systems and Stocks (BIOMASS) programme (Trathan *et al* 1993, El Sayed 1994), to set the first catch limits on the krill fishery in 1991 and 1992 (Nicol & Endo 1997).

The KYM developed into the Generalised Yield Model (GYM) that is now also used to set precautionary catch limits for most other fisheries in the Convention Area (Constable & de la Mare 1995). Precautionary limits are calculated from an estimate of the total biomass of the krill stock in an area obtained from an acoustic survey; an estimate of the rate of natural mortality (including natural predation); a model of how individual krill grow in weight during their lives; and an estimate of the interannual variability in recruitment—replenishment through reproduction (Constable *et al* 2000). This information is used to make a computer population model of a krill stock that can generate a distribution of population sizes, both in the absence of fishing and at various levels of fishing. These distributions are used to determine an estimate of the proportion of the unexploited biomass that can be caught each year ($\gamma$). CCAMLR has also developed a three-part decision rule for determining this proportion that matches the aims of Article II:

1. Choose $\gamma_1$ so that the probability of the spawning biomass dropping below 20% of its pre-exploitation median level over a 20-year harvesting period is 10%.
2. Choose $\gamma_2$ so that the median krill escapement in the spawning biomass over a 20-year period is 75% of the pre-exploitation median level.
3. Select the lower of $\gamma_1$ and $\gamma_2$ as the level of $\gamma$ for the calculation of the krill yield.

The requirements of krill predators are taken into account, to some extent, in the decision rules. Thus, in rule 2, the level of 75% of the pre-exploitation biomass was chosen as an intermediate value between the case of single-species management—where 50% escapement would be acceptable—and the case where no fishing is permitted, which is 100% escapement. This value is subject to refinement as more information becomes available. Additionally, the incorporation of natural krill mortality into the model's formulation should ensure that the needs of predators are taken into account because they are assumed to be the prime cause of mortality in the krill population. This approach to managing the krill fishery, although some way from the desired implementation of an "ecosystem approach," is still novel in its incorporation of precautionary principles into management decision rules. The development of a management feedback procedure for krill, which will incorporate a fully elaborated ecosystem approach, is planned over the next five years (Constable 2002).

The initial catch limits on krill (totalling 1.5 million tonnes per year) were established for the southwest Atlantic near the Weddell Sea in 1991. This was followed by the establishment of a catch limit (650,000 tonnes per year) on the krill fishery in the southwest Indian Ocean (Nicol & Endo 1997). These catch limits utilised all the available krill abundance data from the BIOMASS programme (Trathan *et al* 1993). Other parts of the CCAMLR area were fished through the 1990s, but these areas had not been surveyed for krill so catch limits were not set. In 1996, the first acoustic survey of a CCAMLR area specifically designed for establishing a precautionary catch limit was conducted in a large area of the southeast Indian Ocean (Nicol *et al* 2000). A catch limit (775,000 tonnes per year) established in 1996 on the basis of this survey data was subsequently revised in 2000 and subdivided into two catch limits (277,000 tonnes per year in the west and 163,000 tonnes per year in the east) based on environmental data (Nicol & Foster 2003). In 2000, CCAMLR sponsored a multi-nation survey of the South Atlantic sector to establish new precautionary catch limits (Trathan *et al* 2001, Hewitt *et al* 2002). The results of this survey—carried out by vessels from Japan, Russia, U.K. and the U.S.—were used to establish a new catch limit of four million tonnes for the South Atlantic. The new, higher catch limit reflects more advanced techniques and methods and is not strictly comparable to the older survey results, so the difference certainly cannot be interpreted simply as a change in biomass (Nicol & Foster 2003).

The catch limits are set for enormous areas of ocean, whereas the resource and the fishery naturally operate at much smaller scales. For example, in the southwest Atlantic sector, the krill fishery operates in less than 20% of the available area. In the South Orkney and South Shetland Island areas, the fishery consistently operates in the same locations as the main feeding grounds of breeding penguins and fur seals that feed on krill, creating a significant risk that excessive quantities of krill could be removed from local areas to the detriment of dependent species. To address this problem, CCAMLR has begun the complex task of managing krill fishing at the scales most relevant to many of the natural environment/prey/predator interactions. The first step has been to define smaller and more ecologically realistic management areas—Small Scale Management Units—mainly by using satellite-tracking data to define the foraging ranges of krill-dependent species (CCAMLR 2002a). The next step is to quantify the spatial and temporal demand for krill by predators and fisheries, the nature of potential competition between predators and fisheries and the main fluxes of krill into and out of these areas. While this approach is being developed, the essential complementary step is to start testing potential decision rules and management procedures for minimising, on a precautionary basis, adverse interactions between predators and the fisheries.

While distribution and demographic data relevant to managing the krill fishery were being collected, a parallel initiative addressing ecosystem interactions of krill-dependent

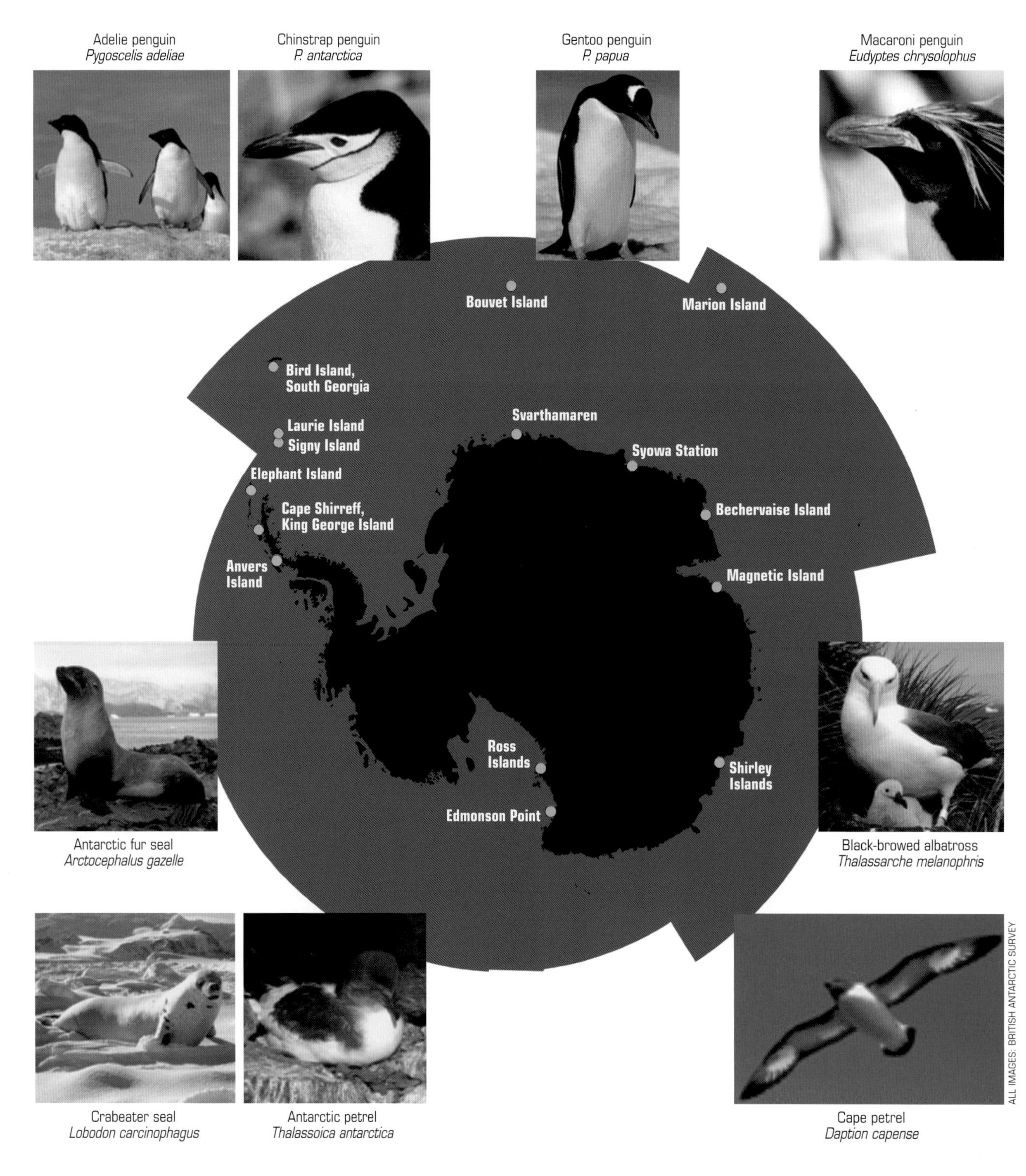


**Figure 3** Monitored sites and monitored species that are dependent on those harvested within the CCAMLR Ecosystem Monitoring Program.

species was also being developed. Concerns for dependent species exist in four main categories:

- Effects of years when krill is in low availability
- The fact that the frequency of these low availability years may be increasing
- The fact that these two factors combined may be having an amplified impact on the predators
- A possible direct impact through competition from the fishery

In order to investigate these issues, CCAMLR established an Ecosystem Monitoring Programme (CEMP) to detect significant changes in critical components of the ecosystem and distinguish between changes due to harvesting and those due to environmental variability. This programme uses standard methods to collect systematic data at a network of sites (Figure 3) on: krill itself; on one krill-related fish species; on a selected group of top predator species that eat mostly krill; and on various environmental variables (Agnew 1997, Croxall *et al* 1988, Croxall 1989, 1994, forthcoming). For the species selected, the programme monitors trends in population size, diet (including condition, meal size and foraging efficiency), various attributes of reproductive performance and survival by means of population studies. The variables were selected to cover a range of temporal scales from short-term (e.g., activity budgets in foraging trips at the scale of days, diet at the scale of weeks) to medium-term (e.g., breeding performance over several months) to long-term (one or more years for the population variables). The variables measured, along with the areas and time scales, are shown in Table 2. This system was designed to represent in space and time an integration of effects on species operating at the top of the ecosystem but which depend on prey lower down the food chain. In particular, the intention was to look at variation in performance within and between years; how performance related to the availability of the prey; how this varied at different spatial and temporal scales; and how the fisheries and dependent predators interacted.

Although distinguishing between natural and harvesting-induced changes has so far proved impossible (CCAMLR 2003a), the CEMP data do provide clear evidence of the effect of years of low krill availability when predator performance is greatly reduced. These data show that availability of prey has an important and almost immediate influence on the predators, initially in terms of productivity (Figure 4) but often also in terms of population effect. If all variables are combined, there is further evidence of a statistically significant change in the system, whereby years of low krill availability have increased in the last decade compared to the 1980s (Reid & Croxall 2001). The causes of this are uncertain. We believe it is unlikely to be anything to do with the level of harvesting but is more likely a result of, and evidence for, system and regime changes within the whole

Table 2

**CCAMLR Ecosystem Monitoring Programme: Variables (and nominal scales of integration) measured for monitored species dependent on krill and harvested species**

| | | Integration scale | |
|---|---|---|---|
| **Category** | **Variable** | **Area (km²)** | **Time** |
| Population size | Breeding count | 1,000–100,000 | Years |
| Survival | Count at age | 1,000–50,000 | Years |
| Adult pre-breeding condition | Mass | 1,000–10,000 | Months |
| Reproductive performance | Breeding success | 100–1,000 | Months |
| Foraging performance | Foraging<br>• Trip duration<br>• Feeding pattern/efficiency<br>• Offspring growth rate<br>• Mass at independence | | Weeks to days |
| Diet | Overall composition<br>• Fish size composition (from otoliths)<br>• Squid size composition (from beaks)<br>• Krill length/size/status | 100–1,000 | Days |

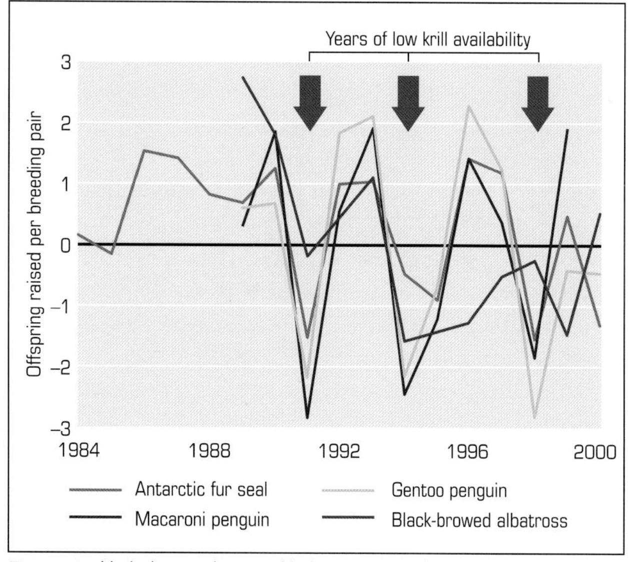


**Figure 4** Variations—above and below average—in reproductive output for krill-dependent top predators breeding at South Georgia Island (extended from Reid & Croxall 2001).

Antarctic system (Croxall 1992, Trathan *et al* forthcoming). Regardless of the cause, the predators appear to have less food, as it is less regularly available to them today; this is reflected, at least in part, by observations of reduced growth and survival, and population decreases in various predator species (Reid & Croxall 2001, Croxall forthcoming).

Furthermore, although the CEMP is one of the world's longest running—and most extensive—programmes monitoring population performance of land-based marine predators in relation to indices of prey and environment, CCAMLR is still trying to develop procedures for using these data to link directly to the management of krill. The recent development of rigorous and objective ways of combining the various types of indicators (Boyd & Murray 2001, de la Mare & Constable 2000) should facilitate this. It is hoped that these performance indicators, whose relationships to krill abundance are becoming increasingly well defined (Barlow *et al* 2002, Reid & Croxall 2001, Trathan *et al* forthcoming, Reid *et al* forthcoming), can now be applied in the context of precautionary management to interactions within the Small Scale Management Units. It is also hoped these changes will lead directly, through new decision rules, to appropriate modifications in fishing practice. One unexpected product of monitoring diet composition of predators has been the ability to provide new insights into krill population dynamics using predator samples (Reid *et al* 1999, 2002, forthcoming, Murphy & Reid 2001), generating the possibility of predicting years of exceptionally low krill availability (Trathan & Murphy 2003, Trathan *et al* 2003)—with obvious implications for predators and fisheries alike.

CCAMLR seems to have succeeded in developing an agreed model—using precautionary approaches and incorporating discount factors for the needs of dependent species—for its basic area-wide management procedures. It has so far failed, however, to develop appropriate ways of managing the fishery at the scales and places most relevant to dependent species, nor has it devised ways of incorporating data from dependent species into its krill management models (Table 3). Much of the credibility of CCAMLR rests on its ability to manage krill stocks, especially in relation to likely future fishery expansion. This makes it exceptionally important to negotiate now the appropriate decision rules and reference points before any new expansion takes place, as well as to acquire economic data to enable independent forecasting of fishery growth.

Table 3

**Some successes and challenges for CCAMLR's management of krill fisheries**

| Successes | Challenges |
|---|---|
| Precautionary Generalised Yield Model (GYM) | Acquisition of economic data to enable independent forecasting of fishery growth |
| Triggered decision rules | Observer programme and vessel monitoring system (VMS) for krill fishing vessels |
| Catch limits in place before fishery expansion | Catch limits defined at appropriate spatio-temporal scale |
| Monitoring (CEMP) programme for dependent species | Incorporation of CEMP data into management advice |
| Recent definition of areas for management at more appropriate scales | Application of Small Scale Management Units |

## Icefish *Champsocephalus gunnari* Fishery

The icefish *Champsocephalus gunnari* is a midwater epipelagic species, meaning it lives in the upper 200 metres of the open ocean that is confined to continental shelves. Many Antarctic species eat icefish but none seem to depend on it totally. It has been subject to massive pulses of fishing using midwater trawls after which the stock has collapsed, recovered and collapsed again from overfishing. In the past decade, it has failed to recover (Figure 5). The ability of CCAMLR's fishery models to predict what the stock size will be the following year has been highly variable (Table 4). The predictions have proved particularly optimistic in years of low availability of krill (an important part of the diet of icefish). Therefore, if krill is becoming consistently less available to species with relatively restricted foraging ranges, this may be contributing to the failure of icefish stocks to recover their former abundance.

Another possible influence on icefish dynamics, however, relates to the role of Antarctic fur seals as predators. The fur seal population recovered from about 100,000 in the 1960s to perhaps over three million nowadays (Boyd 1993, British Antarctic Survey unpublished data). Although icefish are only a very small part of the fur seal diet (perhaps less than 1% by mass), the consumption—particularly by male fur seals (Figure 6) of larger numbers of pre-recruiting (pre-reproductive age) icefish—may have an effect both on recruitment and subsequent productivity (Everson *et al* 1999). Thus a ten-fold increase in the proportion of icefish in the fur seal diet (from about 0.1% to 1%) would result in the consumption of some 40,000 tonnes of icefish, not dissimilar to the differences between the predicted and observed values from the fish surveys. Although this last point remains conjectural, it is highly likely that fur seals are

Table 4

**Abundance of icefish predicted versus observed by fishery surveys, in relation to krill density**

| Year | Predicted (tonnes) | Observed (tonnes) | P/O ratio | Krill density (gm²) |
|---|---|---|---|---|
| 1990 | 14,297 | 27,979 | 0.51 | 45–75 |
| 1991 | 89,697 | 15,805 | 5.68 | 6 |
| 1992 | 17,999 | 21,236 | 0.85 | 95 |
| 1993 | 43,735 | 45,200 | 0.97 | 66 |
| 1994 | 41,618 | 1,383 | 30.1 | 7 |

**Note:** Data from acoustic surveys.
**Source:** Everson *et al* 1994.

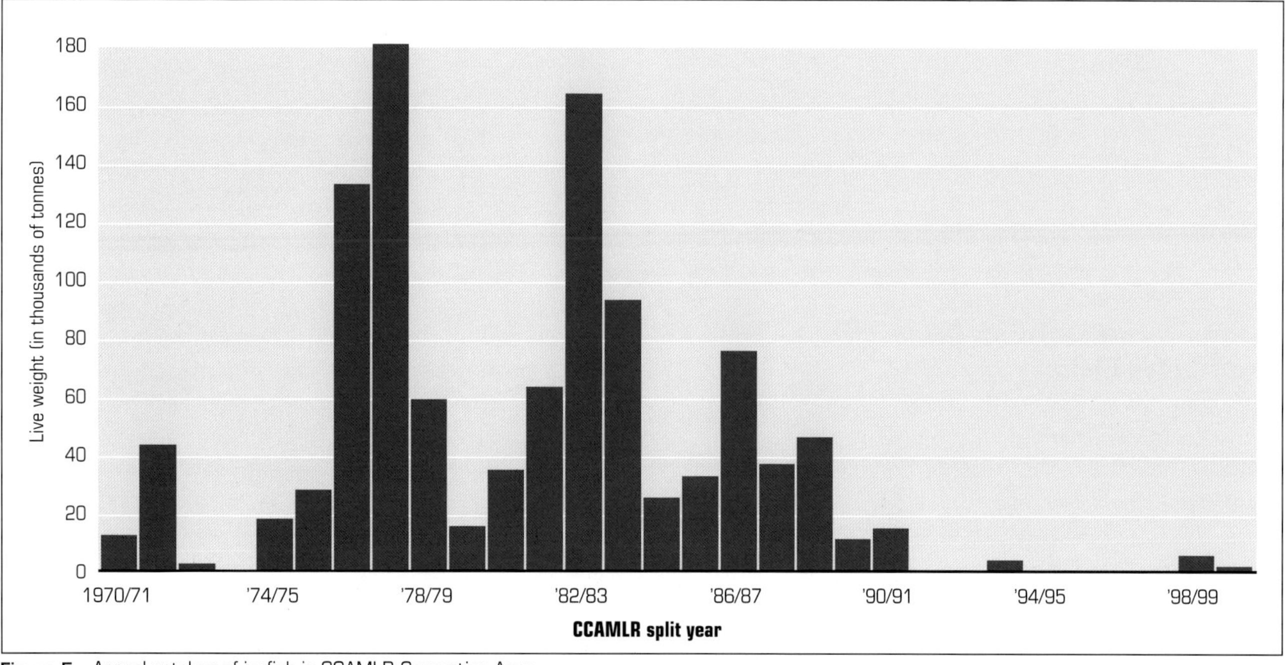


**Figure 5** Annual catches of icefish in CCAMLR Convention Area.

BRITISH ANTARCTIC SURVEY

**Figure 6** Although icefish are a very small part of the diet for Antarctic fur seals, their consumption may affect both recruitment and subsequent productivity of the fish.

now playing a much more important role in icefish population dynamics than two decades ago.

It is somewhat ironic that the recovery of fur seals—from human over-exploitation and possibly from the reduced competition for krill from competitors such as baleen whales that have also been over-exploited—may now be impeding the recovery of icefish from human over-exploitation. These are yet more examples of the reality that perturbed systems cannot readily be restored to previous dynamic states or equilibria.

CCAMLR's experience with the icefish fishery therefore highlights two further fundamental issues. We need to understand tropho-dynamic relationships, where dependent species may switch to alternative prey in years where the biomass of their principal prey is unusually low due to either natural or human-induced causes. We also need to identify management measures where a dependent species like icefish which feeds upon Antarctic krill may also be harvested.

## Patagonian Toothfish *Dissostichus eleginoides* Fishery

In Southern Ocean fisheries, Patagonian toothfish *Dissostichus eleginoides* first appeared as a shallow-water bycatch of juveniles in trawl fisheries around South Georgia in the late 1970s. By the late 1980s, the deep-water (500–1,000 metres) habitat of the larger adults (reaching lengths over two metres and weights in excess of 100 kilograms) could be targeted by using longlines. Typically each longline set involves deploying about 10,000 to 20,000 baited hooks on branch lines from the main longline weighted to sink to the seafloor. Surface buoys mark the ends of the lines that are retrieved 10 to 20 hours later (Figure 7). Fishing directed at Patagonian toothfish started around South Georgia in 1987, but their high market value led to rapid expansion throughout the sub-Antarctic parts of the Convention Area in the 1990s, and a subsequent southward extension of fishing into Antarctic coastal waters, mainly in the Ross Sea, where Antarctic toothfish *D. mawsoni* is the principal target.

Because toothfish are long-lived (30–50 years natural life span), slow growing and slow to reach reproductive age, the rapid expansion of this fishery has caused major concern. This was compounded by data from the fishery being insufficient for conventional stock assessment methods and the adult stocks being largely inaccessible to fishery-independent trawl surveys (Constable *et al* 2000). To take account of the absence of assessments of the unexploited stock, and the fact that fishing was already in progress, CCAMLR developed two approaches for management: a Generalised Yield Model (Constable & de la Mare 1995) and regulations restricting the development of new fisheries (including existing fisheries moving into new areas).

The Generalised Yield Model is very similar in approach to that developed for krill. It uses estimates of the abundance of recruits (from surveys, especially) around South Georgia and projects these forward in the simulations. Known catches are therefore directly discounted from the population and the long-term annual yield is assessed in tonnes, rather than as a proportion of an estimate of standing stock. As a large deep-water predator, toothfish are not key items in the diet of land-based predators; therefore the escapement value was set at 50% (rather than 75% as for krill) but higher than for more conventional fisheries management schemes that typically prescribe escapement levels of 20–30%. For toothfish, the assessment of long-term (20-year) annual yields is a further example of a precautionary approach, taking account of stock fluctuations and incorporating many sources of uncertainty into a single assessment.

These approaches, on their own, are not sufficient to deal with opening new fishing grounds in the absence of survey data, particularly estimates of recruitment. For toothfish, this is particularly exacerbated by the often restricted amount of toothfish habitat in such new areas, making stocks very vulnerable to over-exploitation. CCAMLR has so far dealt with this situation in two ways. First, estimates of recruitment from fished areas have been extrapolated to the new areas in proportion to the ratios of areas of seabed;

the resulting yield estimates are then discounted by 60% to take account of the special uncertainties surrounding the opening of new fisheries. Second, catches in new areas are restricted to 100 tonnes per rectangle of 1° longitude by 0.5° latitude, to prevent all of the allowed catch being taken in a small area. These regulations are to be maintained until sufficient research data, especially on recruitment, are available for a full stock assessment to be carried out, at which time that particular fishing area would be treated in the same way as the existing established and assessed fisheries for toothfish.

Despite enacting these precautionary procedures, CCAMLR's management of toothfish has been severely compromised by difficulties relating to bycatch and the incidental mortality of non-target species, and by high levels of illegal, unregulated and unreported (IUU) fishing.

### Bycatch

Within a few years of the start of longline fishing for toothfish around South Georgia, it became clear there was a potentially substantial bycatch of seabirds (and also of skates and rays). This results from the nature of the longline fishing operation, whereby each day each vessel sets at least 10,000 hooks baited with squid and fish. In the brief interval before the hooks sink, they are accessible to albatrosses and petrels. Although many birds are successful in stealing bait, ensuring that flocks of birds associate with such vessels, typically a few are hooked and drowned on each set. Throughout a fishing season, the scale of the problem becomes substantial, with estimates from observer data indicating these longline fisheries in the CCAMLR area killed thousands of birds annually. These levels had the potential of breaching a fundamental requirement of the Convention, because it was unlikely that the resultant observed population decreases in albatrosses (Figure 8) could be reversed within 20–30 years (particularly as albatrosses have one of the lowest reproductive rates amongst birds and do not breed until, on average, ten years old). Indeed, projections indicated that some albatross populations were liable to become extinct within 50–100 years (Arnold, Brault & Croxall forthcoming).

In 1994, CCAMLR established an *ad hoc* Working Group to advise on the problem, which recommended the use of a variety of measures designed to reduce seabird bycatch. First, CCAMLR developed a suite of compulsory mitigation methods for longline vessels. In essence, these are:

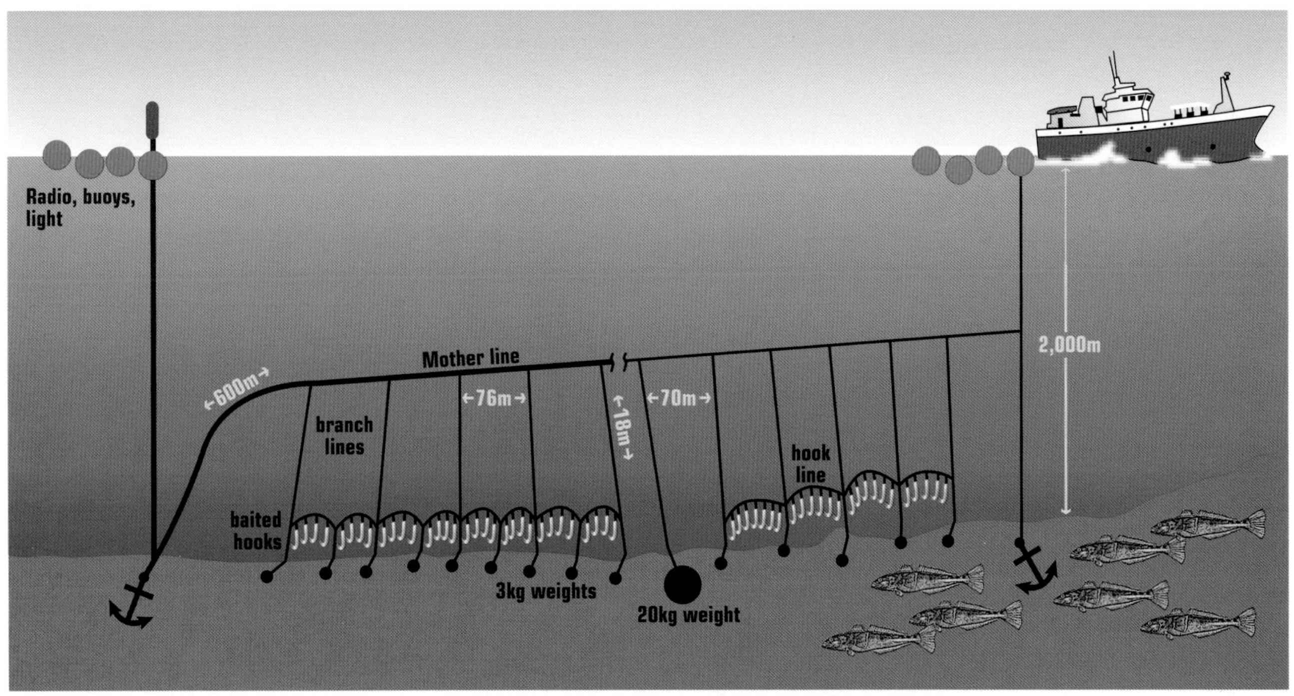


**Figure 7** Longlining for Patagonian toothfish. Every day each vessel sets and hauls 10,000–20,000 baited hooks.

- No discharge of offal, to avoid attracting birds
- Use of streamer lines to keep the birds away from the danger area while the baited hooks are being set
- Increase of weight on the lines to sink them quickly below the reach of albatrosses—which can only dive to about five metres (unfortunately some petrels can dive to 20 metres, so this can require substantial extra weight)
- Setting of lines only at night, albatrosses being largely diurnal, though some petrels can feed at night

Second, using data on albatross distribution and foraging range for each part of the CCAMLR Area, the risk of bycatch was assessed and the use of appropriate combinations of mitigation methods determined. Third, CCAMLR made it obligatory for each vessel to carry an observer from a nation other than that of the vessel's flag-State to collect fishery data (including bycatch) and assist implementation of mitigation measures. Fourth, for the highest risk areas (around the breeding grounds of albatrosses and petrels) fishing was prohibited during the albatross breeding seasons.

These measures had rapid effects at South Georgia where their introduction after 1997 (when nearly 6,000 birds were killed) caused an immediate reduction in seabird bycatch (Figure 9). This was followed by further reductions, once the mitigation measures were used more consistently and correctly, with levels of seabird bycatch being essentially negligible (in terms of effects on the populations concerned) from 2000 onwards.

Seasonal restrictions on longline fishing were not adopted within the South African and French EEZs in the Indian Ocean, however, and seabird mortality levels remained high in those areas. In the South African EEZ, reduction in fishing effort, fishing further away from seabird breeding colonies and more effective use of mitigation measures did reduce bycatch to levels similar to those at South Georgia by 2002. In the French EEZs, however, over 10,000 white-chinned petrels are still being killed annually. New regulations and urgent studies are being implemented to try to rectify the problem (CCAMLR 2003b). With this

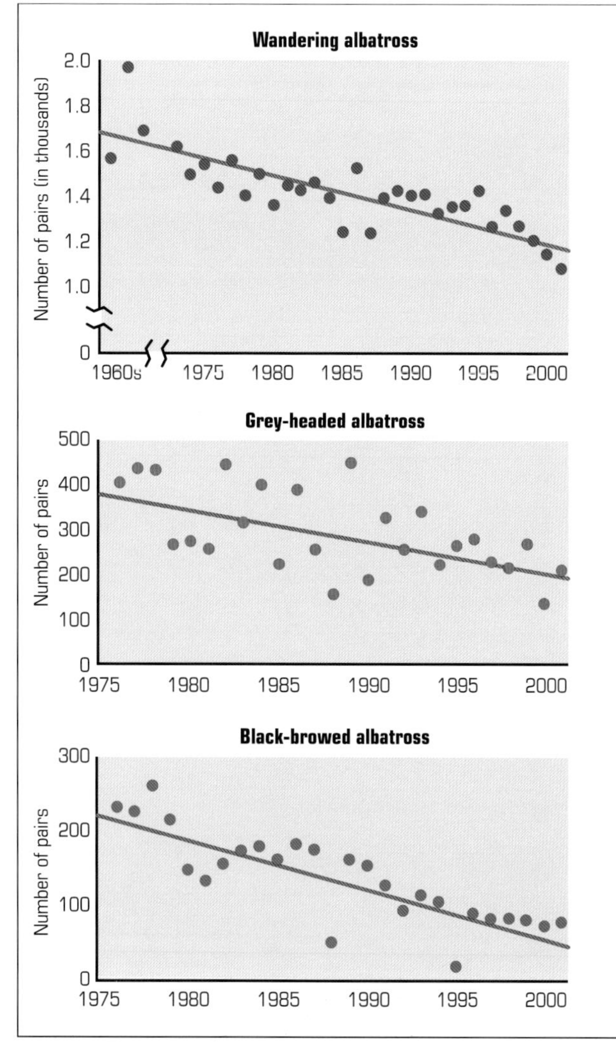


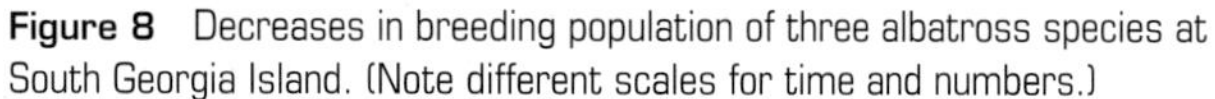
**Figure 8** Decreases in breeding population of three albatross species at South Georgia Island. (Note different scales for time and numbers.)

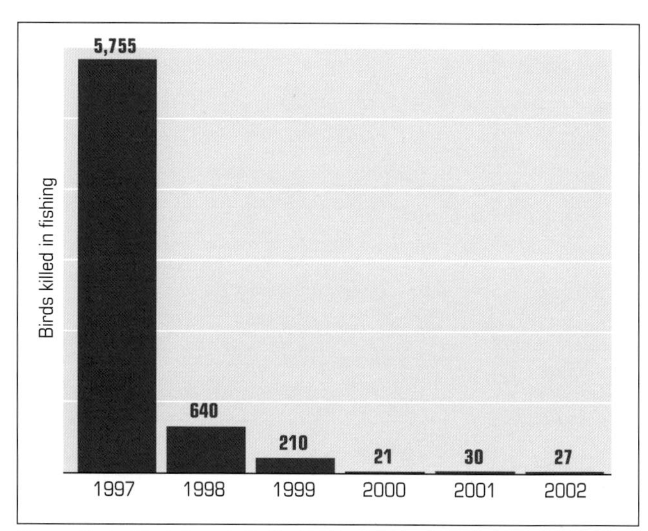


**Figure 9** Incidental mortality (bycatch) of seabirds by longline fishing for toothfish at South Georgia since 1997. Since the first CCAMLR Conservation Measures to reduce seabird bycatch were introduced after 1997, there has been a 100-fold reduction in bird kills (CCAMLR 2003b).

exception, CCAMLR—and notably the fishing interests represented therein—have proved able to address seabird incidental mortality in a manner more effective than in any other Regional Fisheries Management Organisation (Small forthcoming).

Despite CCAMLR's success in addressing this issue, seabird populations are still in peril from longline fisheries for tuna and allied species in waters to the north of the Convention Area and by IUU fisheries in the Convention Area. The latter, where vessels are unlikely to bother to use mitigation measures to avoid catching birds, are estimated to kill tens of thousands of seabirds in the Convention Area annually (CCAMLR 2003b).

CCAMLR's attention to bycatch issues has also extended to fish; many skates and rays are caught and killed or discarded due to toothfish longlining. Because the effect of this bycatch on the populations concerned cannot currently be quantified, CCAMLR has adopted interim bycatch limits for these taxa, together with rules requiring fishing activities to move a specified distance once a threshold level of skate and ray bycatch has been reached (CCAMLR 2003b).

**Illegal, unregulated and unreported (IUU) fishing**

The threat posed by IUU activity to regulated toothfish fisheries is well documented (Agnew 2000, Constable *et al* 2000, CCAMLR 2003b). IUU catches of toothfish were initially confined to the South Georgia region, but by 1996, levels in the Indian Ocean may have reached 100,000 tonnes—well in excess of the aggregate limit for regulated fisheries throughout the whole Convention Area (Figure 10).

To address this situation that was threatening the complete collapse of toothfish stocks in most of the Convention Area, the Commission set up three main initiatives. First, it collaborated operationally and through information exchange to police its area more effectively—resulting in numerous arrests, fines and confiscation of catch. Second, the use of satellite-linked tracking systems called vessel monitoring systems (VMS) was made mandatory for longline vessels fishing in the Convention Area. To be fully effective, however, this system needs to be operated centrally in real time, rather than through various flag-States. Third, it developed a catch documentation scheme whereby only fish with certificates indicating legal capture in the Convention Area can be landed and marketed. This has enjoyed some success but, as with any such trade measure, many difficulties remain to be overcome.

These measures indicate the global nature of problems facing the management of Southern Ocean living resources. Despite these initiatives, especially the resolute policing of the main fishing grounds around sub-Antarctic Islands, only at South Georgia and in the Ross Sea are IUU catches currently at insignificant levels. There is evidence of reduction in IUU catch levels elsewhere, which may augur better for the future. If not, many stocks of toothfish will soon

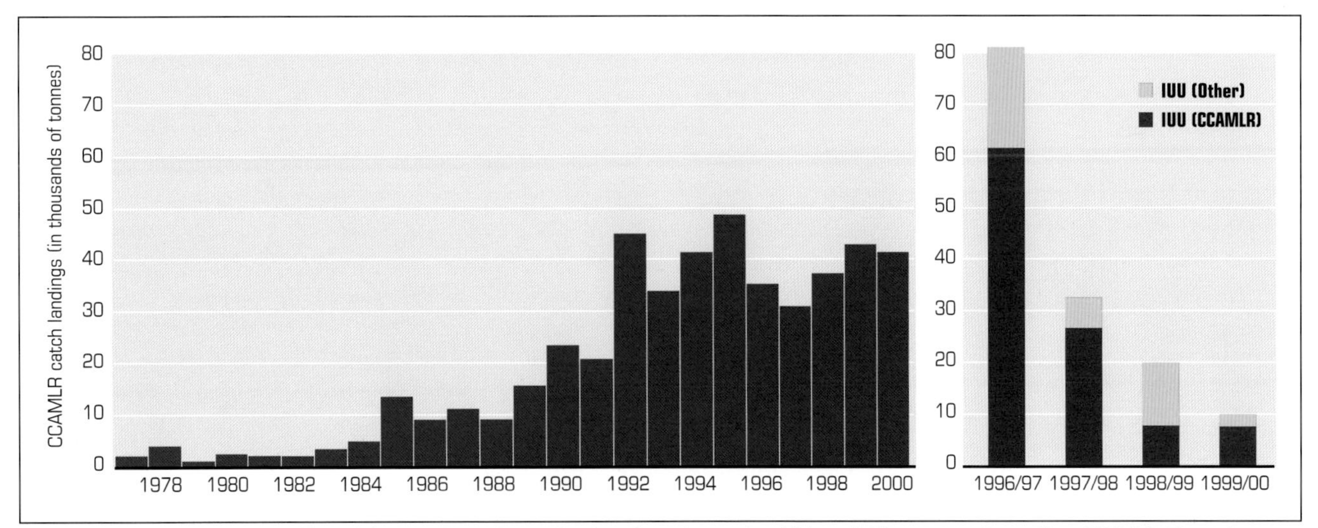


**Figure 10** Catches of toothfish *Dissostichus* spp. in the Convention Area from the regulated fishery (left) as reported annually to CCAMLR. The estimates of IUU (illegal, unregulated and unreported) catches in and around the CCAMLR area (right) are derived by CCAMLR from various sources (CCAMLR 2002b, CCAMLR 2003b).

Table 5

**Some success and challenges for CCAMLR's management of toothfish fisheries**

| Successes | Challenges |
|---|---|
| Within the CCAMLR area, good precautionary management of main stocks | Within the CCAMLR area, limited control of IUU outside South Atlantic |
| Reasonable control on fishery expansion into new areas | Outside CCAMLR area, bycatch measures inadequate to protect "CCAMLR" species |
| Development of catch documentation scheme (CDS) | Improve implementation and efficiency of CDS |
| Compulsory use of vessel monitoring system (VMS) | Convert VMS to centralised system, rather than via flag-State filter |
| Within the CCAMLR area, seabird bycatch largely under control | Within CCAMLR area, skate and ray bycatch still concerns |

become depleted to levels from which they may not be able to recover.

As a summary of its successes and challenges indicates (Table 5), CCAMLR was able to develop a reasonably robust stock assessment model for toothfish, precautionary decision rules and a suite of measures to mitigate bycatch of seabirds. Its ability to control the expansion of the fishery outside areas under national jurisdiction (i.e., into the Ross Sea) has been less successful despite well intentioned procedures for collection of data to enable timely assessment of toothfish fisheries as they expand into new areas. The main challenges are those posed by activities largely outside the control of CCAMLR itself, principally relating to the activities of flags of convenience and IUU fishing. And seabirds breeding in the Convention Area are facing potentially unsustainable levels of mortality in adjacent areas. These problems illustrate the need for similar best-practice standards of governance and management to prevail throughout the world's ocean.

## Summary and Recommendations

Based on CCAMLR experiences, there are three additional general observations relevant to management of marine resources and systems that may provide good lessons for other Regional Fishery Management Organisations (RFMOs).

First, many features of CCAMLR's successes in fishery management have been achieved through its scientific observer scheme. This arose out of the need to collect data to help manage specific fisheries but has now become an invaluable source of data and information on other topics such as bycatch and operational fishing practices. It is greatly to the credit of all parties, especially the fishers, that observers are required—and accepted—on all vessels fishing for finfish in the Convention Area; the reluctance to have observers on krill fishing vessels needs to be rapidly overcome.

Second, there has been a lack of attention to the development of integrated approaches for identifying critical habitats that could be accorded special management protection and/or be included within systems or networks of marine protected areas. There is considerable and growing pressure on those responsible for the governance and management of marine systems and habitats to investigate how best to use marine protected areas—and other related approaches—within the areas of their jurisdiction and responsibilities. Although CCAMLR has been proactive in applying the precautionary principle to the management of marine systems, this has not yet included any systematic review of ways in which concepts like marine protected areas could be used or incorporated into its management systems and approaches. CCAMLR needs to move rapidly to create frameworks and mechanisms for addressing this topic in general and for the development and examination of specific proposals in particular. These proposals should include, but not be limited to:

- Areas within EEZs in the Convention Area
- Areas adjoining existing EEZ marine protected areas (e.g., Heard and MacDonald Islands), perhaps especially in considering the potential for protected corridors
- Areas surrounding archipelagos with exceptional marine biodiversity (e.g., Balleny Islands)
- Seamount and canyon habitats and areas
- Large-scale areas with unique characteristics, perhaps particularly where management of exploitation of marine living resources coexists with extensive scientific

research programmes—the Ross Sea is an area of particular relevance (Ainley 2002)
- Potential enhancement of existing marine protected areas within the Convention Area

Third, CCAMLR management does not yet adequately incorporate information on a range of topics that will have potentially fundamental effects on Southern Ocean fisheries. These include long-term and medium-term changes in the physical environment (Vaughn *et al* 2003, Trathan & Murphy 2003), the impact of scientific and technological developments and the linkage of these to economic forces. Many of the more important drivers, reviewed in some detail in Croxall and Nicol (forthcoming), originate from outside the Southern Ocean system. But these outside forces are likely to produce greater changes in—and pressures on—resources than those events CCAMLR is currently managing.

Based on the CCAMLR experience, there are a number of steps that governments need to take:

- Eliminate illegal, unregulated and unreported (IUU) fishing
- Ensure that all RFMOs adopt best-practice management standards

This would be facilitated by:

- Outlawing "flags of convenience" and improving flag-State controls for fishing
- Transferring some appropriate management responsibilities to RFMOs
- Implementing the FAO draft International Plan of Action for IUU with mandatory as well as voluntary measures

These steps will also require expanding and coordinating intelligence collection, vessel monitoring (VMS), fishery protection, port-State inspection, catch documentation and vessel blacklisting.

To improve the governance and management of fisheries, RFMOs will need to:

- Adopt a precautionary approach to sustainable fishing
- Fully take into account the needs of dependent species
- Implement appropriate mitigation methods to eliminate bycatch of non-targeted species
- Adopt adequate reporting standards of effort, catch and bycatch (using independent observers—and/or video monitoring systems—whenever possible)
- Adopt compulsory (tamper proof) vessel monitoring systems linked to central processing systems (avoiding flag-State filters)
- License only compliant (green-listed) vessels, linked to appropriate incentives and training schemes

## Acknowledgements

*We are grateful to the many colleagues and collaborators who have contributed to the development of our ideas through the numerous stimulating discussions that we have had whilst working with CCAMLR. We also thank the many BAS scientists and field assistants who have helped by contributing data and analyses. Also, particular thanks to both Janet Silk and Liz Dixon for their help preparing the figures and text.*

## Literature Cited & Consulted

Agnew, D. J. 1997. Review: The CCAMLR ecosystem-monitoring programme. *Antarctic Science* 9:235–242.

——. 2000. The illegal and unregulated fishery for toothfish in the Southern Ocean and the CCAMLR catch documentation scheme. *Marine Policy* 24:361–374.

Ainley, D. G. 2002. The Ross Sea: where all ecosystem processes still remain for study, but maybe not for long. *Marine Ornithology* 30:55–62.

Arnold, J. M., S. Brault and J. P. Croxall. Forthcoming. Albatross populations in peril? A population trajectory for black-browed albatrosses at South Georgia. *Ecological Applications.*

Barlow, K. E., I. L. Boyd, J. P. Croxall, K. Reid, I. J. Staniland and A. S. Brierley. 2002. Are penguins and seals in competition for Antarctic krill at South Georgia? *Marine Biology* 140:205–213.

Boyd, I. L. 1993. Pup production and distribution of breeding Antarctic fur seals Arctocephalus gazella at South Georgia. *Antarctic Science* 5:17–24.

Boyd, I. L. and A. W. A. Murray. 2001. Monitoring a marine ecosystem using responses of upper trophic level predators. *Journal of Animal Ecology* 70:747–760.

Butterworth, D. S., A. E. Punt and M. L. Basson. 1991. A simple approach for calculating the potential yield of krill from biomass survey results. *Selected Scientific Papers* 1991 (SC-CAMLR-SSP/8), 207–217. Hobart, Australia: CCAMLR.

CCAMLR. 1989. Report of the Workshop on the Krill CPUE Simulation Study (La Jolla, USA, June 7–13, 1989). Report of the eighth meeting of the Scientific Committee (SC-CAMLR-VIII), Annex 4:81–137. Hobart, Australia: CCAMLR.

——. 1995. Report of the Working Group on Ecosystem Monitoring and Management. Report of the fourteenth meeting of the Scientific Committee. Hobart, Australia: CCAMLR.

——. 2002a. Report of the twenty-first meeting of the Scientific Committee. Appendix D Report of the Workshop on small-scale management units, such as predator units. Hobart, Australia: CCAMLR.

——. 2002b. Report of the twenty-first meeting of the Scientific Committee. Hobart, Australia: CCAMLR.

——. 2003a. Report of the twenty-second meeting of the Scientific Committee. Report of the Working Group on Ecosystem Monitoring and Management, Appendix D, CEMP Review Workshop. Hobart, Australia: CCAMLR.

——. 2003b. Report of the twenty-second meeting of the Scientific Committee. Hobart, Australia: CCAMLR.

Constable, A. J. 2002. CCAMLR ecosystem monitoring and management: Future work. *CCAMLR Science* 9:233–256.

Constable, A. J. and W. K. de la Mare. 1996. A generalised model for evaluating yield and the long-term status of fish stocks under conditions of uncertainty. *CCAMLR Science* 3:31–54.

Constable, A. J., W. K. de la Mare, D. J. Agnew, I. Everson and D. Miller. 2000. Managing fisheries to conserve the Antarctic marine ecosystem: Practical implementation of the Convention on the Conservation of Antarctic Marine Living Resources. *ICES Journal of Marine Science* 57:778–791.

Croxall, J. P. 1989. Use of indices of predator status and performance in CCAMLR fishery management strategies. *CCAMLR Selected Papers,* 353–365.

——. 1992. Southern Ocean environmental change: Effects on seabird, seal and whale populations. *Philosophical Transactions of the Royal Society of London,* Series B, 338, 319–328.

——. 1994. Biomass—CCAMLR relations: past, present and future. In *Southern Ocean Ecology: The BIOMASS Perspective,* ed. S. Z. El-Sayed, 339–353. Cambridge: Cambridge University Press.

——. Forthcoming. Monitoring of predator-prey interactions using multiple predator species: The South Georgia experience. In *Management of Marine Ecosystems: Monitoring Change in Upper Trophic Levels,* ed. I. L. Boyd and S. Wanless. Cambridge: Cambridge University Press.

Croxall, J. P. and S. Nicol. Forthcoming. Management of Southern Ocean resources: A model for global sustainability? *Antarctic Science.*

Croxall, J. P., T. S. McCann, P. A. Prince and P. Rothery. 1988. Reproductive performance of seabirds and seals at South Georgia and Signy Island 1976–1986: Implications for Southern Ocean monitoring studies. In *Antarctic Ocean and Resources Variability,* ed. D. Sahrhage, 261–285. Berlin: Springer-Verlag.

de la Mare, W. K and A. J. Constable. 2000. Utilising data from ecosystem monitoring for managing fisheries: Development of statistical summaries of indices arising from the CCAMLR ecosystem-monitoring programme. *CCAMLR Science* 7:101–117.

El-Sayed, S. Z. 1994. History, organization and accomplishments of the BIOMASS Program. In *Southern Ocean Ecology: The BIOMASS Perspective,* ed. S. Z. El-Sayed, 1–8. Cambridge: Cambridge University Press.

Everson, I., G. Parkes, K. H. Kock And I. L. Boyd. 1999. Large variations in natural mortality rate of marine fish and its effects on multispecies management: the mackerel icefish *(Champsocephalus gunnari)* a practical example. *Journal of Applied Ecology* 36:591–603.

Hewitt, R. P. and E. H. L. Low. 2000. The fishery on Antarctic krill: Defining an ecosystem approach to management. *Reviews in Fisheries Science* 8:235–298.

Hewitt, R. P., J. L. Watkins, M. Naganobu, P. Tshernyshkov, A. S. Brierley, D. A. Demer, S. Kasatkina, Y. Takao, C. Goss, A. Malyshko, M. A. Brandon, S. Kawaguchi, V. Siegel, P. N. Trathan, J. H. Emery, I. Everson and D. G. M. Miller. 2002. Setting a precautionary catch limit for Antarctic krill. *Oceanography* 15:26–33.

Miller, D. G. and D. J. Agnew. 2000. Krill fisheries in the Southern Ocean. In *Fish and Aquatic Resources, Krill: Biology, Ecology and Fisheries,* Series 6, ed. I. Everson, 300–337. Oxford: Blackwell Science.

Murphy, E. J. and K. Reid. 2001. Modelling Southern Ocean krill population dynamics: Biological processes generating fluctuations in the South Georgia ecosystem. *Marine Ecology Progress Series* 217:175–189.

Nicol, S. and Y. Endo. 1997. *Krill fisheries of the world.* Rome: Food and Agriculture Organization of the United Nations.

Nicol, S., T. Pauly, N. L. Bindoff and P. G. Strutton. 2000. "BROKE": A biological/oceanographic survey of the waters off East Antarctica (80-150°E) carried out in January–March 1996. *Deep-Sea Research II* 47:2281–2298.

Nicol, S. and J. Foster. 2003. Recent trends in the fishery for Antarctic krill. *Aquatic Living Resources* 16:42–45.

Reid, K., J. L. Watkins, J. P. Croxall and E. J. Murphy. 1999. Krill population dynamics at South Georgia 1991–1997, based on data from predators and nets. *Marine Ecology Progress Series* 177:103–114.

Reid, K. and J. P. Croxall. 2001. Environmental response of upper trophic level predators reveals a system change in an Antarctic marine ecosystem. *Proceedings of the Royal Society,* Series B 268:377–384.

Reid, K., E. J. Murphy, V. Loeb and R. P. Hewitt. 2002. Krill population dynamics in the Scotia Sea: variability in growth and mortality within a single population. *Journal of Marine Systems* 36 (1–2): 1–10.

Reid, K., J. P. Croxall, E. Murphy and P. N. Trathan. Forthcoming. Monitoring krill population variability using seabirds and seals at South Georgia. In *Management of Marine Ecosystems: Monitoring Change in Upper Trophic Levels,* ed. I. L. Boyd and S. Wanless. Cambridge: Cambridge University Press.

Small, C. Forthcoming. *An evaluation of the environmental performance of southern hemisphere Regional Fishery Management Organisations, with particular attention to bycatch and incidental mortality of seabirds.* Cambridge: BirdLife International.

Trathan, P. N., D. Agnew, D. G. M. Miller, J. L. Watkins, I. Everson, M. R. Thorley, E. J. Murphy, A.W.A. Murray and C. Goss. 1993. *Krill Biomass in Area 48 and Area 58: Recalculation of FIBEX Data.* (SC-CAMLR-SSP/9 WG-KRILL-92/20), 157–181.

Trathan, P. N, J. L. Watkins, A. W. A. Murray, A. S. Brierley, I. Everson, C. Goss, J. Priddle, K. Reid, P. Ward, R. Hewitt,

D. Demer, M. Naganobu, S. Kawaguchi, V. Sushin, S. M. Kasatkina, S. Hedley, S. Kim and T. Pauley. 2001. The CCAMLR-2000 Krill Synoptic Survey: A description of the rationale and design. *CCAMLR Science* 8:1–23.
Trathan P. N., A. S. Brierley, M. A. Brandon, D. G. Bone, C. Goss, S. A. Grant, E. J. Murphy and J. L. Watkins. 2003. Oceanographic variability and changes in Antarctic krill abundance at South Georgia. *Fisheries Oceanography* 12:569–583.
Trathan, P. N. and E. J. Murphy. 2003. Sea surface temperature anomalies near South Georgia: relationships with the Pacific El Niño regions. *Journal of Geophysical Research* 108, no. C4, 8075, doi:10.1029/ 2000jc000299.
Trathan, P. N., E. J. Murphy, J. Forcada, J. P. Croxall, K. Reid and S. E. Thorpe. Forthcoming. Physical forcing in the southwest Atlantic: Ecosystem control. In *Management of Marine Ecosystems: Monitoring Change in Upper Trophic Levels.* ed. I. L. Boyd and S. Wanless. Cambridge: Cambridge University Press.
Vaughan, D. G., G. J. Marshall, W. M. Connolley, C. Parkinson, R. Mulvaney, D. A. Hodgson, J. C. King, C. J. Pudsey and J. Turner. 2003. Recent rapid regional climate warming on the Antarctic Peninsula. *Climatic Change* 60:243–274.

CHAPTER 4

# Coral Triangle

Jamie Bechtel, *Conservation International*
Timothy B. Werner, *Consultant*
Ghislaine Llewellyn, *Ecosafe Consultants*
Rodney V. Salm, *The Nature Conservancy*
Gerald R. Allen, *Conservation International consultant*

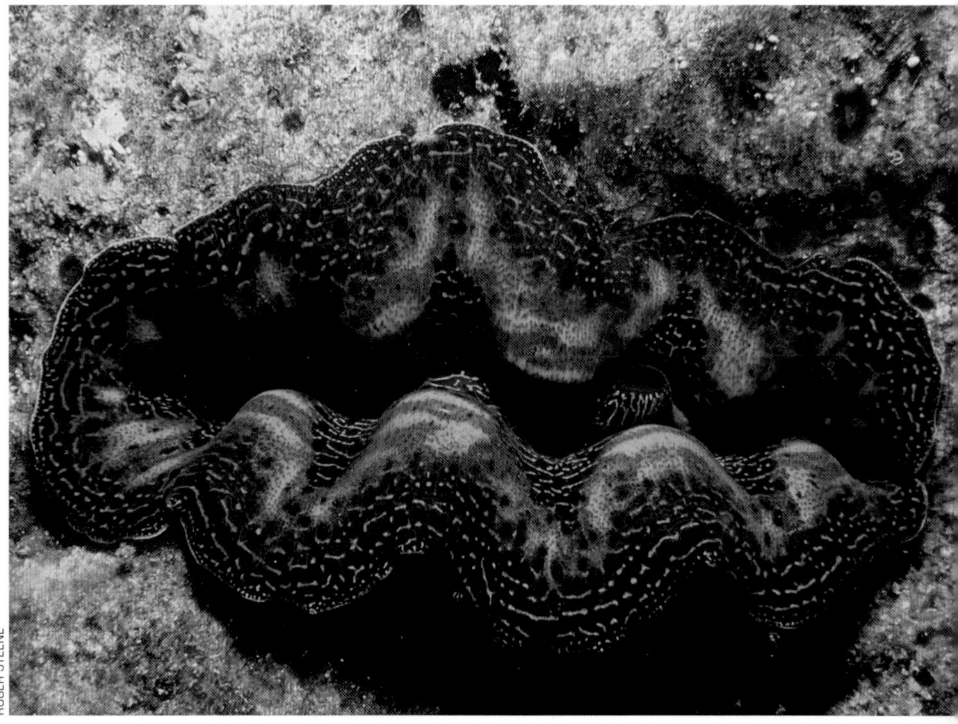
ROGER STEENE

## Geographic History and Biodiversity

The marine biodiversity of the region known as the Coral Triangle is the richest on Earth. The term "Coral Triangle" was first introduced by Werner and Allen (1998), and although research to delineate the area accurately is ongoing, it is generally accepted to be centered on Eastern Indonesia, and to encompass the waters of Indonesia, Philippines, Malaysia and Papua New Guinea (Hoeksema & Putra 2000, Vernon 1995). Recent biodiversity assessments suggest the Coral Triangle may also include the Bismarck/Solomon archipelago.

The earliest reference to the exceptional biodiversity found in the Western Pacific region is likely that of Clark (1912) who described the distribution of crinoids and found an "area of maximum intensity within a triangle whose apices are Luzon, Borneo and New Guinea." (Crinoids are flower-like marine animals related to starfish, with cup-shaped bodies and feathery arms, which attach to the sea floor.) Marine biodiversity in the Coral Triangle today is an extension of the oldest and most predominant center of diversity in the marine system for the past ten million years. New species continue to be discovered, and it is believed that the area has influenced marine community structures as far east as the eastern tropical Pacific and as far west as the Atlantic side of the Cape of Good Hope. While the extraordinary diversity of the region has been evident for some time, theories to explain it are still at issue. Recent paleontology data from the remains of ancient animal communities suggest that current Coral Triangle biodiversity is the result of ancient movements of the continents and corresponding effects on their surrounding seas. It is believed that an eastern migration of species within the Tethys Sea in the late Mesozoic and early Tertiary—roughly 135 to 50 million years ago—was caused by degrading climatic conditions coupled with tectonic movements of the continental and oceanic crustal plates of that period. These processes resulted in the establishment of an Indo-Pacific center of origin for new species.

More recently, two distinct speciation processes (the evolution of new species) have contributed to the vast wealth of biodiversity in this region. First, the large central populations have supported the continued production of successful, wide-ranging species that continue to evolve, expand from their point of origin and establish themselves in new territories (Briggs 1995, 1999). In short, high diversity has begotten higher diversity and this historic center of origin has continued to generate new species for millions of years. Second, small peripheral populations have given rise to new species where local habitats and ecological conditions are favorable, resulting in a high speciation rate throughout the Coral Triangle. This type of speciation is particularly successful in generating new species, including endemics (species found nowhere but in a limited area), when extinction occurs throughout the central portion of a species range (Mayr 1954). Although both types of evolutionary processes occur, the latter process may be extremely important in the Coral Triangle because of the complex array of islands in the Indo-Pacific which provide innumerable isolated habitats that facilitate the creation of unique subpopulations isolated from the

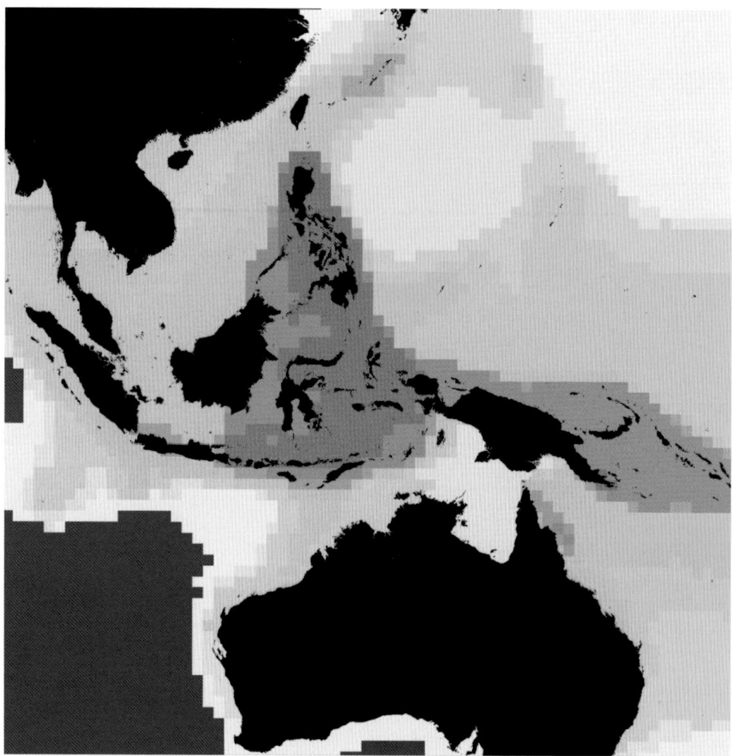

**Figure 1** General outline of the "Coral Triangle" area of high biodiversity is indicated by the combined number of reef fish species from three families—butterflyfishes, damselfishes and angelfishes.

Number of reef fish species identified

- 1–17
- 18–39
- 40–87
- 88–115
- 116–152
- 153–159
- 160–170
- 171–185

rest of a species range. The repeated rise and fall of sealevels in this region has periodically isolated, then reunified, the many islands over geological time, which has also facilitated the expansion and contraction of species ranges and contributed to the remarkable diversity in this area (Huston 1994).

### The Coral Triangle's exceptional diversity

The waters around Indonesia, Malaysia, Papua New Guinea and the Philippines contain 32.9% of all coral reefs in the world, but they include fully 77% of the world's coral reef species—including nearly 70 genera (the biologic class above species, seen as the first name in Latin species names) and almost 500 species of hermatypic corals (reef-building corals that grow in the "photic zone," the sunlit shallow layers of water). In addition, over 50% of the world's reef fish species (Figure 1), 58% of tropical marine molluscs, 40% of the seagrasses and 88% of the mangrove species are found in the Coral Triangle. In Indonesia alone there are approximately 2,057 species of coral reef fishes (Allen & Adrim 2003), 782 species of algae (Gomez *et al* 1990), thirteen seagrass species (Moosa 1999) and 38 mangrove species. These tremendously high levels of biodiversity occur in less than 1% of the global ocean's surface area. The exceptional importance of the region in terms of marine biodiversity extends to deep-water and open-ocean habitats as well. Within the Coral Triangle there is both the largest tuna fishery and the most diverse assemblage in the world of deep-water corals. In addition to the high species diversity, the Coral Triangle exhibits exceptional rates of marine endemism. For example, 97 species of Indonesian reef fishes are endemic (found nowhere else), and there are more than 50 endemic fish species in the Philippines. This amazing phenomenon, in which so many major groups of organisms achieve their maximum biological richness and high endemism appears unrivaled, with no comparable analogues anywhere in the world in either terrestrial or freshwater areas.

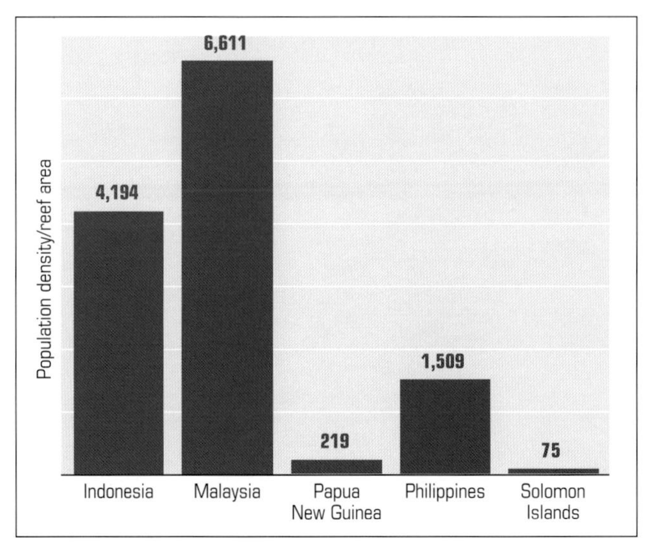


**Figure 2** The concentration of human population on land versus the area of reefs for each nation in the Coral Triangle. This ratio can be used to assess the impact of human land-based activities on the health of reefs surrounding each country.

## Status and Threats

No systematic analysis of the status of the marine environment has been conducted for the whole of the Coral Triangle, though some such analyses have been done for small areas in the region. These studies indicate that, overall, marine biodiversity in the Coral Triangle has not escaped the general trend of loss and degradation occurring around the world. High human population density in the Philippines, Indonesia and Malaysia (Figure 2), especially given the low ratio of land area to coastline (Table 1), exerts overwhelming pressure on coastal marine diversity as does historically high levels of fishing effort for both subsistence and profit. In Papua New Guinea and the Solomon Islands, anthropogenic (human-induced) impacts have historically been lower, but reefs are beginning to show signs of degradation due to overfishing, destructive fishing practices and other threats.

Table 1

**Land, coastline and maritime jurisdiction areas for the Coral Triangle nations**

| Country | Land area (thousands of km²) | Maritime area (thousands of km²) | Coastline (thousands of km) |
|---|---|---|---|
| Indonesia | 1,900 | 6,121 | 95.2 |
| Philippines | 298 | 974 | 33.9 |
| Malaysia | 330 | 450 | 9.3 |
| Papua New Guinea | 462 | 2,366 | 20.2 |
| Solomon Islands | 28 | 1,589 | 9.8 |
| **Total** | **3,018** | **11,500** | **168.4** |

**Note:** Indonesia occupies about two-thirds of the land mass in the Coral Triangle area. The region's maritime area, however, is more than three times the size of the land area in the five countries.
**Source:** World Resources Institute 2004.

One of the most systematic studies of marine habitat status in the Coral Triangle is the "Reefs at Risk" series produced by World Resource Institute and partners (Bryant *et al* 1998, Burke, Selig & Spalding 2002). In the Coral Triangle, this analysis classified nearly all coral reefs in the Philippines (97%), Indonesia (82%) and Malaysia (91%) at either medium or high risk. Less than half of Papua New Guinea and Solomon Islands reefs are at medium or high risk. Other studies have shown that coral reefs in the region are highly degraded, with one report estimating that only 4–5% of coral reefs in the Philippines are in excellent condition.

In addition to the significant degradation recorded for marine habitats, there are numerous species in trouble in the Coral Triangle, many of which are included on the IUCN–The World Conservation Union Red List of threatened and endangered species. In Indonesia, for example, dugongs, all six species of marine turtles, 29 species of whales and dolphins, including the highly endangered blue whale, and twelve species of mollusc have been listed for protection by the government.

### Blast fishing

Blast fishing is the use of explosives to kill fish—which also kills many other species and undermines the structure of the reefs. This practice is prevalent throughout the Coral Triangle, particularly in Indonesia, Malaysia and the Philippines. Blast fishing was introduced to Indonesia by the Japanese during World War II. Over the past 60 years, it has become a traditional fishing practice in the area, conducted predominately in areas where the coastline has significant buildings and human population. The extent of coral reef coverage of the seafloor in Indonesia's Jakarta Bay has been reduced to 2.5%, due largely to fishing by blasters. Small-scale blast fishermen use fertilizer to extract fish from nearby reefs. Medium- to large-scale fishery operations use strong blast devices with detonators and have adequate storage for large catches of up to two tons of fish. Because they remain at sea for nearly two weeks covering expansive areas and reaching deeper reefs with their detonations, the medium- to large-scale fishing operations are much more destructive than small-scale fishing. Subsequent coral damage is found as deep as 20 meters. In addition to damaging habitats, this type of fishing also removes large schools of herbivores, kills large amounts of unintended species (bycatch) and removes from the ecosystem fish of all sizes and stages of their reproductive cycle. The economic costs of these unaccounted-for environmental impacts are in addition to direct economic costs from this type of fishing. The blasting results in lower market quality and subsequent lower market price for the fish; an inability to access distant markets because of damage to the fish, which exhibit a faster decay rate; and lower fish yield on the blasted reefs because of damage to the habitat (Figure 3). It is estimated that in Indonesia alone the total revenue lost over the next 20 years because of blast fishing will be approximately US $3 billion (Pet-Soede, Cesar & Pet 1999).

© LYNN FUNKHAUSER

© WOLCOTT HENRY 2001

**Figure 3** Blast fishing (above) is prevalent throughout the Coral Triangle, particularly in Indonesia, Malaysia and the Philippines. Its impacts range from the indiscriminate removal of fish from the ecosystem to destruction of the coral reef, leaving rubble like that seen at the base of this reef (below).

Blast fishing is also practiced extensively in Malaysia, particularly along the Sabah coast. Migrant fishers are prevalent; they degrade local fishing resources, disrupting established local economies and then move on to other locations. In the Philippines blast fishing has decreased throughout much of the country due to successful awareness campaigns, the establishment of marine protected areas (MPAs) and the presence of alternative income sources. Blast fishing still remains a problem, though, throughout much of the southern Philippines and much of Papua New Guinea, where the reefs are increasingly exhibiting signs of degradation. Although blast fishing is now illegal in most places in the Coral Triangle, it remains a very active fishery where bribery and the lack of enforcement are common.

**Fishing with poisons**

Another destructive extractive method frequently practiced in the western and Indo-Pacific is cyanide fishing. Not only does the application of cyanide to the reef destroy coral tissues leaving the corals susceptible to disease and overgrowth by algae, but the corals themselves are often crushed in order to allow fishermen access to the stunned fish that seek shelter in the protective canopy of the corals. In Indonesia, this practice dates as far back as the 17th century when crushed roots and stems containing poison were used to capture fish (Cesar 1996). In Papua New Guinea, derris root and other traditional poisons have been used for generations to collect fish. The use of poisons to extract fish is closely associated with the Live Reef Fish Trade (LRFT). In 2002, Indonesia was the largest supplier of live reef food fish to Asian markets with nearly 50% of wild-caught fish exported to Singapore and Hong Kong. Of particular concern is the targeting of spawning aggregations as supply sources for this industry. Cyanide fishing is also used to supply ornamental fish to the aquarium industry. It is estimated that 85% of traded aquarium fish have been caught by being stunned with cyanide. As fish populations collapse in Southeast Asia, areas such as Papua New Guinea and the Solomon Islands will become increasingly subject to this destructive fishing process.

**Runoff and sedimentation**

Sedimentation often has a major impact on marine systems in the Coral Triangle, particularly in areas with numerous islands, because of increased land-to-water ratios. Increased rates and amounts of freshwater and sediment pouring into the sea from deforestation, unsustainable agriculture practices and coastal building development disrupt the natural hydrological cycles and place significant pressures on already taxed marine ecological processes. These threats are increasingly problematic in the Coral Triangle. In heavily settled areas of Indonesia, sediment runoff has more than doubled. In Malaysia, the sedimentation load carried by the many rivers of Sarawak has increased a hundred-fold or more in the past 30 years. Nearly half of Papua New Guinea's forests have been allocated for logging. Although little data exists regarding the impact of sediment loads from logging, anecdotal observations on reefs in the Sepik Province and on Manus Island have shown coral decline linked to logging practices (Figure 4). The cultivation of oil palm, copra, rubber, coffee and other agriculture products are also likely to result in increased sedimentation in Papua New Guinea waters, although quantitative studies have not yet been conducted. Edinger *et al* (1998) have demonstrated that reefs subjected to a variety of land-based pollutants show a decrease in coral cover of 30 to 60%. These impacts are exacerbated when both sewage and sediments are released into a reef system, a combination that is frequently found throughout the region.

Extraordinary sedimentation rates associated with mining have a tremendous impact on reef systems and are increasingly a threat within the Coral Triangle due to the highly lucrative nature of the extracted resources. In Papua New Guinea, the Bougainville copper mine discharged 150,000 tons per day of tailings (solid waste materials from mining) into the Jaba River. The tailings eventually reached Empress Augusta Bay where sediments accumulated to a depth of 27 meters, creating a delta of seven square kilometers and negatively impacting 100 square kilometers of seafloor. In addition to the direct disposal of mine tailings, mining operations often allow a general increase in sediment runoff into nearby rivers, which subsequently flows into marine waters and onto the coral reefs.

In some instances, mine tailings are released directly into the marine system. These "submersible tailings" are carried to sea and usually released at a depth greater than 100 meters. In response to the Bougainville mining experience, the government of Papua New Guinea has been careful to ensure that deep tailing disposals do not impact nearshore coral habitats. This avoids socio-economic impacts to local communities, such as decreased tourism because of lower water quality and visibility and reduced fish catches due to ecological degradation. When analyzing the impacts associated with submersible tailings, however, little consideration has been given to the potentially important biodiversity found at depths, the ecological linkages between shallow and deepwater communities and the associated indirect socio-economic issues. For instance, large migratory animals—whales, dolphins and porpoises, turtles, deepsea sharks and some game fish—that rely on deepwater habitats are directly affected by submarine tailing disposal. Furthermore, some scientists argue that in oceanographically complex areas such as the Coral Triangle, the thermocline—the normally sharp density change between cold deep water and warm shallow water—is not an effective barrier against waste being transported up from the depths over the long term, so nearshore habitats could also be affected.

**Figure 4** Deforestation through logging is prevalent throughout the Coral Triangle region. Logging in Indonesia (above) has resulted in a doubling of sediment loads on area reefs (below) which eventually kills the corals and associated marine life.

### Sewage

Sewage outflow is a threat to marine systems throughout the world. In the Coral Triangle, sewage is one of the most predominant pollutants, resulting in decreased overall biodiversity throughout much of the region. The effects of sewage on coral reefs can be highly detrimental and may include:

- Stifled growth from increased sediment loading
- Lowered photosynthesis rates in zooxanthallae (single-celled algae that live inside coral tissues) on which corals depend for nutrients and growth
- Eutrophication, an over-abundant concentration of nutrients that supports dense growths (blooms) of algae that use all of the oxygen, killing off other lifeforms
- Extensive shifts in community structure which threaten the survival of some species
- Accelerated growth rates of corals, resulting in weaker skeletal structures that are more susceptible to fractures and breaks

The extent of sewage input into the Coral Triangle is exemplified by cities such as Makassar, Indonesia. Makassar has a population of over one million people but no primary sewage treatment facilities, and its discharged sewage impacts reefs nearly 35 kilometers away. In Port Moresby, Papua New Guinea, the only primary treatment of sewage consists of its settlement in a natural swamp; eight sewer lines and numerous other discharges empty into the Papuan lagoon near Port Moresby. Consequently, significant health concerns and ecosystem degradation are ongoing problems in the area. Toxic dinoflagellates (minute marine organisms) have caused paralytic shellfish poisoning in the area since the 1970s (Maclean 1973). The Solomon Islands have no sewage treatment at all; raw sewage is put directly into the sea with sewage plumes clearly visible from the air. As development continues, sewage inputs will have an increasingly negative effect on marine habitats within the Coral Triangle.

### Mining of reefs

The Coral Triangle has a significant proportion of the world ocean's shallow coral reefs (Table 2), and reef materials are mined in this region to provide foundations for housing, seawalls and jetties, foundations for roads, linings for shrimp ponds and other construction uses. This practice has obvious and immediate implications for degradation to reef structures, but also results in decreased overall biomass and increased turbidity (cloudiness in the water from suspended sediments). Economic incentives include the need to mine corals for sand material to offset beach erosion for hotels and other tourist industries relying on beach-based activities. In Papua New Guinea and the Solomon Islands, reefs are mined to produce lime (calcium carbonate) to chew with betel nut. An estimated ten million kilograms of live Acroporid corals are destroyed each year in the Solomon Islands to meet the demand for lime, and consumption of lime with betel nut is considered to be one of the most serious threats to coral reefs in this area.

## Regional Policy Development

Regional policy mechanisms are an important aspect of marine conservation in the Coral Triangle. Although some of these mechanisms have been in place for decades, they have only recently considered marine conservation and management as high-priority issues. Consequently, there exists a solid policy framework within which to implement conservation agendas and strategies in the Coral Triangle. Examples of these important regional policy structures include the Pacific Island Forum and the Association of Southeast Asian Nations. These groups are able to address regional issues effectively, and develop and implement international policies and conventions such as the Conservation and Management of Highly Migratory Fish Stocks in the western and central Pacific Ocean as described below.

Table 2
**Importance of the coral reefs of the Coral Triangle area**

| Country | Coral reef area (sq. km) | Percent of world | Percent of Coral Triangle |
|---|---|---|---|
| Indonesia | 51,020 | 17.95 | 51.40 |
| Philippines | 25,060 | 8.81 | 25.24 |
| Papua New Guinea | 13,840 | 4.87 | 13.94 |
| Malaysia | 3,600 | 1.27 | 3.63 |
| Solomon Islands | 5,750 | 2.02 | 5.79 |
| **Total** | **99,270** | **34.92** | **100.00** |

**Source:** Spalding, Ravilious & Green 2001.

### Pacific Islands Forum

The Pacific Islands Forum (PIF)—the only regional grouping of Pacific nations—includes Papua New Guinea and the Solomon Islands among its member States. The Forum assembles annually to discuss issues of regional concern and has observer status at both the U.N. and in the Asia Pacific Economic Cooperation process. The Pacific Islands Forum Secretariat administers the programs and activities that support and implement decisions made by its leaders.

In 2002, the leaders endorsed the Pacific Islands Regional Ocean Policy at the World Summit on Sustainable Development. This unique policy pursues the vision of a healthy ocean that sustains the livelihoods of Pacific island communities. It was developed in recognition of increasing levels of threat to the ocean that jeopardize the long-term development of the region and its ability to use marine resources sustainably. The Council of Regional Organizations of the Pacific (CROP) is developing a plan for implementation of this Regional Ocean Policy, a process initiated in February 2004 at the Pacific Islands Regional Ocean Forum (PIROF). During that conference, regional priorities and actions were established, including articulation of the following strategic steps:

- Improvement of coastal and ocean governance
- Building of local capacity for sustainable development and management
- Fostering of alliances for securing a healthy ocean
- Establishment of high-level leadership on ocean issues and commitments to effective management of ocean resources

With technical and financial support from the Secretariat of the Pacific Community, the plan was finalized and presented to the Pacific Islands Forum for endorsement in August 2004. At that summit, leaders proposed the ocean policy and associated implementing framework be submitted to the ten-year review of the Barbados Program of Action as a major regional initiative for funding and development of partnerships. In addition to the Regional Ocean Policy, the Secretariat of the Pacific Community supports several programs within its Marine Resource Division that have direct bearing on conservation efforts for PIF member States. These include the Oceanic Fisheries Program, the Coastal Fisheries Program and the Aquaculture Initiative.

The Oceanic Fisheries Program was formed in 1980 at the South Pacific Conference. Its mandate is "to provide member countries with the scientific information and advice necessary to rationally manage fisheries exploiting the region's resources of tuna, billfish and related species." With funds from the European Union, Australian Agency for International Development and the Food and Agriculture Organization (FAO) of the United Nations, among others, the Oceanic Fisheries Program works to monitor fish stocks and provide statistical analysis of stock populations and catches, acquire information on the biological and ecological parameters that may impact stock populations and productivity, and make available the information necessary to model stocks and assess population status.

The Reef Fisheries Observatory is a second component of the Marine Resource Division pertinent to the conservation of resources for Pacific island States. The Observatory's mandate is to provide member countries with scientific information concerning reef and lagoonal fisheries. It also provides information to parties involved with Pacific island reef resource management such as local communities, international donors and Pacific island governments requiring information on the management of local fishery resources under their jurisdiction.

Finally, the aquaculture component of the Marine Resource Division provides information regarding economically, socially and environmentally sustainable aquaculture to member States. The program serves as a network for aquaculture practitioners, a mechanism for the exchange of information and a source for researching and sharing aquaculture best practices.

**Association of Southeast Asian Nations**

The Association of Southeast Asian Nations (ASEAN) was established in 1967 in Bangkok with Indonesia, Malaysia, Philippines, Singapore and Thailand as original members. Membership has since expanded and today's ASEAN States represent a total of 173,000 kilometers of coastline and produce a full 14% of the global fish catch (ASEAN 2004). ASEAN's primary mandate is to deliver political and security cooperation while dealing with regional issues and promoting economic development of member States. (Nations are typically referred to as "States" in international law and agreements.) Given the vastness of the coastal and marine areas, however, and the economic importance of their associated resources, marine issues have emerged in recent years as a priority for ASEAN.

Heads of State and government representatives of ASEAN members convene annually with concurrent Ministerial meetings on issues such as the environment, science and technology. The inaugural meeting of Environmental Ministers from ASEAN countries—plus China, Japan and the Republic of Korea (ASEAN+3)—occurred in November 2002. The Ministers agreed at that meeting to foster closer collaboration on priority issues, including global environmental issues, coastal and marine environments, sustainable management of parks and protected areas, public awareness and environmental education, sustainable development and environmental education.

There are several important projects that represent policy decisions made by the ASEAN Ministers and other government officials. The ASEAN Regional Center for Biodiversity Conservation (ARCBC) is one of several flagship projects stemming from the obligations assumed by ASEAN member States in international and regional arenas such as the World Summit on Sustainability and ASEAN+3. The ARCBC project is supported by the European Union and is aimed at strengthening biodiversity conservation through networking, research, database management, training and technical assistance. It has also contributed to marine conservation in the region through its support of technical training in the areas of taxonomy, biodiversity assessment and management, biodiversity awareness and information management. Additionally, ARCBC has published an illustrated map of protected areas in its area and has funded an array of research projects including a study on marine biodiversity loss in Spermonde Islands, the identification of coral reef biodiversity of the Krakatau Islands in Lampung Province, Indonesia and an inventory of scleractinian (hard) corals on the Pacific coast of the Philippines.

In addition to developing and funding the ARCBC, ASEAN has adopted the Marine Water Quality Criteria for the ASEAN Region, the ASEAN Criteria for National Marine Protected Areas and the ASEAN Criteria for the Marine Heritage Areas. These criteria were established to ensure that national action to protect shared marine waters is coordinated throughout the region. ASEAN has also established an information exchange mechanism on issues such as coral reefs, seagrass and mangroves, merchant ship ballast water and tanker sludge, coastal erosion, ecotourism, protected marine areas and clean technologies.

ASEAN has invested considerable resources in exploring the potential implementation of multilateral environmental agreements including research on legal frameworks, information management, monitoring and reporting mechanisms, capacity and funding constraints, public awareness and education. Other projects implemented by ASEAN include the Regional Conservation Program for Marine Mammals, a Workshop on the Creation of the ASEAN Marine Turtle Specialist Network and an Australian Marine National Park Staff Exchange Project funded by the U.S. Agency for International Development. Finally, ASEAN nations worked to develop the Hanoi Plan of Action (1999–2004) to improve regional coordination in the integrated protection and management of coastal zones, the development of a regional action plan for the protection of the marine system from land-based and sea-based activities and regional coordination to protect Marine Heritage Parks and Reserves.

### Convention on the Conservation and Management of Highly Migratory Fish Stocks in the Western and Central Pacific Ocean

In September 2000, the Convention on the Conservation and Management of Highly Migratory Fish Stocks in the Western and Central Pacific Ocean was signed. This important document provides a management mechanism for the largest tuna fishery in the world, currently valued at US $2 billion annually. The convention also provides regulation for what was likely the last remaining unregulated migratory fishery in the world.

This treaty is the first of its kind to conclude formal negotiations since the 1995 adoption of the United Nations Straddling Fish Stocks Agreement, which addresses fish populations that "straddle" boundaries among national jurisdictions and the high seas. The Convention is particularly notable because current estimates indicate stocks in the western and central Pacific remain relatively robust in comparison to other fisheries. The Convention is intended to ensure sustainable use of the fishery as well as employment of a successful conservation regime that will guide development in the region. In countries such as Indonesia, Philippines and Papua New Guinea, tuna is a primary source of income, and the protection of this natural resource is critical to the long-term stability of the countries. Consequently, the precautionary principle is applied in the Convention regarding the conservation of the target species, non-target and associated species, newly explored fish stocks and ecological processes.

This Convention is also unique in that it creates a practical regime for the conservation and management of migratory fish stocks that is applicable throughout the range of the fish—encompassing multiple jurisdictions including the high seas and waters under national jurisdiction—and addressing the activities of both local and distant fishing nations. The Convention provides a detailed framework to regulate fishing and to protect the associated marine environment. Mechanisms established within its framework include an international Commission enabled to act by a supermajority (more than 50%) vote. Also important, the Convention incorporates a strong compliance mechanism, designed to be effective on the open ocean away from traditional national enforcement capabilities (Botet 2001). While the Convention is consistent with international law in granting flag-States exclusive jurisdiction over their vessels operating on the high seas, it incorporates additional legal instruments to augment fishing enforcement efforts (Joyner 1998). For example, the Convention provides for boarding and inspection of fishing vessels on the high seas by international inspectors and for the placement of observers on fishing vessels to collect data and report on compliance. It also calls for a vessel monitoring system (VMS) to be used by vessels fishing on the high seas, with data simultaneously relayed to the flag-State and to the Commission. These mechanisms clearly set a precedent and indicate a willingness to develop innovative new cooperative and enforcement mechanisms to protect the environment within the region.

### National marine protected area policy

Within the Coral Triangle, there is an array of legal and management mechanisms at the national level intended to achieve marine conservation outcomes and sustainable resource use. Although these laws are broad in their approach and structure, one of the most important conservation strategies is the development of new marine protected areas (MPAs) and improving the functionality of existing MPAs. Although laws supporting the implementation and development of MPAs vary from State to State, there are often similarities in the type of constraints that inhibit effective conservation through the MPA model.

MPA activities in the Coral Triangle often remain the mandate of the Minister of Forests or the Minister of the Environment. In instances where a specific Marine Department or Bureau is created, there is significant overlap in authority between Ministers. For instance, in Indonesia there exists a Minister of the Environment, a Directorate for Forest Protection and Nature Conservation and a Minister of Marine Affairs. Similarly, in the Philippines there exists a Protected Areas and Wildlife Bureau, the Department of Environment and Natural Resources, the Bureau of Fisheries and Aquatic Resources and a Department of Agriculture, all of which play a role in MPA development and management. In many cases, a department or bureau lacks the financial and personnel capacity to address marine issues but is reluctant to relinquish authority to another Minister. Matters are further complicated when local governments are involved. There are additional jurisdictional problems when threats to a marine habitat or species arise from terrestrial activities or other activities occurring outside the MPA, or when multinational or transboundary issues are at play. For example, throughout most of Malaysia, the federal government's lack of jurisdiction over coastal lands adjacent to MPAs is the preliminary obstacle to effective MPA management.

In many areas of the Coral Triangle there are a significant number of MPAs, either existing or proposed, but they lack sufficient management plans or implementation protocols. For example, in Indonesia there are over 700 terrestrial and marine protected areas (Figure 5), but only 31 of them have plans and not all of those have been implemented (GOI 1997). Of the six marine national parks, only three have implemented management plans (Hopley & Suharsono 2000). Lack of personnel, training, facilities, funding, information or other local capacity often prohibits the establishment of management protocols.

The Coral Triangle is remarkably diverse in cultures, traditions, systems of property rights, capacity and government structures, further complicating MPA management. For instance, government support is critical to MPA development in Indonesia and Malaysia, whereas community-based conservation is frequently employed in the Philippines, Solomon Islands and Papua New Guinea. Consequently, marine conservation strategies and processes to establish MPAs vary significantly within the region, making the establishment of a regionally based mechanism for development of management protocols and MPA standards difficult. Although both ASEAN and PIF are working to establish MPA criteria, the establishment of MPAs will likely remain the initiative of individual States.

## Conclusions

- The Coral Triangle has the most biologically diverse coastal marine environments on the planet
- Biological communities within the Coral Triangle are not homogenous and there are many pockets of high endemism
- The central region of the Coral Triangle, including eastern Indonesia from the Sulawesi Sea east to and including western New Guinea and the eastern Sunda Islands, appears to be the most biologically diverse
- Threat intensity varies within the region from excessive (most of the Philippines) to relatively light (much of Papua New Guinea and the Solomon Islands)

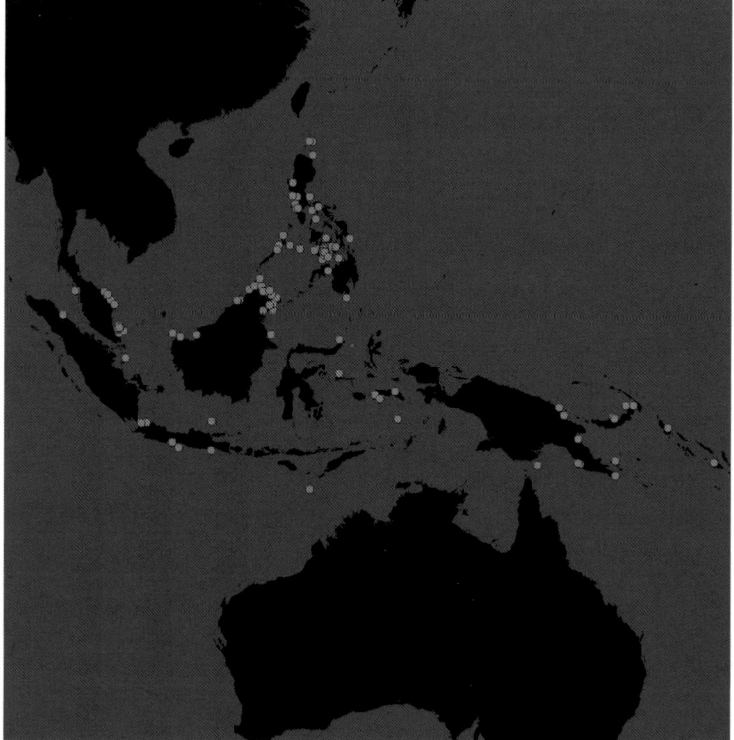

**Figure 5** Marine protected areas (MPAs) within the Coral Triangle (from WDPA Consortium 2004). Actual areas—some of which are very small—are not to scale.

- Marine conservation strategies require a combination of protection and restoration measures
- Most of the existing marine protected areas are small reserves, and these have limited ability to protect biodiversity and maintain fishery yields over the long term; larger areas are needed
- Marine reserves alone will not be enough, but improved management within and outside reserves is a priority, such as:
  - Supportive policy and legislative frameworks
  - Surveillance, enforcement, prosecution of illegal fishing
  - Fishing gear changes
  - Integrated coastal and ocean management
- There are big challenges—lack of data, lack of capacity and the difficulties and expense of surveillance over coastal and ocean areas
- Enhanced fishery yields can be an incentive to local communities for establishing MPAs, but the protected area plans need to address other threats and design criteria, including effective management of surrounding waters
- Conservation is largely perceived within the region as a luxury, except where tangible benefits can be demonstrated

## Marine Conservation Strategies for the Coral Triangle

Implementation of a regional conservation strategy in the Coral Triangle can be broken down into three components: scientific planning; couching the science in socio-economic realities to identify practical targets and approaches that support the achievement of outcomes; and coming up with incentives that would motivate various players to promote marine conservation measures.

While the second and third components of this strategy —understanding socio-economic realities and designing effective incentives—require further definition, the first component—scientific planning—consists of assembling, mapping and analyzing marine species distributions. Table 3 identifies a preliminary assessment of marine taxa that could be analyzed with this objective in mind. A review of the table suggests that fishes—and possibly reef corals—are probably the best groups to focus on initially, though identification of additional taxa for study and mapping should also occur early on. While sources for datasets must be identified, a good many are already available as a result of the *Defying Ocean's End* Conference. Emphasis should be placed on those data layers most readily available and of the greatest utility to marine conservation planning. The geographic scope for biodiversity analyses should extend beyond the Coral Triangle to include all of Melanesia, Palau, northern Australia, western Indonesia, the South China Sea, Taiwan, southern Japan and southern China. This will allow a better understanding of the biogeographic delineation of the Coral Triangle and support further quantification of significant global centers of endemism such as those identified for coral reef biodiversity in Roberts *et al* (2002).

There are three scales at which strategies must be employed to achieve long-term marine conservation success in the Coral Triangle—species, site and seascape.

### Species

A significant number of endangered and threatened species are found within the Coral Triangle. Identifying and determining the distribution and abundance of threatened, endangered and restricted-range species within the Coral Triangle will be an important first step in defining a regional conservation strategy that is both targeted in its approach and measurable in its delivery. This process will necessarily include the compilation and assimilation of existing data and the acquisition of new information in data-deficient areas. There has been some initial consolidation of information, but it is patchy at best. In the Philippines, for example, the Marine Science Institute, the Silliman University Marine Laboratory and the Coral Reef Division of the Bureau of Fisheries and Aquatic Resources have made the Philippines a science leader in the region. Conversely, in eastern Indonesia, there is currently little or no existing scientific capacity. A regional gap analysis will be necessary to identify data needs and ensure that a uniform distribution of data is obtained throughout the region. Identifying gaps and data deficiencies in non-coastal marine habitats, and developing and prioritizing data acquisition strategies with coordinated methodologies, will be a key component of any strategy. The process of data compilation, acquisition and analysis within the Coral Triangle should necessarily involve local as well as international scientists. To the extent that the process engages numerous stakeholders and has broad participation, the results will be representative of the region and will be more useful in future implementation phases.

Table 3

**Preliminary informal assessment of marine taxa for a Coral Triangle scientific analysis**

| Taxon | Availability of data | Utility of data | Preliminary sources of information |
|---|---|---|---|
| **Vertebrates** | | | |
| Mammals | Needs determination | Better for larger scale patterns | IUCN-SSC, Global Mammal Project, Apex International |
| Birds | Good | Good, but not best group for marine biome | BirdLife International |
| Reptiles | Spotty | Needs determination | International Sea Turtle Society, IUCN-SSC, Heatwole for sea snakes |
| Plants | Needs determination | Needs determination | |
| Fishes | Good | Good | G. Allen, FAO, C. Roberts, IUCN-SSC, S. Fowler |
| **Invertebrates** | | | |
| Ascidians | Needs determination | Needs determination | |
| Echinoderms | Spotty, though some families might be useful | Needs determination | D. Pawson, G. Hendler, T. O'Hara |
| Sponges | Needs determination | Taxonomy notoriously problematic | K. Ruetzler |
| Molluscs | Pretty good for some groups (e.g., cowries) | Needs determination | F. Wells, T. Gosliner, FAO |
| Crustaceans | OK for some commercial species, probably good for Indonesia | Needs determination | M. Reaka-Kudla, FAO, Erdman, etc. |
| Polychaetes | Poor | Poor | |
| Reef corals | OK, but coarse | May be good for certain families (e.g., Acroporidae, Fungiidae) | B. Hoeksema, C. Wallace, S. Cairns, J. Veron |
| Other corals | Needs determination | Needs determination | G. Williams |
| Anemones | Needs determination | Needs determination | D. Fautin |
| Bryozoans | Needs determination | Needs determination | |
| Hydrocorals | Needs determination | Needs determination | S. Cairns |

## Site

Marine conservationists have traditionally relied on marine protected areas as the underpinning of a conservation strategy. To ensure optimal effectiveness in meeting conservation goals, however, it is necessary to ensure that MPAs are established and implemented in the Coral Triangle in such a way that they help achieve specific goals of protecting threatened and endangered species. As such, it is first necessary to identify the critical habitats of endangered and threatened, endemic or limited-range species, their aggregation sites and sites that provide essential ecological functions, to further support the long-term sustainability of target species. It is then necessary to conduct a regional analysis to determine potential gaps between existing and needed MPAs, and between existing MPAs with and without suitable management protocols. Once these gaps have been identified, it will be necessary to analyze socio-economic and legal data to determine the feasibility of implementing MPAs in proposed sites. In processing these data, it will be necessary to ensure that the designation and implementation strategies of existing and planned MPAs are appropriate for the users and other stakeholders.

Working within the existing regional organizations such as ASEAN and PIF, it may be possible to create a centralized approach to MPA planning, implementation and management. These regional groups have established technical support mechanisms and data centers that should be used whenever possible to streamline MPA processes and ensure that national MPAs align within a larger regional strategy. In this manner, MPAs can be prioritized across a regional scale and financing can be provided through existing regional mechanisms. Furthermore, to the extent possible, long-term sustainable finance plans should be developed for all priority MPAs. Finally, long-term monitoring and assessment of MPAs—including mapping and use of geographic information system (GIS) tools to facilitate adaptive management of Coral Triangle MPA networks—should be established to ensure that desired marine outcomes are being achieved.

### Seascape

While important conservation outcomes will be achieved through MPAs, additional marine conservation mechanisms must be established at a broader scale to ensure that extinction of target species is prevented. There are several approaches that can be applied to achieve this important goal. First, species-specific strategies should be developed—like those used to protect migratory species, such as turtles and marine mammals—to augment MPAs. For example, these species-specific strategies may pursue changes to law and policy and/or implement outreach and awareness campaigns. Second, broad-scale ecological processes require protection to ensure the stability of marine habitats and systems through the long term. Within the Coral Triangle it is necessary to establish the number and boundaries of "seascapes"—larger areas or marine corridors linking multiple MPA networks—through biological and political analysis. It is then necessary to determine the stakeholders associated with each seascape. Once the stakeholders have been established, a planning and implementation process should be originated, a business plan developed and a determination made on how to operate within the existing regional mechanisms. Finally, new regional approaches must be developed and national and regional policies pursued. Consequently, international law will be amended to reflect the emerging marine conservation priorities of region-wide forums in the Coral Triangle area.

## Summary

The Coral Triangle is an important priority for global marine conservation to avert massive extinction of marine species. It is also a key area for studying the response of the marine environment to global change, including global warming. Conservation in the Coral Triangle must target areas that have high biodiversity, high concentrations of endemic species, areas that are in relatively pristine condition (Figure 6), areas that are critical for the life cycles of endangered species and areas where large reserves can be established. Priority areas identified for biodiversity reasons should also take into consideration the likelihood of conservation success. We need to continue to amass the kinds of information and capacity required to support a region-wide conservation strategy. And more systematic monitoring of key areas is needed to identify changes in the biological composition of communities as well as changes in environmental conditions.

To support the creation of marine reserves that are as large as possible, increased incentives, including private sector approaches, are needed. Some possibilities include:

**Figure 6** The Coral Triangle is extremely rich in colorful and unusual marine life.

- Developing more fiscal incentives, such as conservation concessions
- Responding to government concern for revenue loss from illegal fishing
- Demonstrating fishing benefits to coastal communities
- Establishing higher dive fees since the region has some of the most coveted SCUBA-diving destinations in the world
- Cost-benefit analysis, such as comparing revenues from dive tourism versus revenues from fishing
- Continuing to respect local economic/subsistence needs
- Establishing greater roles outside central governments (such as local governments, communities, NGOs and the private sector)

A central marine conservation strategy for the Coral Triangle should include the creation and strengthening of individual MPAs, but a regional network of marine protected areas should be designed to increase the probability of species survival in the face of global climate change, coral bleaching and increasing negative effects from human population growth and activities in the region. MPAs and MPA networks, and larger seascapes that include multiple MPA networks, should be designed primarily for biodiversity conservation goals, with fisheries enhancement being a secondary goal. Greater capacity in science, communications and funding is needed within the region to achieve conservation goals, and marine conservation strategies in the region need support from communications strategies. MPA networks should be part of a plan for the EEZs of the five nations, including improved fisheries management, better control of pollution, etc., leading to the establishment of large-scale functional seascapes to preserve the unique and important natural richness of this region.

## Literature Cited & Consulted


Allen, G. R. 2000. Indo-Pacific coral-reef fishes as indicators of conservation hotspots. In *Proceedings of the Ninth International Coral Reef Symposium, Bali, Indonesia, 23–27 October 2000,* ed. M. K. Kasim Moosa, S. Soemodihardjo, A. Nontji, A. Soegiarto, K. Romimohtarto, Sukarno and Suharsono, 921–926. Ministry of Environment, the Indonesian Institute of Sciences and the International Society for Reef Studies.

Allen, G. R. and M. Adrim. 2003. Coral reef fishes of Indonesia. *Zoological Studies* 42 (1): 1–72.

ASEAN. 2004. The Official Website of the Association of Southeast Asian Nations Secretariat. *http://www.aseansec.org/home.htm* (accessed August 25, 2004).

Botet, V. 2001. Filling in one of the last pieces of the ocean: Regulating tuna in the Western and Central Pacific Ocean. *Virginia Journal of International Law* 41:221–237.

Briggs, J. C. 1995. *Global Biogeography.* Amsterdam: Elsevier Science.

——. 1999. Coincident biogeographic patterns: Indo-west Pacific Ocean. *Evolution* 53:326–335.

——. 2003. Marine centers of origin as evolutionary engines. *Journal of Biogeography* 30:1–18.

Bryant, D., L. Burke, J. W. McManus and M. Spalding. 1998. *Reefs at Risk: A map-based indicator of threats the world's coral reefs.* World Resource Institute (WRI), International Center for Living Aquatic Resources Management (ICLARM), World Conservation Monitoring Centre (WCMC), United Nations Environment Programme (UNEP).

Burke, L., L. Selig and M. Spalding. 2002. *Reefs at risk in Southeast Asia.* Washington, D.C.: World Resources Institute.

Cesar, H. 1996. *Economic Analysis of Indonesian Coral Reefs.* Environment Department, Coral Reef Rehabilitation and Management. Washington, D.C.: World Bank.

Clark, A. H. 1912. The crinoids of the Indian Ocean. *Echinoderms of the Indian Museum, Part VII Crinoidea.* Calcutta: The Indian Museum.

Convention on the Conservation and Management of Highly Migratory Fish Stocks in the Western and Central Pacific Ocean, opened for signature Sept. 4, 2000.

Edinger, E. N., J. Jompa, G. V. Limmon, W. Widjatmoko and M. J. Risk. 1998. Reef degradation and coral biodiversity in Indonesia: Effects of land-based pollution, destructive fishing practices and changes over time. *Marine Pollution Bulletin* 36 (8): 617–630.

GOI. 1997. *Implementation of Agenda 21: Review of Progress Made Since the United Nations Conference on Environment and Development, 1992. I.* Government of Indonesia, Indonesia Country Report. *http://www.un.org/esa/agenda21/natlinfo/countr/indonesa/natur.htm* (accessed August 30, 2004).

Gomez, E. D., E. Deocadiz, M. Hungspreugs, A. A. Jothy, Kuan Kwee Jee, A. Soegiarto and R. S. S. Wu. 1990. *The State of the Marine Environment in the East Asian Seas Region.* UNEP Regional Seas Reports and Studies No. 126. United Nations Environment Programme.

Green, A. and P. J. Mous. 2004. *Delineating the Coral Triangle, its ecoregions and functional seascapes.* Bali, Indonesia: The Nature Conservancy, Southeast Asia Center for Marine Protected Areas.

Hoeksema, B. W. and K. S. Putra. 2000. The reef coral fauna of Bali in the center of marine diversity. In *Proceedings of the Ninth International Coral Reef Symposium, Bali, Indonesia, October 23–27 2000,* ed. M. K. Kasim Moosa, S. Soemodihardjo, A. Nontji, A. Soegiarto, K. Romimohtarto, Sukarno and Suharsono, 173–178. Ministry of Environment, the Indonesian Institute of Sciences and the International Society for Reef Studies.

Hopley, D. and Suharsono, eds. 2000. *The status of coral reefs in Eastern Indonesia.* Townsville, Australia: Australian Institute of Marine Science.

Huber, M. and J. Opu. 2000. Assessment of Natural Disturbances and Anthropogenic Threats to Coral Reef Biodiversity in PNG. Chapter 5 in *The Status of Coral Reefs in Papua New Guinea,* ed. P. L. Munday. Global Coral Reef Monitoring Network (GCRMN) Report.

Huston, M. A. 1994. *Biological Diversity: The Coexistence of Species on Changing Landscapes.* Cambridge, England: Cambridge University Press.

Joyner, C. C. 1998. Compliance and enforcement in new international fisheries law, 12. *Temple International and Comparative Law Journal,* 327–355.

Maclean, J. L. 1973. Red Tide and paralytic shellfish poisoning in Papua New Guinea. *Papua New Guinea Agricultural Journal* 24:131–138.

——. 1977. Observations on Pyrodinium bahamense Plate, a toxic dinoflagellate, in Papua New Guinea. *Limnology and Oceanography* 22 (2): 234–254.

Mayr, E. 1954. Change of genetic environment and evolution. In *Evolution as a Process,* ed. J. Huxley, A. C. Hardy and E. B. Ford, 157–180. London: Allen and Unwin.

Moosa, M. K. 1999. The extent of knowledge about marine biodiversity in Indonesia. In *Integrated Coastal and Marine Resource Management, Proceedings of International Symposium, Malang, 1998,* ed. J. Rasi, I. M. Duttong, L. Pantimena, J. Plouffe and R. Dahuri R, 126–153. Malang: Batu, ITN, BAKOSURTANAL and Proyek Pesisir.

Pet-Soede, C., H. S. J. Cesar and J. S. Pet. 1999. An economic analysis of blast fishing on Indonesian coral reefs. *Environmental Conservation* 26 (2): 83–93.

Polunin, N.C. 1983. The marine resources of Indonesia. *Oceanography Marine Biological Annual Review* 21:455.

Roberts, C. M., J. P. Hawkins, D. E. McAllister, C. J. McClean and T. Werner. 2000. Global Priority Regions for Coral Reef Conservation. In *Proceedings of the Ninth International Coral Reef Symposium, Bali, Indonesia, 23–27 October 2000,* ed. M. K. Kasim Moosa, S. Soemodihardjo, A. Nontji, A. Soegiarto, K. Romimohtarto, Sukarno and Suharsono. Ministry of Environment, the Indonesian Institute of Sciences and the International Society for Reef Studies.

Roberts, C. M., C. J. McClean, J. E. N. Veron, J. P. Hawkins, G. R. Allen, D. E. McAllister, C. G. Mittermeier, F. W. Schueler, M. Spalding, F. Wells, C. Vynne and T. B. Werner. 2002. Marine biodiversity hotspots and conservation priorities for tropical reefs. *Science* 295:1280–1284

Secretariat of the Pacific Community. 2004. *The Secretariat of the Pacific Community: Serving the Pacific Islands. http://www.spc.org.nc* (accessed August 25, 2004).

Spalding, M., C. Ravilious and E. P. Green. 2001. *World Atlas of Coral Reefs.* Berkeley: University of California Press.

Veron, J. E. N. 1995. *Corals in Space and Time: The Biogeography and Evolution of the Scleractinia.* Sydney, Australia: UNSW Press.

——. 2000. *Corals of the World.* 3 Volumes. Australian Institute of Marine Science.

Wallace, C. C., G. Paulay, B. W. Hoeksema, D. R. Bellwood, P. A. Hutchings, P. H. Barr, M. Erdmann and J. Wolstenholme. 2000. Nature and origins of unique high diversity reef faunas in the Bay of Tomini, Central Sulawesi: The ultimate "center of biodiversity"? In *Proceedings of the Ninth International Coral Reef Symposium, Bali, Indonesia, 23–27 October 2000,* ed. M. K. Kasim Moosa, S. Soemodihardjo, A. Nontji, A. Soegiarto, K. Romimohtarto, Sukarno and Suharsono, 185–192. Ministry of Environment, the Indonesian Institute of Sciences and the International Society for Reef Studies.

Werner, T. B. and G. R. Allen, eds. 1998. *A Rapid Biodiversity Assessment of the Coral Reefs of Milne Bay Province, Papua New Guinea.* Washington, D.C.: Conservation International.

WDPA Consortium. 2004. *World Database on Protected Areas 2004.* Copyright World Conservation Union (IUCN) and UNEP–World Conservation Monitoring Centre (UNEP-WCMC), 2004. *http://sea.unep-wcmc.org/wdbpa/download/wdpa2004/index.html* (accessed October 25, 2004).

World Resources Institute. 2004. EarthTrends: The Environmental Information Portal. World Resources Institute. *http://earthtrends.wri.org* (accessed October 4, 2004).

CHAPTER 5

# *The Gulf of California: Natural Resource Concerns and the Pursuit of a Vision*

María de los Ángeles Carvajal, *Conservation International*
Exequiel Ezcurra, *Instituto Nacional de Ecología, México*
Alejandro Robles, *Conservation International*

DOC WHITE/OCEAN MAGIC PHOTOGRAPHY

## Introduction: A Unique Marine Ecosystem

The Gulf of California in northwest Mexico (also known as the Sea of Cortés) covers an area of 375,000 square kilometers. It has been called a "caricature of oceanography" because its oceanographic features seem dramatically exaggerated, with deep basins in its central and lower portions and some of the greatest tides in the world in its upper reaches. It contains over a hundred islands and offshore rocks, and strong upwellings of cold nutrient-rich waters are evident along both its coasts. The great diversity of topographic and bathymetric features has produced a great variety of habitats for marine life. The Gulf's high biodiversity levels, biological productivity, 770 endemic (unique to this area) species (Findley forthcoming) and its 39 marine species listed on the IUCN Red List (IUCN 2003) as threatened or vulnerable make this one of the large coastal ecosystem conservation priorities on the planet.

The natural history of the Gulf of California and the Baja California peninsula is one of evolution in isolation. All through the region, the driving theme is insularity. During the last six million years, the Sea of Cortés has kept the long and dry peninsula separated from the Mexican mainland, and the peninsula of Baja California has kept the Sea of Cortés sequestered from the Pacific Ocean. Within this landscape of sea and land that mutually enclose each other, preserving the genetic uniqueness of their life forms, some patches of insularity are superimposed at even smaller scales. Marine islands surround the peninsula on all sides, and rich seamounts, true underwater islands, act as a major factor in endemism (species unique to a small area) and speciation (evolutions of new species). On land, palm oases in deep, disjunctive canyons form thousands of small, isolated wetlands within the rocky matrix of the peninsular ranges. The seacoast is fringed by coastal lagoons that repeat the isolation theme in smaller and smaller bodies of water (Bourillón *et al* 1988, Case, Cody & Ezcurra 2002). These patches of spatial segregation are the driving force of biological speciation, adaptation to local conditions and specialization in particular isolated environments.

In the Sea of Cortés, fragmentation has not only yielded unique life forms resulting in the myriad endemic species, but has also created unique human cultures. Separated from the rest of Mesoamerica, the Cochimí Indians developed one of the most incredible assemblages of cave paintings in the world. Later, under Spanish rule, the Jesuit fathers founded a system of missions here that evolved in complete indepen-

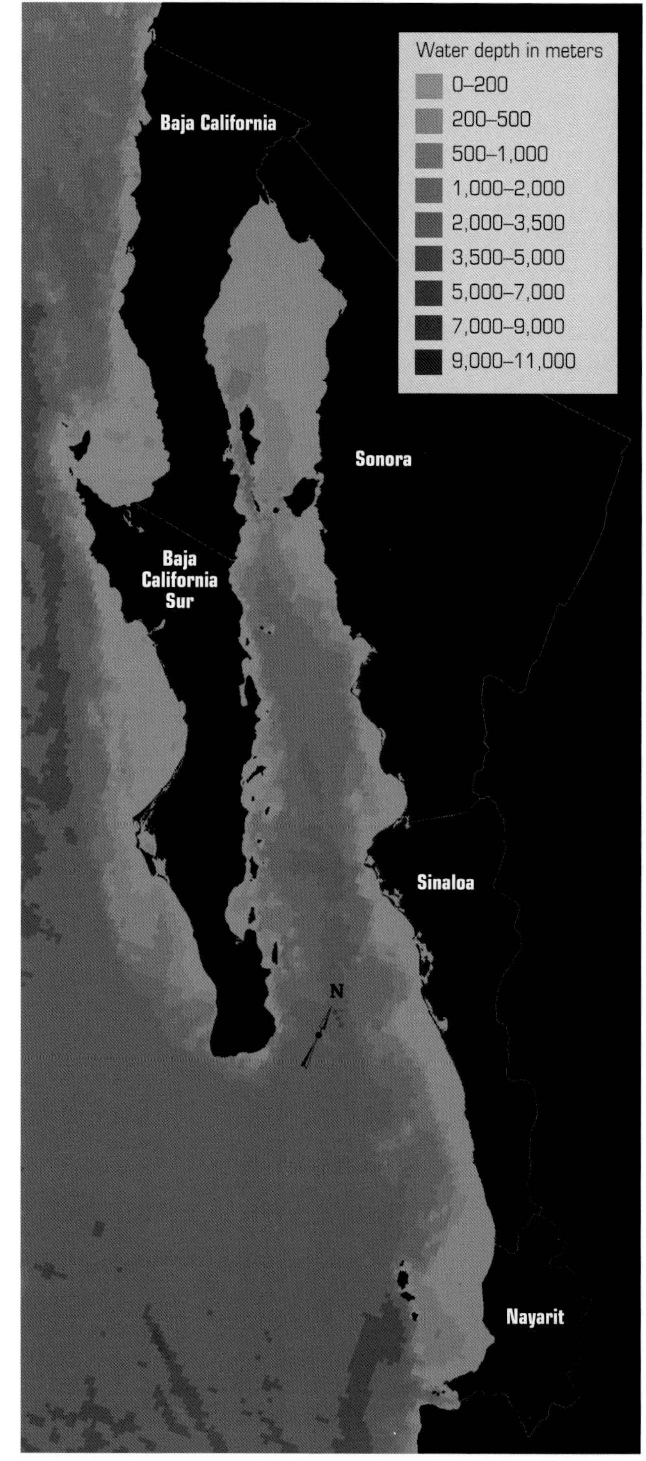


**Figure 1** The deep, narrow Gulf of California is a unique marine environment. The bathymetry (water depth) scale is the same as all other maps in this volume.

dence from the harsh rule of the mainland *conquistadores*. Other coastal indigenous groups, such as the Seri, the Yaqui or the Cucapá, also developed unique lifestyles as fishermen and sailors, with cultures finely adapted to their ocean and coastal resources (Bowen 2000, Bahre & Bourillón 2002, Felger & Moser 1985, Nabhan 2002). True to the etymology of the word, the peninsula has indeed been almost an island, isolated from the rest of Mexico, and the Gulf has been a secluded interior sea, deeply isolated from the larger Pacific (Figure 1). Even in recent decades, journalists such as Fernando Jordán (1951) have referred to this region as *el otro México* (the other Mexico). It has always been a region of fantasy and adventure, a territory of surprising, often bizarre, growth-forms and immense natural beauty.

Few places exhibit the extraordinary environmental heterogeneity seen in the peninsula of Baja California and the Sea of Cortés (Robles-Gil, Ezcurra & Millink 2001, Garcillán & Ezcurra 2003). The regional climates vary from Mediterranean-type winter rains in the north to monsoon-type summer rains in the south. The steep slopes of the mountain ranges, both on land and under water, generate some of the most dramatic environmental gradients on Earth. The northern part of the peninsula on the Pacific Coast harbors rare remnants of Californian coastal scrub, chaparrals, temperate forests and dry deserts. A rare form of tropical deciduous forest occupies the lowlands of the Cape Region on the southern part of the peninsula. Similar areas of geographic isolation and biological rarity are found in the coastal lagoons and wetlands, on the Gulf islands and in the deep-water reefs of the underwater seamounts.

The Sea of Cortés is a sort of "marine peninsula," isolated from the rest of the Pacific by the 1,500-kilometer extension of Baja California. Biologically, it is one of the most productive and diverse seas in the world (Álvarez-Borrego 1983, 2000). The high biodiversity in the Gulf of California is largely due to two phenomena: the great variety of habitats, including mangrove swamps, coastal lagoons, coral reefs, shallow and deepsea basins, hydrothermal vents and a diverse array of shore and subtidal areas (Figure 2); and the complex geological and oceanographic history of the Gulf, including past invasions of animals immigrating from tropical South America, the Caribbean Sea (before Earth's tectonic forces sealed the Panamanian seaway), the cold shores of California (during past glacial periods) and across the vast stretch of the Pacific Ocean from the tropical West Pacific. The Gulf is not only biologically important, it also

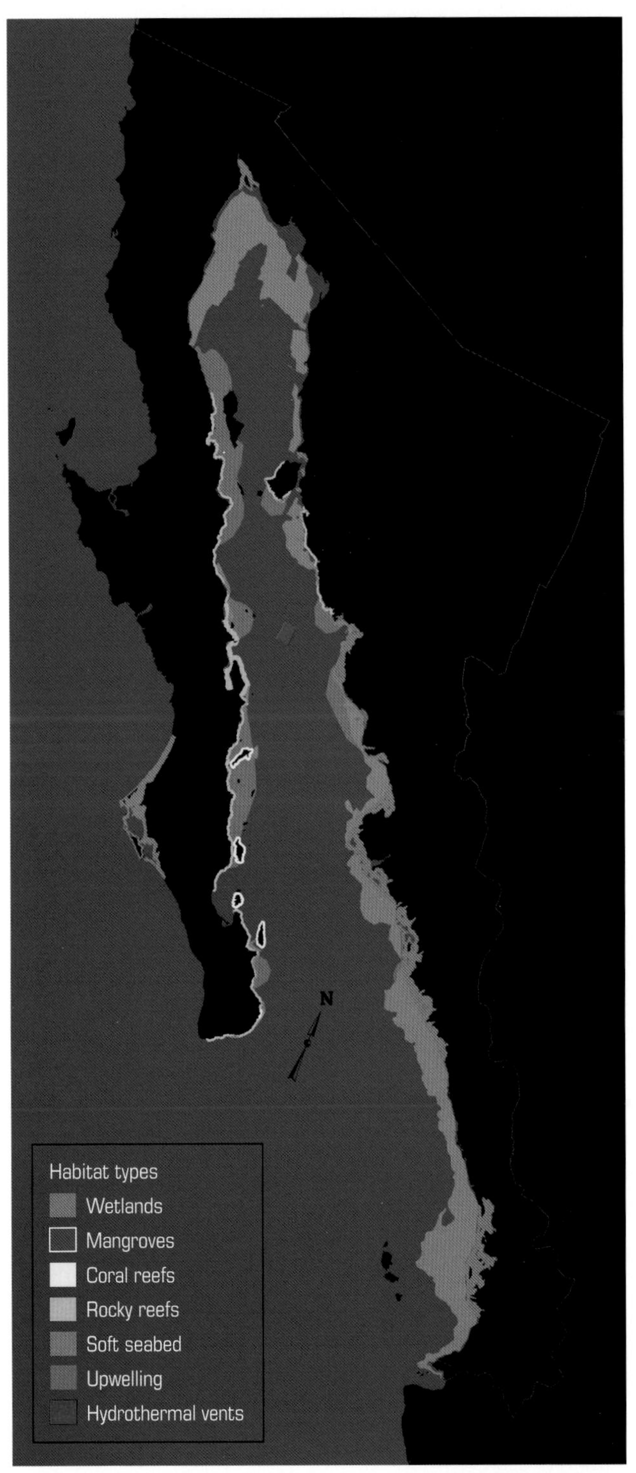


**Figure 2** Coastal and marine habitats in the Gulf of California (modified from National Commission for the Knowledge and Use of Biodiversity–CONABIO).

has immense economic worth; it houses an inordinately high proportion of the marine species richness of Mexico and yields some 30 to 60% of the national fisheries catch. The sustainable use and conservation of the Sea of Cortés are critical issues from both points of view.

The Gulf of California—in both its marine and surrounding terrestrial ecosystems—is in a transition zone between bioregions identified as critically important for conservation. The Sea of Cortés is one of the most important marine ecosystems on the planet. And the Sonoran and Baja Californian deserts are part of one of the world's top five wilderness areas, making up significant portions of two terrestrial hotspots: the southern part of the California Biotic Region and the northern portion of the Mesoamerican Dry Tropical Forests (Ezcurra *et al* 2002, Mittermeier *et al* 1999). The region covers an area of 890,000 square kilometers, of which 57% (515,000 square km) is terrestrial and 43% (375,000 square km) is marine. The terrestrial portion comprises thirteen ecoregions, which include the lowland dry tropical forests of Sinaloa, the Sonoran and Baja Californian deserts, Mediterranean ecosystems of the California Biotic Province and the mountain pine and oak forests. The marine area of the Gulf stretches some 1,200 kilometers in a southeast-to-northwest direction—from the tropical waters of the southern Gulf at 22°N with clear Panamanian biogeographic influences to the colder waters of the upper Gulf at 32°N.

Fourteen of the world's 32 marine phyla are represented in the Gulf (Brusca 1980, Gotshall 1998, Thomson & Gilligan 2002, Thomson & Eger 1966, Thomson, Findley & Kerstitch 2000). An estimated half of its faunal diversity is composed of nearly 6,000 known macrofaunal (larger than microscopic animal) species. The Gulf is home to 907 fish species (more than 55% of which are species from marine families), 240 marine birds, 35 marine mammals (82% of marine mammals found in the northeastern Pacific) and 4,818 known marine macroinvertebrates (Findley forthcoming). Some authors estimate that more than 4,000 invertebrate species remain undescribed in this extraordinarily rich environment. Of all these species, 770 are endemic to the region, including the totoaba *Totoaba macdonaldi*, a giant sea bass, and the vaquita *Phocoena sinus*, the Gulf of California's unique harbor porpoise (Findley forthcoming). The Gulf has the highest diversity of large whales in the world, all of which are legally protected in Mexico. Furthermore, its marine seabed, together with 6,000 square kilometers of coastal lagoons and 2,560 square kilometers of mangrove forests, serve as reproductive, nesting and nursing sites for hundreds of resident and migratory species. The complex archipelago of the Gulf's islands—containing 922 islands and smaller islets—harbors 90 endemic species, five of which are critically endangered and 60 of which are reptilian (Case, Cody & Ezcurra 2002). Such endemism significantly contributes to making Mexico second in the world in terms of reptile biodiversity.

This vast natural wealth is not only of biological and conservation interest; it also provides the socioeconomic sustenance of the inhabitants of the region who have developed systems of natural resource use that often put the long-term sustainability of the resources in peril. The most significant threats to biodiversity are driven by the growth of economic activities in the region, which has caused the deterioration of coastal marine ecosystems due to decreasing freshwater flows, pollution by agrichemicals and urban waste, sedimentation and the use of inappropriate fishing technologies such as bottom-trawling. Critical mangrove habitat is being lost at an annual rate of 9% from the construction of shrimp ponds, marinas, inland channels and deforestation leading to sedimentation, eutrophication and changes in water flows. In addition, the invasion of exotic plant and animal species is putting at risk the native and endemic species of the Gulf's islands and the Sonoran and Baja California deserts.

## The Shaping of a Singular Landscape

While most of the continental mainland of Mexico and the U.S. are attached to the North American plate, Baja California is a sliver of continental crust that is affixed to the Pacific plate. The peninsula is being very slowly torn away from the Mexican mainland in a series of deep trenches and rifts that are gradually moving the whole Pacific plate—Baja California—toward the northwest and widening the Gulf of California. The ridges and fractures that have been formed by this very active tectonic interface are the main causes of the regional topography, both on land and under water. The heterogeneous topography and bathymetry drive the local climates and oceanographic processes and ultimately are a major causal force of the region's unique biological diversity (Carreño & Helenes 2002).

In the same way the geological processes shaped the abrupt geology of the Sea of Cortés, the circulation of winds and oceanic currents is the underlying mechanism that created the region's incredible ecological variability and is the force that maintains it. On the Pacific coast of Baja the cold California current, which flows north-to-south along the shoreline, is deflected westwards by the rotational movement of the Earth. The deflected surface waters are replaced by an upwelling of cold, nutrient-rich water transported up from the ocean floor, bringing fertility to the surface. Inside the Gulf, however, the high productivity is derived mostly from mixing of the water column caused by the tidal waves moving from its mouth toward the cul-de-sac in its northern reaches. The tide-generated upwelling is especially high in the Gulf's mid-region, where a series of islands and islets produce a tight narrowing of the sea, and in the shallow waters of the upper Gulf where the tidal flow is dissipated in 8-meter high waves that provide a strong vertical mixing of the waters. This turbulent upwelling in the mid-Gulf and tidal mixing in the upper Gulf bring nutrients to the surface and make the Sea of Cortés one of the most productive regions in the world ocean (Lavín & Marinone 2003, Lavín, Palacios-Hernández & Cabrera 2003, Marinone & Lavín 2003).

Cold seas are also the major cause of aridity on the land, as the cold waters generate low-pressure centers and a stable atmosphere that forces moisture-laden northwesterly winds from the cold Pacific onto the warm peninsula where they warm up and become drier. Only in the mountain ranges does ascending air cool sufficiently to generate significant rainfall and sustain temperate pine and oak forests. In the northern part of Baja California, the more temperate land cools sufficiently in winter to cause atmospheric condensation and induce winter rains, forming the chaparrals of the California Biotic Province. In the rest of the region, the rich upwelling of the Sea of Cortés coexists side by side with desert on the land.

During El Niño years, random variations in the airflow weaken the trade winds and the westward deflection of surface ocean currents decreases. Warm oceanic waters accumulate along the coast of the North American Continent, the upwelling of nutrient-rich waters decreases, and the Gulf waters become warmer. Thus, the natural cycle becomes inverted. As the currents slow down and the ocean warms up, the sea becomes less productive, and the land is soaked by abundant rainfall that originates from the now-warm marine waters (Holmgren *et al* 2001, Polis *et al* 1997, Velarde & Ezcurra 2002, Velarde *et al* 2004). In the Sea of Cortés region, pulses of productivity in the dry deserts are seen in opposite years from the productivity highs in the Gulf and Pacific waters.

## The Evolution of Regional History

Archeological evidence of the first humans in the Gulf of California region is dated around 14,000 years ago, toward the end of the last glacial period. The land, now occupied by deserts, was at that time covered by grasslands where mammoths, horses, camels and giant sloths roamed in large numbers. In a few thousand years, as the glaciers retreated northward and the Gulf lands became hotter and drier, the changing climate and the lethal efficiency of the newly arrived nomadic hunters drove most of these large herbivores into extinction (Martin 1984). Human populations had to adapt to the absence of large game, learning to gather plants from the desert and fish from the sea. Most of them remained nomadic gatherers until the arrival of the first Europeans, but in Nayarit and southern Sinaloa, small agricultural villages developed, depending primarily on farming and on the plentiful resources provided by the extensive mangrove swamps of local lagoons.

When the Spaniards arrived in the first half of the 16th century, they found about sixteen different ethnic groups inhabiting the Gulf coast. The inhospitable climate and isolated conditions of the Baja California peninsula and the Sonoran desert made it impossible at first for the *conquistadores* to establish themselves permanently along the coast (Robles, Ezcurra & León 1999). It was not until 1667 that the Jesuits first established a mission in Baja California, and soon after an impressive chain of agricultural settlements was set up in the upland sierras of Sinaloa and Sonora and throughout the peninsula. Despite the lack of farming experience, the Cochimí natives of the central peninsula made their transition to sedentary life with apparently little effort and, in doing so, seem to have abandoned their fishing traditions. During the 18th century the missions flourished, developed a prosperous system of irrigated agriculture and served as strategic points of access to the coastal lagoons of the peninsula and to the fertile summer pastures of the cool sierras (Clavijero 1789, del Barco 1973). During the 18th century a prolonged and severe drought, coupled with the expulsion of the Jesuits from the New World, resulted in the collapse of the mission system and drove the Cochimí people into cultural extinction.

For thousands of years, the population of the Gulf of California has benefited from the marine resources of the region, exploiting the bays, coastal lagoons, river mouths, salt marshes and estuaries for food. Until the early 20th century fishing was practiced only in the inshore areas, as vessels were small and powered by oars and sail. The main fishing techniques were lines and hooks (Robles & Carvajal 2001). After the long drought and decline of the missions, the discovery of pearls led to the exploitation of one of the most valuable resources and export items for the Spanish in the Gulf of California. The largest pearl beds in the world were located on the east coast of Baja California Sur, and in the 19th century, the city of La Paz was recognized worldwide for its pearls. By 1892, however, diving for pearl extraction had declined significantly because of the drop in oyster numbers, possibly as a result of over-exploitation. Riverine areas and lower-basin agricultural valleys have always been favored by the region's population for settlement and development. In particular, intensive settlement occurred during the 20th century in the Fuerte, Mayo and Yaqui valleys of Sinaloa and Sonora and in the Colorado River Valley on the Mexico/U.S. border. Until the 1930s the Colorado was the largest river flowing into the Gulf of California with a vast delta covering 300 square kilometers of wetlands (Sykes 1937, Fradkin 1984, Ezcurra *et al* 1988, Felger 2000). The development of steamboat traffic on the Colorado River during the 19th century was made possible partly because of the dense cottonwood forests on the riverbanks, which fed the high demand for charcoal and wood needed by the steamboats. This caused devastation of the forest—the first significant environmental impact on the oasis—but was only a prelude to the devastation the estuary had yet to see. In 1935 the Colorado River was dammed after an international water treaty was signed between the United States and Mexico, starting the development of the Imperial and Mexicali agricultural valleys but bringing about the demise of the great delta. In the same manner, dams and channel works were initiated in the 1940s on the Fuerte, Mayo and Yaqui rivers as part of a regional project for agricultural development. Economically, all these irrigation projects were extremely successful, but they left behind a substantial ecological footprint through the loss of native vegetation in the coastal zones and a reduction in the supply of freshwater and ensuing degradation of the riverine ecosystems, estuaries and associated coastal lagoons.

In the 1930s a technological revolution permanently transformed the Gulf of California; outboard motors and gillnets came into use. As a result, the totoaba—a fish endemic to the Sea of Cortés and highly appreciated for export—saw significant population declines in the Guaymas area of Sonora. Additionally, inshore fisheries in the estuaries and lagoons started to increase steadily and became a highly profitable activity. In 1933, the shrimp fishery developed the use of trawling boats over soft seabed (Figure 3). Since then, the seabed of the Gulf of California has been swept clean every year in search of "pink gold." Everything in the path of the trawling dragnet—fish, octopus, conch, sponges and starfish—is destroyed. Judged to be of lesser economic value, this unintended bycatch is returned dead to the sea. As the 20th century progressed, the shrimp fishery became the most important activity in the fishing sector with the largest contribution to regional income, jobs, supporting infrastructure and foreign income. In 1997 the five states surrounding the Gulf of California produced 57,000 tons of shrimp—approximately 70% of the national shrimp production and 90% of Mexico's Pacific coast production. The use of bottom trawlers by the shrimp industrial fleet also became one of the greatest threats to biodiversity in the Gulf, however, producing high volumes of bycatch and destroying the Gulf's soft seabed.

In the 1950s, fishing became a major driving force in regional development. The success of the fishing industry attracted financial resources to the sector and led to the establishment of freezing and packing companies and shipyards. This period was characterized by apparently inexhaustible abundance. In reality the decline of the fishing sector was already underway, driven by unsustainable harvests, over-capitalization of the fishing fleet and wasteful fishing technologies (Robles & Carvajal 2001). In the 1960s, the sardine fishery developed in the Sea of Cortés using purse seiners. These boats are still used today, and their

RUSSELL A. MITTERMEIER, CONSERVATION INTERNATIONAL

**Figure 3** Rows of shrimp trawlers in port in the Gulf of California. The use of bottom trawls is one of the greatest threats to the Gulf's marine biodiversity, resulting in high volumes of bycatch and destruction of the seabed.

technology represents one of the most selective fishing gears in the region. In the second half of the 20th century, the yellowfin tuna fishery developed in the mouth of the Gulf of California, and in the 1980s the Humboldt squid fishery started and is now the second most important in the area in terms of catch.

A marked increase in the price of beef in the 1980s led to the clearing of substantial areas of native vegetation using heavy machinery and replacing pristine native desert scrub and dry tropical forests with the non-native buffel grass *Pennisetum ciliare* of African origin to improve forage productivity for cattle in desert environments. Highly adapted to the hot and dry tropical environment, buffel is rapidly invading some overgrazed desert lands and, although it supports an increased carrying capacity of cattle, it is not used by the native fauna, and consequently the habitat for native species has been severely diminished. Once invaded by the rapidly growing and leafy buffel, the accumulated biomass burns easily during the dry season, turning the Gulf deserts into a fire-maintained ecosystem that burns seasonally and prevents the re-establishment of the original biologically rich scrub. Apart from its direct impact on the terrestrial ecosystems, the replacement of the native vegetation by buffel grass also represents a threat for the coastal environments, as the fire-prone new grasslands generate more sediment and silt runoff into the coastal areas and lagoons during intense monsoon rains.

## Socioeconomic Overview

With some 26% of Mexico's land area and 8.8% of its population in the year 2000, the states surrounding the Gulf of California produced 9.1% of the country's gross domestic product. The Gulf is a large, still sparsely populated area with human densities of only one-third the national average. It is also a relatively wealthy region within Mexico: the *per capita* contribution of Gulf inhabitants to the country's GDP is 5% above the national average. This productive advantage is even higher in the Baja California peninsula and the state of Sonora, where the *per capita* income is about 22% higher than the national average. The region is a major contributor to the national fisheries sector, producing approximately 50% of the landings and 70% of the value of national fisheries in Mexico. The coastal plains of Sonora and Sinaloa are also major agricultural producers. Approximately 40% of the national agricultural production comes from this region, mainly supported by high-technology irrigation.

The region is not only one of Mexico's richest in terms of natural resources; it also holds one of Mexico's fastest growing regional economies. The *maquiladora* industries (corporations with special foreign investment and import/export privileges) in Tijuana, the high-input crops and associated agro-industries in the agricultural valleys (Mexicali, Tecate, San Quintín) and the booming tourism industry are all powerful driving forces in economic and demographic growth. Some selected indicators of economic development (education, housing and human fertility) show values that suggest a relatively high economic development compared to the rest of Mexico (Ezcurra 1998). For example, the peninsula of Baja California has levels of illiteracy of less than 4%; the number of houses with electricity (including rural dwellings) approaches 90%; and the mean number of live births per woman over twelve years old was around 2.4 in the mid-1990s and has been steadily decreasing since (for comparison purposes, the state of Oaxaca in southern Mexico has 17% of illiteracy, only 73% of its houses have access to electricity, and the mean number of live births per woman over twelve years of age is 3.1).

In spite of the low fertility rates, the success of the peninsular economy has brought a large demographic increase to the region, chiefly derived from immigration. While the demographic growth rates in Mexico have decreased considerably during the last decades from a national average of more than 3% to less than 2%, the growth rates in the northern states of the region—and especially in the peninsula of Baja California—still remain high. Population growth rates in the whole Gulf region averaged 2.4% for 1990–2000 while the national average was around 1.8%. Between 1980 and 2000, the state of Baja California grew at an annual rate of 5.6%, while the neighboring Baja California Sur grew at a rate of 4.9% (Table 1). The cities with the most dynamic and active economies have been growing even more rapidly: Tijuana, fueled by the immigration magnet of the *maquiladora* industry, grew at a rate of 6.5%, while the population of Los Cabos, because of a rapidly growing tourism boom, grew at the extraordinary rate of 9.7%. If these rates are maintained, Tijuana would double in population size every eleven years, while Los Cabos would double every seven years. These remarkably rapid growth rates are not due to reproductive habits; they are chiefly the result of internal migration within Mexico from the impoverished southern states into the more dynamic economy of the Gulf's border region.

Table 1
**Selected demographic and economic indicators for the five states that surround the Sea of Cortés**

| State | Population density/sq km 2000 | Population growth (%) 1980–2000 | Marginalization | Per capita GDP (US$) 2001 | GDP growth (%) 1980–2000 |
|---|---|---|---|---|---|
| Baja California | 34.8 | 5.6 | very low | 7,646 | 4.6 |
| Baja California Sur | 5.7 | 4.9 | low | 7,498 | 4.1 |
| Nayarit | 33.0 | 1.3 | high | 3,809 | 1.1 |
| Sinaloa | 44.2 | 1.9 | intermediate | 4,763 | 2.6 |
| Sonora | 12.3 | 2.3 | low | 7,838 | 3.3 |
| **Regional average** | **20.9** | **2.8** | | **6,311** | **3.4** |

**Sources:** *Instituto Nacional de Geografía e Informática* (INEGI), *Censos de Población y Vivienda X, XI, XII* and *Banco de Información Económica.*

All these indicators show that, in terms of wellbeing, the region as a whole is significantly better off than others in Mexico—and hence its magnet-like attraction for new immigration—but there is also considerable differentiation among the five states, with only some specific zones driving the growth of the regional economy. Furthermore, most of the population is concentrated in the coastal zone where the pressure on natural resources is greatest.

Historically, the economy of the Gulf region has been based mainly on the primary sector—agriculture, fisheries and mining—and the southern continental states have maintained this economic structure. In recent decades, however, declining natural resources and new opportunities in the Gulf states have led to major shifts in the economic structure. Over-exploitation of water and soil resources by agriculture (especially in the drier parts of the region) has encouraged a reorientation toward high-value horticultural crops which provide a better return for water and soil inputs than more traditional field crops. In fisheries, over-capitalization of the shrimp trawling and inshore fishing fleets means that limited resources are shared among too many boats, and profits are low. At the same time macro-economic policy change, new trade agreements and globalization have created new opportunities in export-oriented manufacturing and industry (especially in the border states) and in services such as tourism. This new economic environment provides opportunities for conservation as well as increased threats associated with population growth and increasing conflicts over scarce natural resources.

As natural resources become depleted and new opportunities stimulate demand for these resources and for the environmental services they provide, there is increasing risk of conflict between and within different sectors over resource use. This is especially apparent in the case of common-access resources and in the absence of clear property rights or strongly enforced regulations. In the Gulf, different sectors respond to the demands of different export markets, both national and international, and the region has never been integrated economically as a single, joint economy. The current disintegration of the economy contrasts strongly with the strong local economic linkages among sectors that a vision of integrated coastal management demands.

The implications of this economic change for biodiversity conservation are significant. Decision-makers have an important opportunity to reduce the unsustainable pressure on natural resources in the region by supporting reorientation of the economy away from primary sector activities associated with the over-exploitation of natural resources. Increasing population pressures and conflicts over natural resources need to be addressed through the strengthening of property rights for fisheries and water resources and better enforcement of existing regulations. Urbanization and tourism development must be based on careful land-use planning and environmental management. Support to the development of manufacturing and industry must be balanced by increased attention to environmental standards and the adoption of environmental technologies and best practices. Hydropower development must be accompanied by increased commitment to watershed management. In many parts of the Gulf region, depletion of ocean and water resources is already driving investment and economic opportunities away. The maintenance of fundamental ecological processes and ecosystem functions is critically needed to protect economic investments as well as biodiversity in the Gulf of California region.

There are significant and growing socioeconomic differences among the five states that compose the Gulf of California region. In recent decades, states that continue to specialize in primary products (Sinaloa and Nayarit in the southern continental region) have lost economic preeminence, while those linked to the modern export sectors (in the Baja California and Sonora border belts) have seen growth. Baja California Sur, in the southern part of the peninsula, has seen significant growth through tourism, especially in the Cape Region. Sonora, Baja California and Baja California Sur have per capita GDP 24% higher than the national average, whereas that of Sinaloa and Nayarit is lower. This pattern is paralleled by economic growth rates—with Baja California, Baja California Sur and Sonora enjoying significantly higher growth than Sinaloa and Nayarit. The differentiation among Gulf states is also reflected in levels of marginalization of their populations, meaning the limits on access to education and economic and cultural growth. While Baja California, Baja California Sur and Sonora enjoy a low or very low level of marginalization, Sinaloa and Nayarit show medium and high levels, respectively. In the period 1980–2000, economic growth in Nayarit was outpaced by population growth, indicating a reduction in economic welfare and increased marginalization of its people.

Population density reflects economic history. It is highest in Baja California (mainly in the border zone, as the rest of the state is very sparsely populated) and in Sinaloa and Nayarit; and it is significantly lower in Sonora and Baja California Sur, the two desert states. Population growth rates are higher than the national average, however, in Baja California, Baja California Sur and Sonora, and lower than the national average in Sinaloa and Nayarit, reflecting current economic opportunities.

The primary sector of the economy (agriculture, livestock and fisheries) is showing by far the most sluggish growth rates (Table 2). While the annual growth rate of the primary sector between 1980 and 2000 was 0.7%, the rest of the regional economy grew at an average combined rate of 4.7%. This indicator highlights that further economic growth through new or intensified use of natural resources seems to have reached a limit, and the areas within the region showing the highest dependence on natural resource use are rapidly lagging behind in economic development. Again, the two southern states (Nayarit and Sinaloa) are the ones with the highest dependence on agriculture, livestock and—especially important for this analysis—fisheries. Their continued dependence on primary sector activities, combined with a slowing of economic growth and increasing marginalization, raises concerns in these southern states for the overexploitation of natural resources and increased threats to biodiversity.

These economic indicators highlight some of the most pressing environmental problems of the region. On the one hand, open-access, extractive use of natural resources seems to have reached a limit and little can be expected from this sector for future development. On the other hand, the rapid growth of the manufacturing and services sectors is putting an additional strain on the regional resources. It is extremely difficult to keep supplying services such as running water and sewage to cities that double in size every ten years. Rapid demographic growth means, almost by definition, an increasing demand for and pressure on the regional resources, especially water which is scarce in the peninsula. It also means an increase in pollutants that result from the uncontrolled urban growth and from the growing pressures on the deficient sanitary infrastructure, including poor drainage and lack of water-treatment facilities. Thus, the rapid expansion of the population linked to the more successful sectors of the regional economy is mostly done at the expense of depleting underground aquifers and destroying the natural ecosystems and watersheds that surround the large urban centers.

Table 2

**Economic dynamics (in terms of relative contribution of regional GDP and growth rates) of different sectors of the regional economy of the Gulf of California (in %)**

| Sector | Contribution to regional GDP | Growth rate 1980–2000 |
|---|---|---|
| Agriculture, livestock, fisheries | 10.3 | 0.7 |
| Manufacturing | 15.8 | 4.7 |
| Electricity | 2.4 | 5.8 |
| Commerce, restaurants, hotels | 22.1 | 3.3 |
| Transport, storage, communications | 11.7 | 5.6 |
| Financial services | 15.9 | 5.8 |

**Sources:** *Instituto Nacional de Geografía e Informática* (INEGI), *Banco de Información Económica.*

## Regional Environmental Challenges

The ecology and biodiversity of the Gulf of California have already suffered considerable degradation. Some 39 species that live in the marine zone and on the Gulf's islands are classified as endangered or threatened by the Red List of IUCN–The World Conservation Union (IUCN 2000). Of these, the endemic vaquita porpoise *P. sinus* and the totoaba *T. macdonaldi* (Figure 4) are near extinction, while populations of five species of sea turtles have all but disappeared from the Gulf. Recent estimates indicate there are approximately 567 vaquita individuals, and their mortality is above twelve deaths per year, making the vaquita the most endangered marine cetacean in the world.

Over-exploitation of the fishing stocks is rapidly becoming a strongly limiting factor for the success of the regional fisheries. Twenty years ago there was a correlation between catch and effort in many of the regional fisheries: the more days the fleets fished, the more they caught. Now, that correlation is largely gone. The total landings in most fisheries are chiefly independent of fishing effort, and the catch per unit effort has decreased severely for many species. In short, the fishermen of the Sea of Cortés are often over-exploiting and, in some cases, even depleting their stocks (Sala *et al* 2004, Velarde *et al* 2004). There is also clear evidence that coastal food webs in the Gulf of California have been "fished down" during the last 30 years (i.e., fisheries shifted from large, long-lived species belonging high on the marine food chain to small, short-lived species from lower trophic levels), and that the maximum individual length of fish landed has decreased significantly (about 45cm) in only 20 years (Sala *et al* 2004).

© RICHARD HERRMANN

**Figure 4** Found only in the Gulf of California, the totoaba—a species of large croaker—is in danger of extinction.

In some fisheries, the tragedy of common-access resources has hit the Gulf very hard. For example, 30 years ago the shrimp trawling fleet in the Gulf was around 700 boats, each of which captured about 50 tons of shrimp per season. Now the fleet is almost 1,500 boats and the annual catch scarcely surpasses ten tons per boat. Despite government subsidies of around US $30 million each year provided in the form of cheap fuel, many boats of the fleet are facing economic collapse.

Environmentally the situation is also discouraging, especially in the case of the shrimp trawlers. The bottom trawlers kill some 200,000 tons of unintended bycatch every year for a meager annual catch of about 30,000 tons of shrimp. In so doing, the dragnets also destroy some 30,000 to 60,000 square kilometers of seabed, much of which lies within the Upper Gulf Biosphere Reserve. The boats also collectively emit some 30,000 to 40,000 tons of greenhouse gases derived from the government-subsidized cheap fuel that keeps their inefficient business going. The seabed has been so depleted in some parts that the local artisanal fishermen in places such as Loreto Bay and Bahía de los Ángeles have been demanding the establishment of no-take zones and marine protected areas. In open conflict with the local communities, the larger fleets oppose the establishment of protected areas and demand permits to trawl inside the already established reserves to increase their scanty earnings. In the Gulf, conflict between sectors and particular interests has been the rule.

Not all stories of common resource use in the Sea of Cortés are despairing tales of unsustainability and collapse, however. There are also a number of success stories, and understanding these stories is fundamental for future conservation efforts. For example, local artisanal fishermen have started to work with local researchers in the Sea of Cortés to understand the phenomenon of spawning aggregations to identify and protect reproductive areas. As a result of the pressures from these local resource users, the Loreto Bay is now a marine park, and the fishermen of Bahía de Los Ángeles are supporting the creation of a similar marine protected area. In the San Ignacio Lagoon, fishermen who previously engaged in unsustainable practices have organized to preserve the environment and are training their people in basic natural history to organize whale-watching tours.

The abalone and lobster cooperatives of the Pacific coast of Baja provide yet another example of long-term sustainable use. With no support from the federal government, they have established strict rules for resource extraction and have developed their own law enforcement system. Many generate their own electricity, run their own canneries and finance their own schools. More than 40 years after their establishment, productivity remains high and their natural resources seem to be in fairly good shape.

It is not only small communities and conservationists that are critical of some of the region's unsustainable modes of development; a growing number of entrepreneurs and business people are also becoming committed supporters of the environmental cause. Even large fishing fleets can be sustainable when their operators work in cooperation. In contrast to the failure of the shrimp bottom-trawling fleet, the sardine fishery has been capable of controlling its own fishing effort, and—after a past collapse—their fishery is now healthy and sustainable (Cisneros-Mata, Nevárez-Martínez & Hammann 1995, 1996, Lluch-Belda, Magallón & Schartzlose 1986). As a result of growing concerns about these issues, a cluster of environmentally concerned business leaders working with environmental non-governmental organizations (NGOs) has organized an action and opinion group called *Noroeste Sustentable Iniciativa* (Sustainable Northwest Initiative), or *Iniciativa Nos,* to promote the sustainable use of the resources in the Sea of Cortés. In short, although the Sea of Cortés is undergoing extreme pressures from overfishing in many areas, with the consequent collapse of some of its resources, it also harbors a number of successful and encouraging experiences from communities that are trying to maintain their resources—healthy and productive—for the future.

## Institutional Capacities

Viewed from the center of Mexico, the Gulf of California appears to be a uniformly distinct region formed by five coastal states with a hot, dry climate. From inside the region, however, it is evident that the different states face very different problems and there is little internal cohesion in the policies and views on natural resource management.

For many decades following the 1920s, the political and economic stability of the Gulf of California region has been based on three things:

- A political system dominated by a single party that handled the main negotiations between federal and state governments
- A cooperative structure of government, both at the federal and state levels, with party-controlled unions of industrial workers and farmers
- The involvement of major private interests in the region

This system started to change slowly in the 1980s, when the federal government decreased its support to the previously dominant unions and the union system started to dismantle. In the Gulf of California this meant the disarticulation of the main cooperative structures that operated within the fishing and agricultural sectors. This change allowed the private sector to gain a central role in these activities, while the federal and state governments retreated from direct oversight of production to assume a new role as regulator, financier and referee in the resolution of regional conflicts.

Later, in the 1990s, the democratization of Mexico brought on further change. One-party rule is no longer the norm in the region, and multi-party politics result in debates over governance, sovereignty and decision-making for natural resources. With three different parties governing the different states around the Gulf, there is now a more diverse and complex outlook. And, as the states have demanded more independence from federal rule, local governments now represent more freely the demands and necessities of their people, often in conflict with neighboring states or even with the previously all-powerful federal government.

Institutional change is still underway and more reforms are being demanded by local governments. The federal authorities still retain a high degree of political, fiscal and economical centralization, collecting most of the tax revenue and managing the national budget. The state governments are almost completely dependent on the federal government in fiscal terms, and an almost absolute dependency on the state in economic and regulatory terms persists for the municipal authorities. All this has lead to a growing complexity in the definition of priorities for regional development among the three levels of government, with poor coordination among authorities and increasing dispersion of resources and actions. Although political change has swiftly taken place around the Sea of Cortés, the reformed institutions have not yet been capable of tackling the urgent issues of resource degradation in a coordinated fashion.

One of the key difficulties faced in reversing the trend of environmental and resource degradation has been the notorious absence of a well-developed, consolidated relationship between environmental scientists and decision-makers.

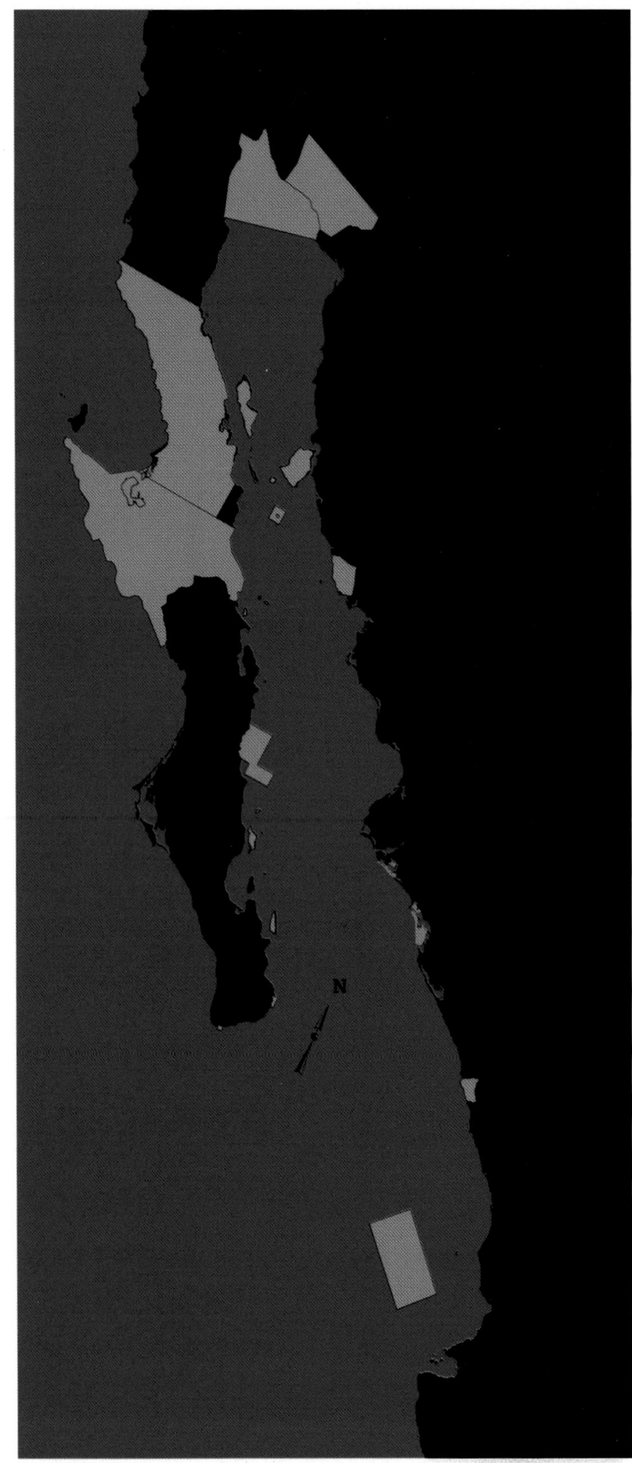


**Figure 5** Protected areas with marine components established by Mexico in the Gulf of California Region (modified from National Institute of Ecology).

The Sea of Cortés region harbors some of the most outstanding environmental research centers in Mexico, but most of this research rarely reaches the sphere of decision-makers. Another element contributing to the failures to preserve the natural resources of the Sea of Cortés is the sheer intricacy of the endeavor. Although the region is one complex, interconnected set of biological communities, the governance is fragmented and most of the research done has been based on narrow, specific aspects of conservation rather than addressing the complex functioning of the regional ecosystem. Few of the regional research and conservation efforts have concentrated on a multidisciplinary approach, with little effort placed on large-scale, ecosystem-based research.

Changes over the past decades, however, have brought some positive transformations. As described in the previous section, globalization of trade and the new open markets have produced a shift in the economy from unsustainable resource extraction to services and manufacturing. In response to this, some of the pressures on the Gulf's open-access natural resources have declined. In contrast, the development of the region has not offered solutions to the old problems of social inequality and territorial isolation. And, despite the common interests among the majority of environmental conservation institutions in the region, the links of cooperation between them are scarce. The region has not been able to develop a common societal vision which would allow planning for a sustainable future and collectively working to ensure that future development also allows conservation of the region's rich natural heritage.

## Conservation Efforts

In the midst of all the political and economic changes the region has undergone, both the Mexican government and conservation NGOs have developed actions to protect the incredibly rich and increasingly endangered ecosystems of the Sea of Cortés, Baja California and the neighboring mainland (Zavala *et al* 2001, Ezcurra *et al* 2002b, Ezcurra 2003). In total, 29 Protected Natural Areas have been established—ten in the U.S. and nineteen in Mexico—for conservation of the natural wealth of this bioregion. Of these, 23 are terrestrial, two are marine and four have both marine and terrestrial portions (Figure 5). The total surface under protection is 98,755 square kilometers (11% of the regional area), of which 14,404 square kilometers are marine (4% of the region's marine area) and 86,350 square kilometers are terrestrial (17% of the region's terrestrial area). Importantly, the capacity to manage protected areas in Mexico in general, and in the Sea of Cortés in particular, has improved during the last decade as a result of additional financial resources from the Mexican federal government and increased support from private groups.

Since 1993 there has been an immense effort in Mexico to protect new areas in and around the Sea of Cortés. Indeed, between 1993 and 1998, there were six new protected areas decreed, totaling more than two million (2,612,126) hectares under some category of protection. In 1993, the Mexican government issued a decree under the Biosphere Reserve protecting the upper Gulf of California and the delta of the Colorado River along a coastal strip of desert lands and coastal ecosystems that join the Sonoran desert with the Baja California peninsula. The creation of this wilderness protected area was achieved largely thanks to the initiative and support of local conservation groups and academic institutions. The new upper Gulf reserve protected two highly endangered marine species: the vaquita porpoise *P. sinus* and the totoaba *T. macdonaldi*. The most remarkable aspect of the project, however, was the wide cooperation it involved. The creation of these desert and coastal reserves was based on the participation of many groups, including indigenous peoples like the Cucapá and the Tohono O'odham, conservation groups like Pronatura, Conservation International, The Nature Conservancy and the Audubon Society, and many academic and research organizations. Some of these organizations eventually coalesced into a conservation bloc called the Sonoran Desert Alliance.

During 1993, the Mexican government prepared a series of documents for presentation to UNESCO to dedicate the Vizcaíno Biosphere Reserve as a World Heritage Site. Shortly afterwards, new decrees followed:

- With cooperation from researchers of the *Instituto de Ecología* (in Jalapa), the California Academy of Sciences and the University of California at Los Angeles, the Mexican government issued a decree for the protection of the Revillagigedo Archipelago in the Pacific.
- Six months later in June 1994, a decree was issued to protect the Sierra de la Laguna in Baja's Cape Region.
- Following a 1995 initiative from Pronatura Peninsula de Baja California, a Mexican NGO, the greatest and most diverse reef in the Sea of Cortés became officially protected under the name of *Parque Marino de Cabo Pulmo.*
- Finally, in July 1996, a decree was issued to protect the Loreto Bay as a marine park. It is remarkable that the creation of this last park was totally the result of a grassroots initiative from the local small fishermen, who were concerned about the continuing decrease of their catch and about the degradation of the hatching grounds of their fisheries.

In 1997, working to promote a conservation agenda, a group of scientists and conservationists—including 30 academic, government and conservation organizations—teamed up in a project known as the Coalition for the Sustainability of the Gulf of California. Among other activities, the Coalition produced scientific information for the region that became a milestone in regional planning. Indeed, the most relevant initiative of the federal government for the Gulf is the *Ordenamiento Ecológico,* i.e., the Regional Marine and Coastal Use Plan. To support the development of this initiative with the most up-to-date scientific information and analysis, the Coalition organized a region-wide biodiversity conservation priority-setting workshop. During the workshop, 180 experts identified 22 marine and 20 terrestrial areas of high biodiversity importance (Figure 6). The final document included a comprehensive conservation priorities map, which in turn provided important information for a range of institutions to use in coordinating and developing conservation strategies in the region.

A complementary process developed through a regional alliance of NGOs—the Alliance for the Sustainability of the Mexican Northwest Coastline, or ALCOSTA, which is composed of 20 conservation groups. ALCOSTA defined a regional vision to unify conservation efforts. Through their rich network of cooperation, ALCOSTA was capable of bringing a voice of alarm and concern into the *Escalera Náutica* (Nautical Ladder) tourism development project, thereby initiating an important effort to diminish the environmental impacts of the project and transforming the initiative forever. Partly as a result of these efforts, President Vicente Fox recently announced that the Gulf of California is a joint priority for both tourism development *and* conservation. As a result, current conservation initiatives are being directed toward enlarging the protected areas in the Gulf's islands to include surrounding waters.

The Sustainable Northwest Initiative *(Iniciativa NOS)* was recently formed by a cluster of environmentally concerned business leaders. Working under the model of the United States' Chesapeake Bay program, these leaders are developing a common vision for sustainable use of the living resources of the Sea of Cortés through high-level regional agreements between government and business. As entrepreneurs, they perceive there is value in conservation in order for their businesses to survive in the long term. As concerned persons, they also feel a responsibility for future generations—a historic imperative that must be met.

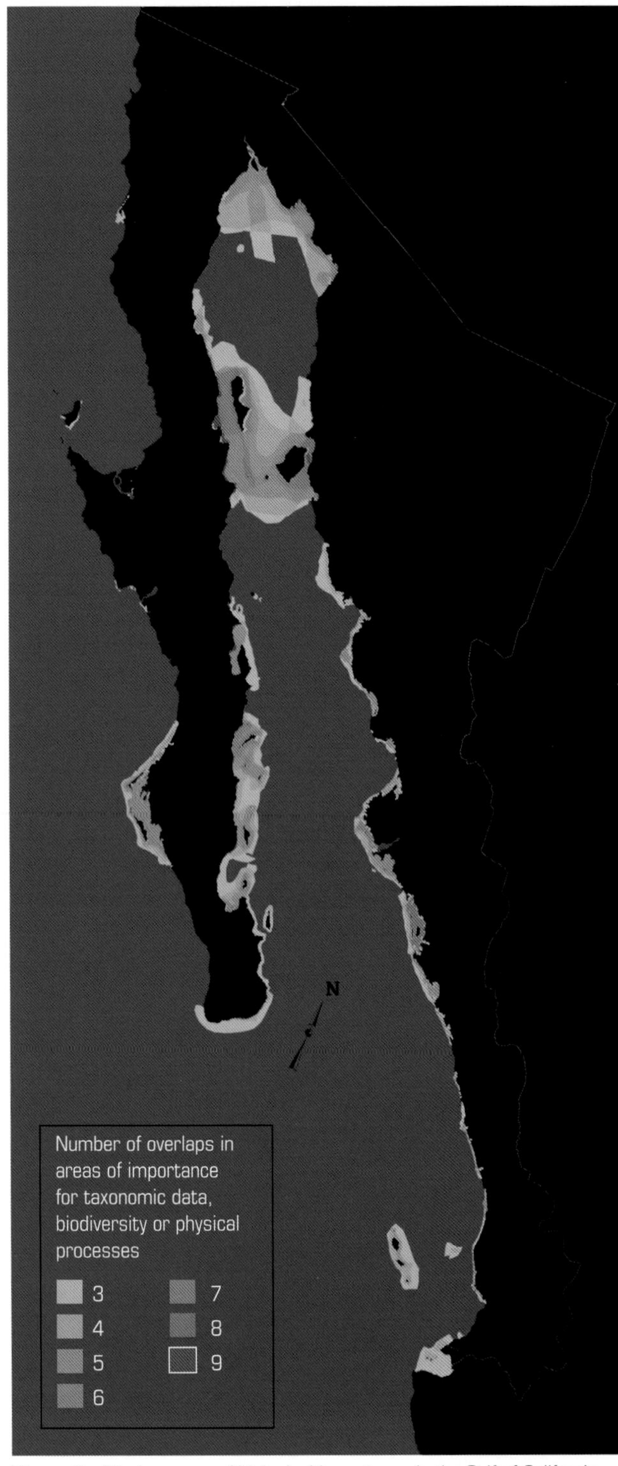

**Figure 6** Marine areas of biological importance in the Gulf of California.

## Toward a Regional Conservation Agenda

In spite of difficulties in maintaining its resource productivity and the past collapses of its fisheries, the Gulf of California region still presents great economic potential, but governance is a critical factor for regional development. The region is governed under a complex mix of authority emanating from the federal government, the surrounding five states and 40 coastal municipalities, and there is often little synergy among them. Poorly planned economies, often the result of the lack of coordination among authorities, are one of the greatest threats to ecosystems in the Gulf of California.

On the other hand, regional cooperation among nongovernmental agencies has yielded impressive results in the past. Through their rich network of cooperation, ALCOSTA, the regional alliance of NGOs, for example, was able to make the *Escalera Náutica* development project much more open to environmental conservation issues. Likewise, the Coalition for the Sustainability of the Gulf of California has produced some of the most notable conservation milestones for the region. And cooperation between regional NGOs and research groups allowed the presentation of a regional agenda at the *Defying Ocean's End* Conference in Los Cabos in 2003. This cooperative agenda established seven specific objectives to approach sustainability in the Sea of Cortés.

### 1. Improve the management of regional marine and coastal protected areas

Although impressive progress has been made during the last decade by the Mexican government in the funding and management of its protected natural areas, many of them still exist as "paper parks," with inadequate funding and little effective management. If the regional protected areas are to be effective in their conservation goals, they must improve in their level of funding, equipment and staffing.

### 2. Enlarge the system of marine and coastal protected areas

Although some marine protected areas have been created in the Gulf (namely, the Upper Gulf Biosphere Reserve and the Cabo Pulmo and Loreto Bay National Parks), these cover less than 4% of the Gulf's marine area. If effective conservation in the region is to be achieved, a significant increase in the marine protected areas must be obtained, reaching at least 15% of the Gulf's waters. This would allow the protection of spawning aggregation areas and critically endangered ecosystems such as seamounts, coastal lagoons, coral and rocky reefs, estuaries and marine mammal habitats (Sala *et al* 2002a, b).

**3. Develop a comprehensive plan to manage and protect priority coastal wetlands**

The degradation of coastal wetlands is one of the Gulf's most serious threats. With little consideration for the ecological services they provide, mangrove forests are being cut for the development of aquaculture—mostly shrimp farms—and tourism projects. Furthermore, coastal wetlands in general are threatened by increased water consumption upstream and by pollution of rivers and waterways. The ecological services provided by estuaries and lagoons are critical for the survival of the Sea of Cortés fisheries and for the health of the large marine ecosystem as a whole. A comprehensive plan to protect coastal wetlands, stop mangrove deforestation and maintain the ecological services of coastal lagoons and estuaries must be developed and its actions implemented immediately.

**4. Reduce the shrimp trawling fleet and improve its fishing technology**

Many of the strongest issues of unsustainability in the Gulf stem from the destructive effect and the economic inefficiency of the current shrimp bottom-trawling fleet. The only alternative to solve this growing problem is to reduce the fleet by at least 50% through a legal buyout. If effective legal means are put in place to ensure that no new fishing permits are issued in future—and hence not allowing the fleet to grow again to unsustainable levels—an action of this sort would allow the negotiation of effective enforcement within existing no-take zones and the introduction of better fishing gear with more efficient excluder devices.

To address this challenge, Conservation International and other partners in the region have developed a strategy with three major objectives:

- Reduce the fishing effort to optimal economic levels, with buyout of 500 shrimp vessels currently operating
- Promote the use of modified bottom-trawling equipment that is less harmful to the environment
- Restrict bottom-trawling in the critical areas for biodiversity conservation that were identified by the Coalition for the Sustainability of the Gulf of California in its regional exercise on conservation priorities

**5. Develop a regional plan to regulate the use of land, coasts and waters**

The main instrument in Mexican legislation to regulate the use of space within environmental guidelines is the *Ordenamiento Ecológico,* or Ecological Planning of the Territory. This plan demands full and comprehensive hearings and negotiations with local governments, local businesses and non-governmental organizations. Because of its complexity, effective territorial planning has been difficult to achieve in the Sea of Cortés but is now one of the most urgent objectives to reach. For this purpose, the participation of the general public and local conservation alliances is of critical importance.

**6. Reorient regional tourism toward low-impact, environmentally sustainable resource use**

The *Escalera Náutica* has become one of the most debated projects in the region. On the one hand, most environmentalists agree that in the Gulf the primary sector of the economy has reached its limits and, in some cases, is even facing collapse. Thus, a shift of the economy from unsustainable fisheries and water-intensive agriculture to the services sector (including tourism) seems a desirable move. However, experiences in Mexico with unsustainable tourism (and its failed and abandoned projects, dredged mangrove swamps and exhausted aquifers) have left a deep scar. The big challenge in the Sea of Cortés seems to be how to promote environmentally sustainable tourism while ensuring the preservation of the natural beauty and the biodiversity of the region, the very attributes that have triggered tourism.

**7. Articulate a common regional vision for development and build capacities for regional management**

Regional conservation will be successful if, in collaboration with local business and political leaders, a regional development vision based on the long-term protection of the Gulf and its resources can be pieced together collectively and agreed upon. *Iniciativa NOS* has been created as an *ad hoc* vehicle to promote and articulate the common regional vision and help develop the regional capacities for manage-

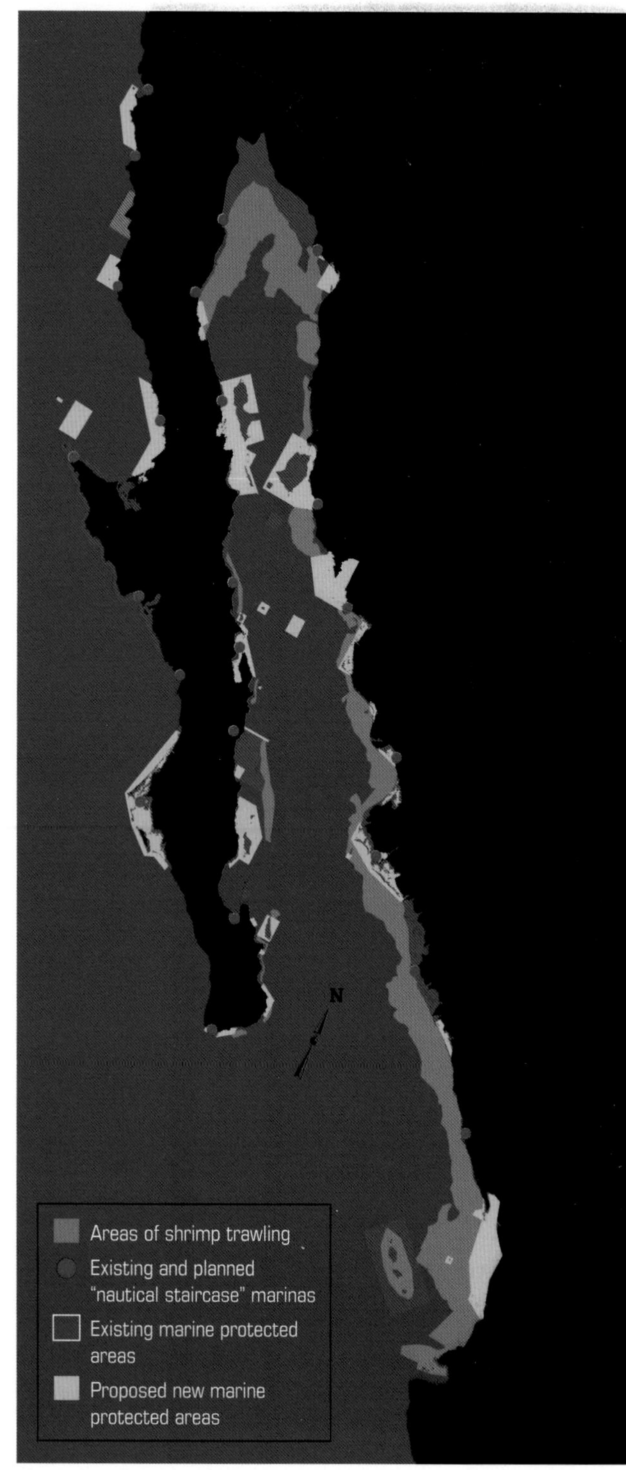


**Figure 7** Marine conservation priorities for the Gulf of California.

ment in close relationship with major stakeholders and governments.

This seven-point agenda, catalyzed by the *DOE* meeting, is now a conservation roadmap for the years to come (Figure 7). To facilitate well-planned sustainable development, it is important to build a common regional development vision and establish a highly motivated and committed group of leaders from businesses, environmental organizations, civil society and government to work together on common regional goals. The creation of a regional development vision with well-defined scenarios for the future has been among the top priorities of NGOs, government, academics and business leaders in the region to define a workable and viable agenda. This agenda is based on five fundamental tenets:

1. Collaborative inter-sectoral efforts to define solutions to threats are critical for success at a regional scale.
2. Improved governance structures are key to effective natural resource management.
3. Adequately managed marine protected areas are fundamental to control open access and support fisheries management.
4. Behavioral changes promoting negotiation as a major mechanism for conflict resolution produce more long-lasting results than imposing the values of one group on another.
5. Sustained progress toward measurable goals should be made at the scale of the regional large coastal ecosystem.

The rich technical information generated by the Coalition for the Sustainability of the Gulf of California is now being used by a group of the Gulf's key stakeholders congregated around *Iniciativa NOS*. The aim of this group is to analyze the environmental issues the region is facing and address them through a region-wide agreement. This agreement should have clear, ambitious and measurable long-term goals for key areas (including Natural Protected Areas, fisheries, tourism and water) and should include a comprehensive implementation plan. To put this vision into practice a permanent structure needs to be created, including a policy board, a scientific board, technical staff and a permanent financial commitment from regional governments and funding agencies for the core funding needed to operate the initiative.

The participatory process initiated with the establishment of *Iniciativa NOS* has already yielded a better understanding of the context necessary to develop a successful regional agreement that considers biodiversity conservation priorities as it builds economic opportunity in the region. The final regional vision must be designed to work with several other plans being implemented in the region, supporting efforts rather than duplicating them. The next step in this initiative is to attract business leaders and government agencies to support the agreement and build broad-based support. If successful, this important initiative will encourage the governments of the five states that surround the Gulf of California and the federal government to sign a regional agreement for sustainable development and commit to its implementation.

## Summary

Hopefully, the increasing pace of conservation efforts will stall the environmental degradation the Sea of Cortés has been suffering and diminish the threats to its long-term sustainability. There seems to be a growing awareness in the region, as never before, of the need to take urgent action to protect the environment. Conservation groups, research institutions, federal and state governments, conscientious businesspersons and ecotourism operators have all been contributing to the growing appreciation of the environment.

It was not by accident that the first Mexican marine protected area—the Upper Gulf Biosphere Reserve—was created in the Sea of Cortés. A number of institutions and organizations in both Mexico and the U.S. had been working together for years promoting the conservation of the Upper Gulf and the Colorado River delta. Research institutions, conservation groups and governmental officials had all been teaming for years in discussion forums, preparing plans and proposals for the first marine protected area in Mexico. The time finally came when a proposal to protect the Upper Gulf and the coasts of the Colorado River estuary was ready and had the consensus and support of hundreds of environmental leaders (Ezcurra *et al* 2001).

It is now time to develop a new vision—a social agreement among sectors that will drive regional development for years to come, with ever-increasing consideration for the Gulf's environment and natural resources, and their sustainability. The Sea of Cortés receives what remains of the discharges of the Colorado River basin, and the survival of the Upper Gulf is a challenge for both Mexico and the United States. Thus, its larger basin is part of a binational wilderness, for which both Mexico and the U.S. share the responsibility of protecting their joint natural heritage, and both countries need to develop further and continuing efforts to promote truly cooperative work. Through the growth of alliances we may, in the end, preserve the Gulf of California. The region is a single large continuum with shared watersheds and estuaries, species and natural resources. The protection of these unique environments is of the utmost importance for the survival and wellbeing of all of us, now and for generations to come.


## Acknowledgements

*The authors would like to acknowledge the many people who contributed to the* Defying Ocean's End *process, especially in the development of this case study. In particular, we wish to thank Juan Bezaury of The Nature Conservancy–Mexico; Ernesto Bolado, Nohemi Camacho, Mauricio Cervantes and Roberto Gonzalez of Conservation International–Mexico; Charlotte Boyd of Conservation International–USA; and Stephen Olsen of the Coastal Resource Center, University of Rhode Island.*


## Literature Cited & Consulted


Álvarez-Borrego, S. 1983. Gulf of California. In *Estuaries and Enclosed Seas,* ed. B. H. Ketchum, 427–449. Amsterdam: Elsevier Science Publishing Co.

——. 2002. Physical oceanography. In *A New Island Biogeography of the Sea of Cortés,* ed. T. Case, M. Cody and E. Ezcurra, 41–60. Oxford: Oxford University Press.

Bahre, C. J. and L. Bourillón. 2002. Human impact in the Midriff Islands. In *A New Island Biogeography of the Sea of Cortés,* ed. T. Case, M. Cody and E. Ezcurra, 383–406. Oxford: Oxford University Press.

Bourillón, L., A. Cantú, F. Eccardi, E. Lira, J. Ramírez, E. Velarde and A. Zavala. 1988. *Islas del Golfo de California.* México, D. F.: Secretaría de Gobernación–Universidad Nacional Autónoma de México.

Bowen, T. 2000. *Unknown Island: Seri Indians, Europeans, and San Esteban Island in the Gulf of California.* Albuquerque: University of New Mexico Press.

Brusca, R. C. 1980. *Common Intertidal Invertebrates of the Gulf of California,* 2nd ed. Tucson: University of Arizona Press.

Carreño, A. L. and J. Helenes. 2002. Geology and ages of the islands. In *A New Island Biogeography of the Sea of Cortés,* ed. T. Case, M. Cody and E. Ezcurra, 14–40. Oxford: Oxford University Press.

Case, T. J., M. Cody and E. Ezcurra, eds. 2002. *A New Island Biogeography of the Sea of Cortés.* Oxford: Oxford University Press.

Cisneros-Mata, M. A., M. O. Nevárez-Martínez and M. G. Hammann. 1995. The rise and fall of the Pacific sardine, *Sardinops sagax caeruleus Girard,* in the Gulf of California, México. *California Cooperative Oceanic Fisheries Investigation Report* 36:136–143.

Cisneros-Mata, M. A., G. Montemayor-López and M. O. Nevárez-Martínez. 1996. Modeling deterministic effects of age structure, density dependence, environmental forcing and fishing in the population dynamics of the Pacific sardine *(Sardinops sagax caeruleus)* stock of the Gulf of California. *California Cooperative Oceanic Fisheries Investigation Report* 37:201–208.

Clavijero, Francisco Xavier. 1789. *Historia de la Antigua o Baja California.* México, D. F.: Republished by Imprenta del Museo Nacional de Arqueología, Historia y Etnografía, 1933.

del Barco, M. 1973. *Historia natural y crónica de la Antigua California. Adiciones y correcciones a la Noticia de Miguel Venegas.* Original manuscript aprox. 1780, edited with a preliminary study by Miguel León-Portilla. México, D. F.: Universidad Nacional Autónoma de México.

Ezcurra, E. 1998. Conservation and Sustainable Use of Natural Resources in Baja California: An Overview. Briefing Paper prepared for San Diego Dialogue's Forum Fronterizo, October 1998. University of California, San Diego (UCSD), Division of Extended Studies and Public Programs, San Diego. *http://www.sandiegodialogue.org/cb_research.htm* (accessed August 13, 2004).

——. 2003. Conservation and sustainable use of natural resources in Baja California. In *Protected Areas and the Regional Planning Imperative in North America,* ed. J. G. Nelson, J. C. Day and L. Sportza, 279–296. Calgary, Alberta: University of Calgary Press & Michigan State University Press.

Ezcurra, E., R. S. Felger, A. D. Russell and M. Equihua. 1988. Freshwater islands in a desert sand sea: The hydrology, flora, and phytogeography of the Gran Desierto oases of northwestern Mexico. *Desert Plants* 9 (2): 1–17.

Ezcurra, E., L. Bourillón, A. Cantú, M. E. Martínez and A. Robles. 2002a. Ecological Conservation. In *A New Island Biogeography of the Sea of Cortés,* ed. T. Case, M. Cody and E. Ezcurra, 417–444. Oxford: Oxford University Press.

Ezcurra, E., E. Peters, A. Búrquez and E. Mellink. 2002b. The Sonoran and Baja Californian Deserts. In *Wilderness: Earth's Last Wild Places,* ed. R. A. Mittermeier, C. G. Mittermeier, P. Robles-Gil, J. Pilgrim, G. A. B. da Fonseca, T. Brooks and W. R. Konstanteds, 315–333. Monterey: CEMEX, Washington D.C.: Conservation International and México, D. F.: Agrupación Sierra Madre.

Felger, R. S. 2000. *Flora of the Gran Desierto and Río Colorado Delta.* Southwest Center Series, Tucson: University of Arizona Press.

Felger, R. S. and M. B. Moser. 1985. *People of the Desert and Sea: Ethnobotany of the Seri Indians.* Tucson: University of Arizona Press.

Findley, L. T., M. E. Hendrickx, R. C. Brusca, A. M. Van der Heiden, P. A. Hasting and J. Torre. Forthcoming. Macrofauna del Golfo de California. CD-Rom Version 1.0. Macrofauna Golfo Project. Center for Applied Biodiversity Science, Conservation International, Washington, D.C., and Conservation International –Programa Golfo de California, Guaymas, Sonora, Mexico.

Fradkin, P. L. 1984. *A River No More: The Colorado River and the West.* Tucson: University of Arizona Press.

Garcillán, P. P. and E. Ezcurra. 2003. Biogeographic regions and beta-diversity of woody dry land legumes in the Baja California peninsula. *Journal of Vegetation Science* 14 (6): 859–868.

Gotshall, D. W. 1998. *Sea of Cortez Marine Animals.* Monterey: Sea Challengers.

Holmgren, M., M. Scheffer, E. Ezcurra, J. R. Gutiérrez and G. M. J. Mohren. 2001. El Niño effects on the dynamics of terrestrial ecosystems. *Trends in Ecology & Evolution* 16 (2): 59–112.

INEGI. 2004. *Instituto Nacional de Geografía e Informática.* México. *http://www.inegi.gob.mx/inegi/default.asp* (accessed September 30, 2004).

IUCN. 2003. *2003 IUCN Red List of Threatened Species. http://www.redlist.org* (accessed August 23, 2004).

Jordán, F. 1951. *El otro México: Biografía de Baja California.* México, D.F.: Biografías Gandesa.

Lavín, M. F. and S. G. Marinone. 2003. An overview of the physical oceanography of the Gulf of California. In *Nonlinear Processes in Geophysical Fluid Dynamics: A Tribute to the Scientific Work of Pedro Ripa,* ed. O. U. Velasco Fuentes, J. Sheinbaum and J. Ochoa, 173–204. Dordrecht: Kluwer Academic Publishers.

Lavín, M. F., E. Palacios-Hernández and C. Cabrera. 2003. Sea surface temperature anomalies in the Gulf of California. *Geofísica Internacional* (Special Volume: Effects of El Niño in Mexico) 42 (3): 363–376.

Lluch-Belda, D., B. F. J. Magallón and R. A. Schartzlose. 1986. Large fluctuations in sardine fishery in the Gulf of California: Possible causes. *California Cooperative Oceanic Fisheries Investigation Report* 37:201–208.

Marinone, S. G. and M. F. Lavín. 2003. Residual flow and mixing in the large islands region of the central Gulf of California. In Nonlinear Processes. In *Geophysical Fluid Dynamics: A Tribute to the Scientific Work of Pedro Ripa,* ed. O. U. Velasco Fuentes, J. Sheinbaum and J. Ochoa, 213–236. Dordrecht: Kluwer Academic Publishers.

Martin, P. S. 1984. Prehistoric overkill: the global model. In *Quaternary Extinctions: A Prehistoric Revolution,* ed. P. S. Martin and R. G. Klein, 354–403. Tucson: University of Arizona Press.

Mittermeier, R. A., N. Myers, P. Robles-Gil and C. G. Mittermeier. 1999. *Hotspots: Earth's Biologically Richest and Most Endangered Terrestrial Ecoregions.* Monterey, Mexico: CEMEX, Conservation International, Agrupacion Sierra Madre.

Nabhan, G. 2002. Cultural dispersal of plants and reptiles. In *A New Island Biogeography of the Sea of Cortés,* ed. T. Case, M. Cody and E. Ezcurra, 407–416. Oxford: Oxford University Press.

Polis, G. A., S. D. Hurd, C. T. Jackson and F. Sánchez-Piñero. 1997. El Niño effects on the dynamics and control of an island ecosystem in the Gulf of California. *Ecology* 78 (6): 1884–1897.

Robles, A., E. Ezcurra and C. León. 1999. *The Sea of Cortés.* México, D. F.: Editorial Jilguero/México Desconocido.

Robles, A. and M. A. Carvajal. 2001. The sea and fishing. In *The Gulf of California: A World Apart,* ed. P. Robles-Gil, E. Ezcurra and E. Mellink, 293–300. México, D. F.: Agrupación Sierra Madre.

Robles-Gil, P., E. Ezcurra and E. Mellink, eds. 2001. *The Gulf of California: A World Apart.* México, D. F.: Agrupación Sierra Madre.

Sala, E., O. Aburto-Oropeza, G. Paredes and G. Thompson. 2002a. Spawning aggregations and reproductive behavior of reef fishes in the Gulf of California. *Bulletin of Marine Sciences* 72 (1): 103–121.

Sala, E., O. Aburto-Oropeza, G. Paredes, I. Parra, J. C. Barrera and P. K. Dayton. 2002b. A general model for designing networks of marine reserves. *Science* 298:1991–1993.

Sala, E., O. Aburto-Oropeza, M. Reza, G. Paredes and L. G. López-Lemus. 2004. Fishing down coastal food webs in the Gulf of California. *Fisheries* 28 (3):19–25.

Sykes, G. 1937. *The Colorado Delta.* Carnegie Institution of Washington, Publ. No. 460.

Thomson, D. A. and W. H. Eger. 1966. *Guide to the Families of the Common Fishes of the Gulf of California.* Tucson: University of Arizona Press.

Thomson, D. A., L. T. Findley and A. N. Kerstitch. 2000. *Reef Fishes of the Sea of Cortez: The Rocky-Shore Fishes of the Gulf of California* (revised edition). Austin: University of Texas Press.

Thomson, D. A. and M. R. Gilligan. 2002. Rocky-shore fishes. In *A New Island Biogeography of the Sea of Cortés,* ed. T. Case, M. Cody and E. Ezcurra, 154–180. Oxford: Oxford University Press.

Velarde, E. and E. Ezcurra. 2002. Breeding dynamics of Hermann's Gulls. In *A New Island Biogeography of the Sea of Cortés,* ed. T. Case, M. Cody and E. Ezcurra, 313–325. Oxford: Oxford University Press.

Velarde, E., E. Ezcurra, M. A. Cisneros-Mata and Miguel F. Lavín. 2004. Seabird ecology, El Niño anomalies, and prediction of sardine fisheries in the Gulf of California. *Ecological Applications* 14 (2): 607–615.

Zavala, A., G. Anaya, C. Godínez and E. Ezcurra. 2001. El camino andado. In *The Gulf of California: A World Apart,* ed. P. Robles-Gil, E. Ezcurra and E. Mellink, 223–241. México, D. F.: Agrupación Sierra Madre.

CHAPTER 6

# Lines on the Water: Ocean-Use Planning in Large Marine Ecosystems

Dee Boersma, *University of Washington*
John Ogden, *Florida Institute of Oceanography, University of South Florida*
George Branch, *University of Cape Town, South Africa*
Rodrigo Bustamante, *Commonwealth Scientific & Industrial Research Organisation, Australia*
Claudio Campagna, *Centro Nacional Patagónico, Argentina*
Graham Harris, *Fundación Patagónica Natural, Argentina*
Ellen K. Pikitch, *Pew Institute for Ocean Sciences, University of Miami*

GREG FOSTER

## Introduction

In the terrestrial realm, both ecological and ownership boundaries are discrete and easily observed. Satellite and remote sensing images can be used to map habitat types, readily distinguishing among deserts, prairies, forests and lakes at a great distance. Furthermore, land uses that are codified by land ownership and zoning are often manifest in visible boundaries, such as cities and housing developments, industrial areas, roads and highways, power line corridors, agricultural areas and buffer strips surrounding streams. On land, changes in human use patterns and the impact of human activities are obvious and may easily be assessed over time. These changes, suitably presented in map formats, are powerful demonstrations of the need for management and zoning.

We do not as easily see lines on the water. Most habitat types are not as distinct. It is difficult to distinguish functional ecological boundaries and impossible to see legal jurisdictions. The edge of ocean currents or underwater continental shelves are both invisible without special equipment, and we cannot easily monitor the migrations of fishes, sharks, marine turtles, penguins or albatross. The measures that mark state and national borders are also not obvious in the ocean. There are no easily observed ocean-use changes and few obvious physical signs to indicate ecological or jurisdictional boundaries in the ocean.

Over the past 50 years, principally through rapid development of satellite and remote sensing technology, we have begun to penetrate and categorize areas within the uniform blue of the ocean. Satellite sensors can distinguish the physical and chemical characteristics of the ocean, and satellite tracking is beginning to detect the movements of migratory species in the ocean. New global positioning technology makes it possible to determine locations within a few centimeters. By adding precise positioning to multibeam acoustic technology, we can now conduct large-scale three-dimensional (3D) surveys of the seafloor to map and classify the elusive benthic (deep seafloor) habitats. Technology is now making it possible to put lines on the water.

Over 70% of the Earth's surface is ocean and it contains most of the phyla (higher order biodiversity) represented on land, plus many more phyla unique to the ocean (Table 1). Yet we know far less about biodiversity of the ocean, the distribution and abundance of its marine species, and how humans impact the ocean than we do for the land. Threats to biodiversity for both the land and the sea are quite similar, though, including over-exploitation of resources, habitat loss and degradation, pollution, invasive species and climate change. As they do for land activities, governments subsidize many ocean exploitation efforts from fishing to oil and gas drilling, with tax incentives and cheap fuel that promote over-use and degradation.

The 2002 World Summit on Sustainable Development (WSSD) emphasized the crisis of ocean resources exploitation and habitat destruction—mostly from large-scale commercial fishing—and urged the implementation of ecosystem-based management and conservation, including networks of marine protected areas (MPAs). In the United States, the Pew Oceans Commission and the U.S. Commission on Ocean Policy featured larger ecosystem-

Table 1

**Animal phyla in marine and non-marine ecosystems**

Biological life is identified and categorized in a hierarchy of taxonomic orders—kingdom, phylum, class, order, family, genus and species. There are many phyla having representatives living in both land/freshwater environments and in saltwater habitats. But many phyla are found only in saltwater areas.

| Both marine and non-marine | Exclusively non-marine | Exclusively marine |
|---|---|---|
| Porifera* | Onychophora | Placozoa |
| Cnidaria* | | Ctenophora |
| Platyhelminthes | | Mesozoa |
| Nemertina* | | Gnathostomulida |
| Gastrotricha | | Kinorhyncha |
| Rotifera | | Loricifera |
| Acanthocephala | | Phoronida |
| Entoprocta* | | Brachiopoda |
| Nematoda | | Priapulida |
| Nematomorpha | | Sipuncula |
| Ectoprocta* | | Echiura |
| Mollusca | | Pogonophora |
| Annelida | | Echinodermata |
| Tardigrada | | Chaetognatha |
| Pentastoma | | Hemichordata |
| Arthropoda | | |
| Chordata | | |
| **17 phyla** | **1 phylum** | **15 phyla** |

**Note:** For phyla marked * more than 95% of the species are marine.
**Source:** Adapted from Norse 1993.

based management as the best alternative to the current piecemeal system of managing ocean areas and resources. Recommendations from these groups have created a window of opportunity for implementing new ocean governance and management structures within coastal nations and on the high seas.

In this chapter we present the discussions and recommendations of the Ocean-Use Planning theme group at the *Defying Ocean's End* Conference. Our theme group began with the premise that a comprehensive ecosystem framework for management, conservation and scientific understanding of the ocean was needed. Concentrating on the similarities rather than the differences between land and sea, we determined that the tools of land-use planning and zoning that separate and govern exploitive human activities can be used also in the ocean. Within a framework of marine ecosystem management, these land-based tools have the potential to protect ecosystem services and allow sustainable use of the ocean resources upon which we depend.

## Continental Shelves and Open Ocean

Two major realms are recognized in the ocean. The coastal areas and continental shelves are the most productive ocean areas and support most of the world's fisheries yield. Over 50% of the world's population lives within 100 miles of the ocean and, not surprisingly, this is where habitat modification and pollution are most pronounced. Under the 1982 United Nations Convention on the Law of the Sea (UNCLOS), most countries in the world have declared an Exclusive Economic Zone (EEZ) extending from twelve to 200 nautical miles from their coastlines and have asserted their sovereign jurisdiction over the resources in their EEZ. These rights are currently being expanded as nations make their claims, under UNCLOS rules, for continental shelf and slope extensions beyond 200 nautical miles (Ray & McCormick-Ray 2004).

Vast areas of the ocean beyond the EEZs of coastal nations—called the high seas—cover 64% of the ocean's surface. Average depth of the high seas is 4,000 meters and these waters comprise more than 80% of the global biosphere, the area inhabited by plants and animals. The sea contains most of the higher order biological diversity on Earth and some scientists suggest that the sediments of the deep seafloor contain as many as 100 million species, most of them not yet discovered. The open ocean and deepsea contain hydrothermal vents, cold seeps, deep corals and seamounts—still largely unexplored areas of our planet. Just a few liters of seawater from the Sargasso Sea in the northwestern Atlantic, thought to be an "ocean desert" without many species, was recently shown to have 1,800 species of microbes new to science (Venter *et al* 2004).

Among the most pressing threats to the high seas is industrial-scale fishing. Many species, particularly large predatory fish like tuna, sharks and marlin (Figure 1), have experienced reductions in abundance of about 90% (Myers & Worm 2003). Bycatch (unintended catch during fishing operations) of albatross, other seabirds, sea turtles and marine mammals has caused many other marine species to plummet. Today's modern fishing technologies allow new regions and depths to be exploited and leave almost nothing beyond human reach (Myers & Worm 2003).

Taken together, the continental shelves and the open ocean contribute almost half of the planet's primary production—the creation of food by converting sunlight to plant life—and support commercial fisheries. The lack of a universally applied scheme of ocean planning and governance has led to the "tragedy of the commons" where the ocean is available to all but the responsibility of none, and there is global over-exploitation of the ocean's critical resources and ecosystem services.

## Background and Rationale for Ocean-Use Planning

In many countries of the world, significant areas of land are government owned and under some form of management, including fully protected zones in national, provincial

**Figure 1** Overfishing on a global scale has led to a 90% reduction in the abundance of large predatory fish such as tuna, sharks and marlin, like this striped marlin in the eastern Pacific.

and State parks. In the ocean, less than 1% of coastal nations' territorial seas and EEZs are under comparable management—and virtually none of these areas are fully protected—in spite of manifest and growing human impacts of marine ecosystems (Norse forthcoming).

The general failure of fisheries management policy and ubiquitous overfishing has created an intense global interest in marine protected areas (MPAs), particularly fishing prohibition zones or "no-take" marine reserves. But in spite of increasing scientific justification and outreach efforts over several decades, there has not been significant progress in implementing MPAs. The reasons for this are many, but significant resistance by fishing interests is a major cause (Murray *et al* 1999, Sobel & Dahlgren 2004).

MPAs are necessary but not sufficient. Human-induced disturbances to ocean areas are not limited to fishing. There are numerous other factors involved. The growing problem of coastal "dead zones," for example, arises from the application of fertilizers, pesticides and herbicides in agricultural areas sometimes thousands of miles upstream from the coast. We need a more comprehensive framework for scientific input, management and conservation.

Ocean-use planning recognizes that we must use the ocean, but we can't afford to use it up. Ocean-use planning broadens the stakeholders from just fishing interests to society as a whole. There is sufficient information about the EEZs of most coastal nations to begin a decade-long national marine conservation planning process. We recom-

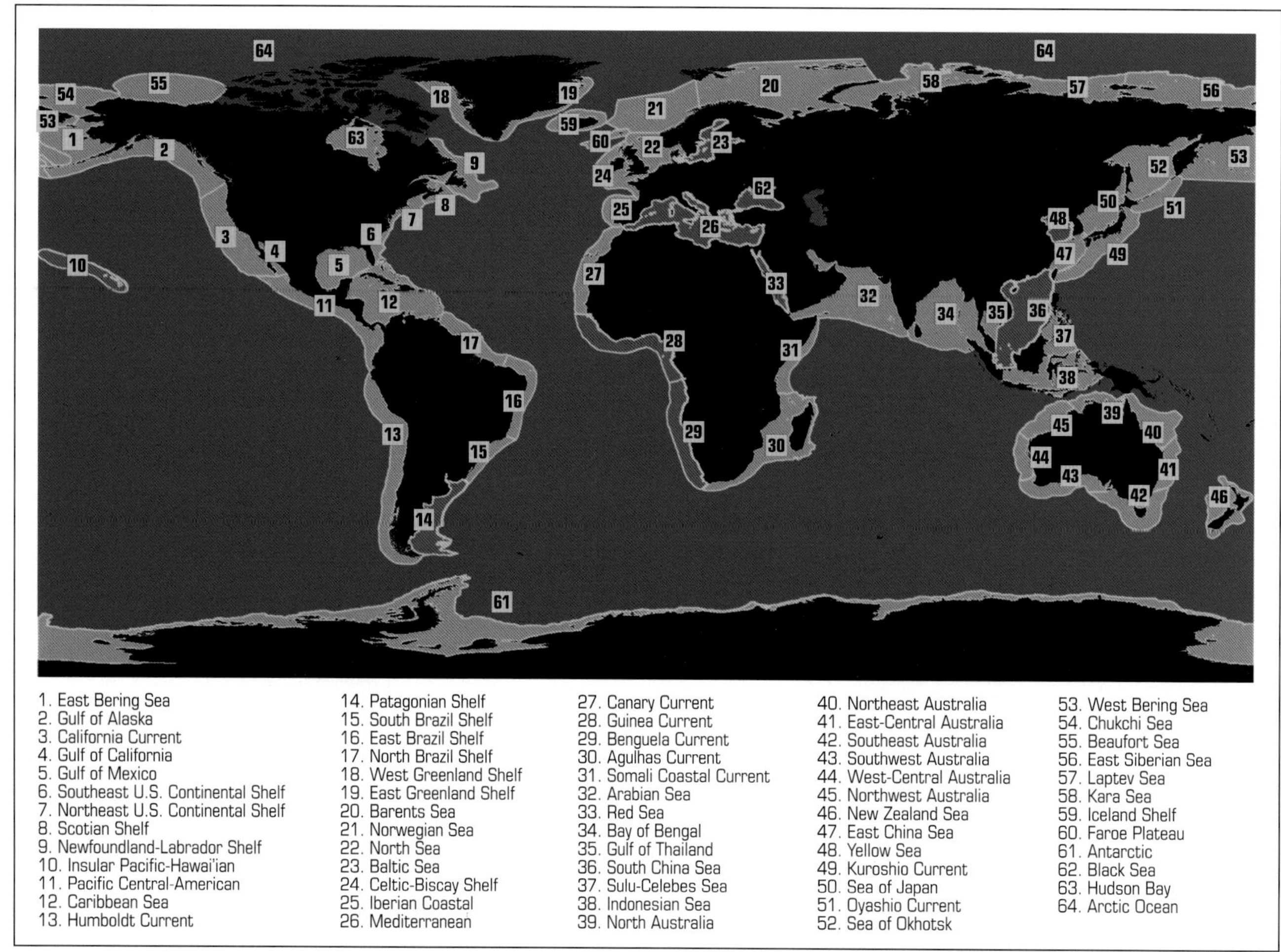


**Figure 2** Large marine ecosystems (LMEs) of the world as defined by the U.S. National Oceanic and Atmospheric Administration (Sherman, Alexander & Gold 1990). LMEs in darker green have active conservation programs funded by the Global Environment Facility of the World Bank.

mend as a first step mapping human-use patterns; collecting geolocated multibeam sonar data to show the terrain in underwater habitats; collecting fisheries data and monitoring data; conducting benthic (seafloor) habitat surveys; and using remote sensing data to detail current patterns, temperatures, tides, chlorophyll and other parameters.

Ocean-use planning should be implemented within defined large marine ecosystems (LMEs) spanning waters both under national jurisdiction and on the high seas if a natural ecosystem crosses these boundaries. Using an ecosystem approach and geographic information systems (GIS) technology can assist in visualization of problems by the stakeholders and assist in gaining the necessary political support. The need for MPAs will become more apparent within this context of multiple use, and the usual contention over their implementation may be mitigated.

**Ocean ecoregions and large marine ecosystems**

Our *DOE* theme group decided to concentrate on coastal and shelf regions for ocean-use planning. Productivity is concentrated in the coastal region and human disturbances are most intense there. We also have the best chance of gathering the political support needed for a new approach to ocean governance and management in these coastal areas.

The search for biological and ecological boundaries in the ocean has a long and rich tradition. Biogeographic patterns have fascinated naturalists for hundreds of years (Briggs 1987, 1995) and the coastal ecoregions of the world have been defined in various schemes. Longhurst (1998) in his *Ecological Geography of the Sea* did an exhaustive analysis of biogeographic patterns and oceanographic characteristics and constructed a global ecoregional scheme based on key habitats, mostly in the open ocean. WWF and other nongovernmental conservation organizations have defined global coastal ecoregions based on biology, ecology and their particular conservation agendas. The U.S. National Oceanic and Atmospheric Administration (NOAA) has divided coastal regions of the world into 64 large marine ecosystems (LMEs) and has provided a global map (Figure 2) of these LMEs (Sherman 1990, Sherman, Alexander & Gold 1990, Sherman 1994). Within these boundaries, 90% of the global fisheries catch is taken and habitat degradation and coastal pollution are the highest (Watson *et al* 2003).

The Sherman LME concept has proven very useful to coastal nations and international funding agencies concerned with increasing human impacts on the decline of coastal ocean resources. Since 1995, the Global Environment Facility (GEF) of the World Bank has provided approximately US $650 million to countries for planning and implementation of ecosystem-based management of marine resources and environments focused on LMEs (Table 2). The GEF programs follow a "country-driven" approach, wherein governments come together to assess the problems, define action plans, build awareness and local capacity to address the problems and set priorities for management. The World Summit on Sustainable Development (WSSD) in 2002 set target dates for achieving marine conservation goals: 2006 for substantial reductions in coastal pollution, 2010 for introduction of ecosystem-based management, 2012 for designation of a network of marine protected areas within global LMEs and 2015 for the restoration of fish stocks to maximum sustainable yield. More recently, a consensus statement of seventeen internationally recognized experts published in *Science* (Pikitch *et al* 2004) supported the urgent need for, and the feasibility of, moving forward with an ecosystem-based approach to fisheries management that would rely on measures such as ocean zoning.

**Selection of a case study for ocean-use planning**

Based on the existing schemes for definition of global ecoregions and LMEs, our group concluded that we did not have adequate time to merge the most mature schemes (e.g., Longhurst 1998, Sherman 1994, Olson & Dinerstein 1998). We agreed that doing this should be a priority for the future and that it could be accomplished by a group of knowledgeable scientists in a relatively short time.

There has been agreement since the *DOE* Conference on integration of the Longhurst (1998) biogeochemical open-ocean provinces with the coastal LMEs (Sherman, Alexander & Gold 2003) in a geographic global representation of ecologically determined ocean units (Watson *et al* 2003). Based on our personal expertise and experience, our theme group—at the conference—identified a preliminary list of large marine ecosystems of extreme importance including: Benguela Current, Bering Sea, Baja California System (Baja and Sea of Cortés), Caribbean Sea, Central West Pacific (Palau to Tuvalu), CCAMLR Region (Convention for the Conservation of Antarctic Marine Living Resources), Coral Triangle (Tropical Indo-West Pacific), Eastern Tropical Pacific Seascape (from Costa Rica to Ecuador), Humboldt (Chilean-Peruvian) System, Patagonian System and Western Indian Ocean/East Africa region (Figure 3).

Table 2

**Countries participating in Global Environmental Facility (GEF)–funded projects in large marine ecosystems (LMEs) worldwide**

| LME | Countries |
|---|---|
| **Approved GEF projects** | |
| Gulf of Guinea | Benin, Cameroon, Côte d'Ivoire, Ghana, Nigeria, Togo |
| Yellow Sea | China, Korea |
| Patagonia Shelf/Maritime Front | Argentina, Uruguay |
| Baltic | Denmark, Estonia, Finland, Germany, Latvia, Lithuania, Poland, Russia, Sweden |
| Benguela Current | Angola, Namibia, South Africa |
| South China Sea | Cambodia, China, Indonesia, Malaysia, Philippines, Thailand, Vietnam |
| Black Sea | Bulgaria, Georgia, Romania, Russia, Turkey, Ukraine |
| Mediterranean | Albania, Algeria, Bosnia-Herzegovina, Croatia, Egypt, France, Greece, Israel, Italy, Lebanon, Libya, Morocco, Slovenia, Spain, Syria, Tunisia, Turkey, Yugoslavia, Portugal |
| Red Sea | Djibouti, Egypt, Jordan, Saudi Arabia, Somalia, Sudan, Yemen |
| Insular Pacific Provinces* | Cook Islands, Micronesia, Fiji, Kiribati, Marshall Islands, Nauru, Niue, Papua New Guinea, Samoa, Solomon Islands, Tonga, Tuvalu, Vanuatu |
| **GEF projects in the preparation stage** | |
| Canary Current | Cape Verde, Gambia, Guinea, Guinea-Bissau, Mauritania, Morocco, Senegal |
| Bay of Bengal | Bangladesh, India, Indonesia, Malaysia, Maldives, Myanmar, Sri Lanka, Thailand |
| Humboldt Current | Chile, Peru |
| Guinea Current | Angola, Benin, Cameroon, Congo, Democratic Republic of the Congo, Côte d'Ivoire, Gabon, Ghana, Equatorial Guinea, Guinea, Guinea-Bissau, Liberia, Nigeria, São Tomé and Principe, Sierra Leone, Togo |
| Gulf of Mexico | Cuba, Mexico, United States |
| Agulhus/Somali Currents | Comoros, Kenya, Madagascar, Mauritius, Mozambique, Seychelles, South Africa, Tanzania |
| Caribbean LME | Antigua and Barbuda, The Bahamas, Barbados, Belize, Colombia, Costa Rica, Cuba, Grenada, Dominica, Dominican Republic, Guatemala, Haiti, Honduras, Jamaica, Mexico, Nicaragua, Panama, Puerto Rico, Saint Kitts and Nevis, Saint Lucia, Saint Vincent and the Grenadines, Trinidad and Tobago, Venezuela |

**Note:** * Provisionally classified as Insular Pacific Provinces in the global hierarchy of LMEs (Watson *et al* 2003).
**Source:** Adapted from Duda & Sherman 2002.

At the *DOE* Conference we chose one of these regions, Patagonia, for special focus to develop detailed recommendations for action and costs.

## Ocean-Use Planning for the Patagonian Large Marine Ecosystem

We focused our attention on a case study of ocean-use planning for the Patagonia shelf and shelf-break ecosystem, which we refer to as the Patagonian large marine ecosystem (PLME). Our *DOE* theme group had substantial expertise in the southwest Atlantic, so we believed we could build a plan for ocean-use planning and zoning with more confidence in this area than in others. In addition the PLME has attributes that make an ocean-use plan here a potentially practical and successful model for other large ocean areas of exceptional productivity.

The Wildlife Conservation Society has focused on the PLME area for decades, and together with the government of Argentina and local organizations like the Fundación Patagonia Natural, has developed the first GEF-funded initiative of ecosystem conservation in coastal Patagonia. Today, there are several ocean conservation initiatives implemented by the Argentine government and by non-governmental organizations (NGOs). The *Sea & Sky Program* is promoting an ecosystem approach under the "precautionary principle" for the continental shelf and slope areas of

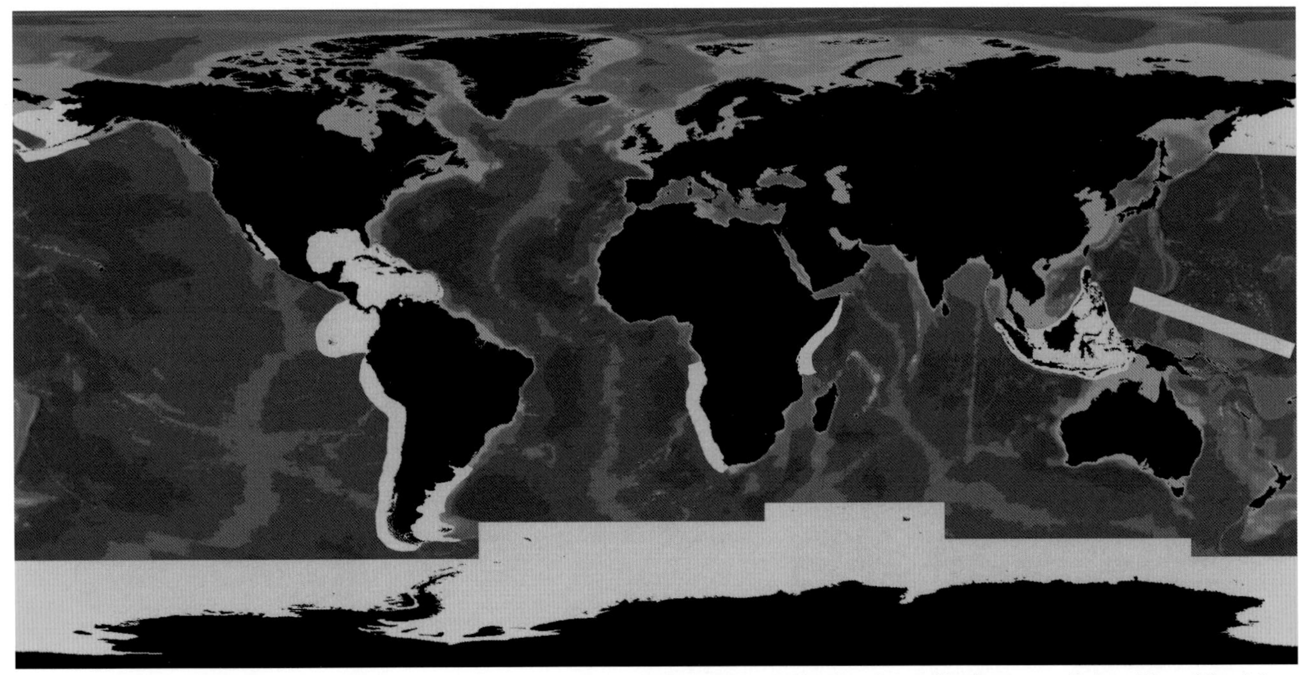

**Figure 3** Large marine ecosystems (LMEs) proposed for high-priority attention: Benguela, Bering Sea, Baja Californian system (Baja and Sea of Cortés), Caribbean, central west Pacific Islands (Palau to Tuvalu), Convention for the Conservation of Antarctic Marine Living Resources (CCAMLR), Coral Triangle (Tropical Indo-Pacific), eastern tropical Pacific (Costa Rica Basin), Humboldt (Chilean/Peruvian) System, Patagonian and western Indian Ocean/East Africa.

Argentina. Priority-use zones and sectors under particularly precautionary management—including corridors and permanent as well as seasonally defined marine protected areas—are designed based on regional oceanographic regimes to meet the requirements of critical species.

### The Patagonian large marine ecosystem

The PLME is one of the largest and most important coastal shelf areas on Earth (Esteves *et al* 2000). The largest continental shelf in the southern hemisphere, it measures more than two million square kilometers—70% of the size of the land area of Argentina. It has some of the richest resources south of the equator, sustained by the nutrient transport systems of the Falklands-Malvinas and Brazil Currents (Figure 4).

#### *Oceanographic regimes*

Western South Atlantic waters can be classified into regions or oceanographic regimes according to their physical characteristics and water masses and by boundaries called "fronts" between them. Subtropical waters reach the region by the southward flowing Brazil Current—the western limb of the counterclockwise South Atlantic subtropical gyre. The boundary between the subtropical and subpolar regions occupies a wide area from 38° to 47°S and extending eastward to the collision of the western boundary currents. This area is marked by moderate surface chlorophyll-a concentrations. Subpolar waters enter the region from the south as part of the Malvinas/Falkland current that extends northward into the western Argentine Basin to about 38°S. The polar water regime lies close to the Antarctic Polar Front. The front at the shelf break (the continental shelf drops off steeply to the continental slope) is a narrow transition region between subpolar and shelf waters. Waters on the shelf are sub-Antarctic waters diluted by the influence of continental runoff, such as the outflow of the Rio Plata.

#### *Biological and jurisdictional aspects*

Approximately 75% of the global population of black-browed albatross live and breed in these waters. More than a million breeding pairs of Magellanic penguins, 72,000 southern sealions and 60,000 elephant seals depend on the Patagonia shelf ecosystem for breeding and feeding. Together these species constitute one of the greatest wildlife concentrations and spectacles on Earth. The local breeding populations of these species are augmented seasonally by migrants from

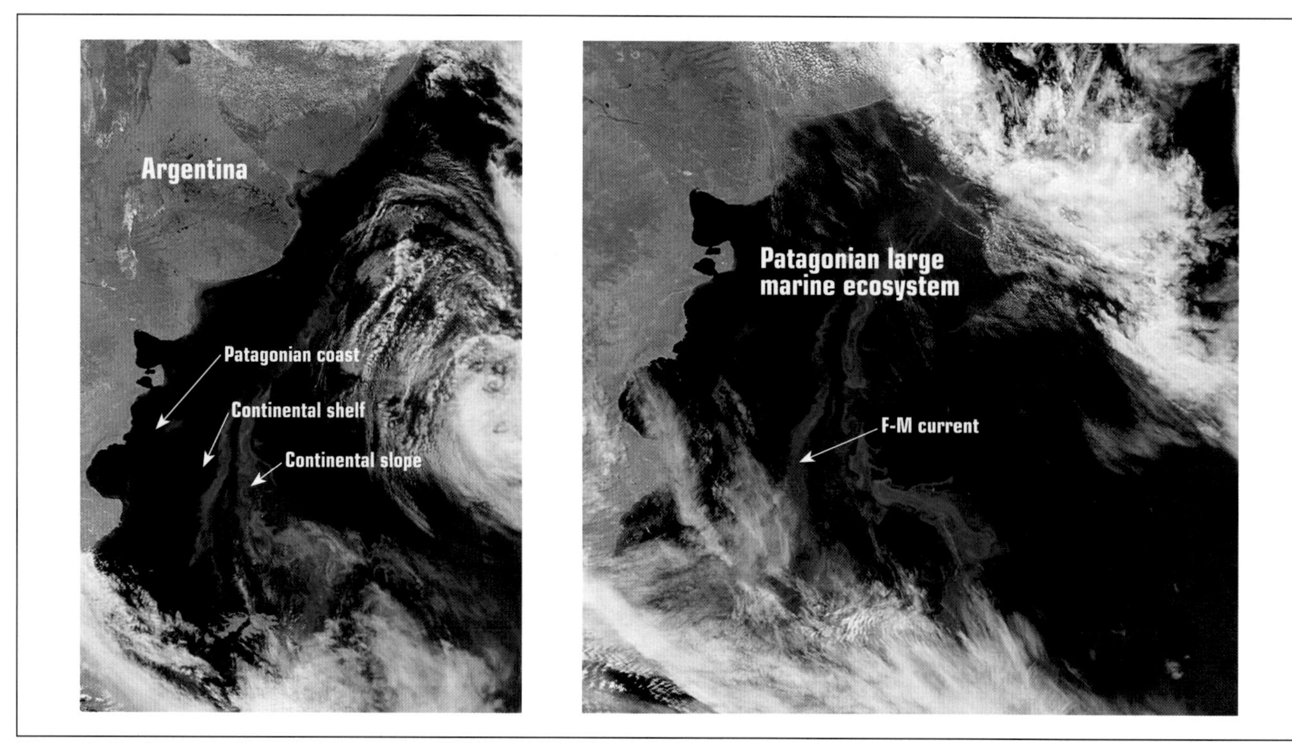


**Figure 4** The Patagonian large marine ecosystem (PLME) and the Falklands-Malvinas current on the Patagonian Shelf.

the Antarctic and South Georgia and as far away as New Zealand and Australia.

The PLME also supports some of the world's major fisheries for squid and finfish. Some of these fisheries are clearly over-exploited and in decline. Hydrocarbon resources are actively exploited along the mainland coasts and potential major offshore resources are currently under exploration license. The PLME is unquestionably of global significance because of its location, linking Antarctic waters to the south with tropical waters to the north.

Argentina is the dominant country in the region with a coastline of more than 4,000 kilometers, but its jurisdiction over the South Atlantic is limited. Sovereignty of the Falkland Islands (Malvinas) is still in dispute between Argentina and the United Kingdom. Uruguay and Brazil to the north are important stakeholders as many PLME species use their coastal waters for parts of the year.

Argentina has fifteen coastal marine protected areas (MPAs), all under provincial jurisdiction, but Argentina and Uruguay have no national marine parks. In Brazil, there are two national marine parks and two more proposed. Although MPAs are virtually absent, there are extensive areas within the EEZ under fishery management. The need for increased protection and conservation is urgent.

*Costs of ocean-use planning and implementation for the PLME*

Our theme group considered what research, monitoring and institutional structure would be needed to implement a management plan for the PLME and estimated its cost over ten years. We developed a spreadsheet outlining the required actions, timing and costs over ten years to develop an ocean-use plan for the PLME. Some planning activities apply only in the first year and other activities would begin in years three to five. Our estimates for PLME served as the basis for our estimates of costs for implementing ocean zoning in high-priority coastal areas around the world. Effective management in the Patagonia area would require an investment of US $64 million for the first three years and a total of US $615 million over ten years (Table 3). We also scaled the costs to cover a global system: for fifteen high-priority large marine ecosystems, we estimated the cost for ten years of zoning and monitoring to be more than US $9 billion. For the world to implement this kind of zoning for ten years, the estimated cost would be approximately US $40 billion (Table 4).

Table 3

**An ocean-use plan for the Patagonian large marine ecosystem (PLME)—activities and costs are shown for year 1, years 2–5 and years 6–10 (in US $)**

| Recommended actions | Year 1 | Years 2–5 | Years 6–10 |
|---|---|---|---|
| **Initial planning: Vision and mission development** | | | **Total $27.6M** |
| Establish task force with key planning people (NGOs, governments or multinational) | Involve stakeholders in process **$0.5M** | | |
| Collate GIS-based data on the following: *Activities:* capture fisheries, shipping and ports, aquaculture, mining, cable laying, energy production, protected areas, tourism, defense *Pressures/threats:* pollution, habitat modification, introduced species, threatened species, human population *Environment:* biota, habitats and physical parameters | GIS database produced **$0.5M** | | |
| Survey who benefits and who is responsible | Conduct initial survey **$0.2M** | Responsibilities established, benefited sectors identified | |
| Determine economic flow | Ongoing **$0.2M** | Ongoing **$0.8M** | |
| Ground-truth GIS data | Ongoing **$0.2M** | Ongoing **$0.8M** | Ongoing **$0.2M** |
| *Fill data gaps:* benthic sampling, seasonal changes, refinement of habitat characterization. Support research cruises. | 120 days of mapping ($20K/day) **$2.4M** | Update base layers with new information **$9.6M** | Annual updating of database **$12.0M** |
| Map existing MPAs | Produce maps **$0.2M** | | |
| **Subtotals** | **$4.2M** | **$11.2M** | **$12.2M** |

Although substantial new funding is required, some of the estimated total US $64 million for first steps in the PLME project can be supplemented by re-directing existing funds. New funds will be needed for satellite monitoring of vessel activity in the protected areas, scientific data collection on bycatch and implementation of the project.

The group also reached consensus that conservation organizations and governments should focus on the overarching goal of implementing ecosystem-scale management globally. Although this is an ambitious goal that has political, economic and management ramifications, the group feels the PLME case study can serve as a model of how management efforts at the large marine ecosystem scale can begin.

## Recommendations

- Governments in high-priority coastal areas should implement a geographic information system (GIS) assessment of the coastal marine resources of their EEZ, including coastal lands, benthic (seafloor) and water column communities, features such as river drainages and outflow plumes (Figure 5), estuaries, currents, gyres, upwelling areas, ocean productivity patterns, migration routes,

Table 3 (continued)

**An ocean-use plan for the Patagonian large marine ecosystem (PLME)—activities and costs are shown for year 1, years 2–5 and years 6–10 (in US $)**

| Recommended actions | Year 1 | Years 2–5 | Years 6–10 |
|---|---|---|---|
| **Decision-making process** | | | **Total $92.55M** |
| Develop and implement communications strategy | *First year communications efforts:* defining messages and targeting stakeholders (15% of total cost) **$8.0M** | Implement and update communications strategy **$32.0M** | **$40.0M** |
| Establish long-term steering committee | | *Meetings:* two per year ($0.2M/year) **$0.8M** | Continued meetings and oversee plan implementation **$1.0M** |
| Convene scoping meetings; discuss options | Meetings with interested parties and public **$0.2M** | Meetings with interested parties and public; include mid-term plan revision ($0.2M/year) **$0.8M** | |
| Secure endorsement by national governments and stakeholders | Meetings, presentations (10 at $50K each) **$0.5M** | Meetings, presentations **$4.0M** | Meetings, hearings, lobbying **$5.0M** |
| Draft Ocean-Use Plan | | Prepare the plan **$0.25M** | |
| **Subtotals** | **$8.7M** | **$37.85M** | **$46.0M** |
| **Implementation** | | | **Total $494.5M** |
| Establish ecologically responsible aquaculture practices | Monitoring and control **$0.2M** | Ongoing **$0.8M** | Ongoing **$1.0M** |
| Establish zoning for cable laying | Coordinate with industry **No cost** | | |
| Establish training programs and student support | Training programs and student support **$1.0M** | Ongoing **$4.0M** | Ongoing **$5.0M** |
| Identify the conservation priorities | Study multiple-use needs and survey potential MPA sites **$1.0M** | Determine where and how MPAs will be established and begin implementation process **$4.0M** | Regional network of MPAs implemented **$5.0M** |
| Prevent ecologically damaging activities | Monitoring and ensuring compliance **$0.3M** | Ongoing **$1.2M** | Ongoing **$1.5M** |
| Prevent invasive species | Monitoring and control **$0.2M** | Ongoing **$0.8M** | Ongoing **$1.0M** |

Table 3 (continued)

**An ocean-use plan for the Patagonian large marine ecosystem (PLME)—activities and costs are shown for year 1, years 2–5 and years 6–10 (in US $)**

| Recommended actions | Year 1 | Years 2–5 | Years 6–10 |
|---|---|---|---|
| **Implementation (continued)** | | | |
| Control ballast water to control introductions of invasive species | Monitoring and control **$0.2M** | Ongoing **$0.8M** | Ongoing **$1.0M** |
| Include defense needs in zoning schemes | Coordinate with defense departments **No cost** | | |
| Determine optimal energy sources | Desktop study; other costs covered by industry **$0.2M** | | |
| Develop monitoring, control, surveillance and enforcement plan, including Vessel Monitoring Systems (VMS) | Monitoring, control and surveillance ($1.5M/yr); conference costs **$20.0M** | Monitoring, control and surveillance ($1M–10M/yr) **$80.0M** | Ongoing monitoring ($1M–10M/yr) **$100.0M** |
| Develop and implement best fishing practices | Technological advances, gear improvements, mitigation and threat abatement plans **$10.0M** | Ongoing **$40.0M** | Ongoing **$50.0M** |
| Fund conservation-oriented research | Scientific monitoring **$6.0M** | Ongoing **$24.0M** | Ongoing **$30.0M** |
| Secure endorsement of relevant IPOAs (international plans of action) | Work with the regional governments **$1.0M** | | |
| Control pollution from mining | Desktop study; other costs covered by industry **$0.2M** | | |
| Identify and implement international and other agreement options | | Establish conservation fund; governments provide additional funding **$1.0M** | |
| Implement best available practices for pollution prevention | Monitoring and control; other costs covered by industry **$0.3M** | Ongoing **$1.2M** | Ongoing **$1.5M** |
| Establish contingency plans for emergencies and disasters | Costs to be recovered from polluter **$1.0M** | Ongoing **$4.0M** | Ongoing **$5.0M** |

Table 3 (continued)

**An ocean-use plan for the Patagonian large marine ecosystem (PLME)—activities and costs are shown for year 1, years 2–5 and years 6–10 (in US $)**

| Recommended actions | Year 1 | Years 2–5 | Years 6–10 |
|---|---|---|---|
| **Implementation (continued)** | | | |
| Establish remote sensing program for monitoring, tracking | Investigate feasibility and cost savings of a dedicated satellite **$2.0M** | Establish remote sensing monitoring program **$8.0M** | Continue remote sensing monitoring program **$10.0M** |
| Study and assess population and ecosystem impacts | Status and trends **$6.0M** | Ongoing **$24.0M** | Ongoing **$30.0M** |
| Address discharge, shipping lanes, ship condition, products transported and offloaded | Monitoring and control; other costs borne by industry **$0.5M** | Ongoing **$2.0M** | Ongoing **$2.5M** |
| Establish zoning | Assess new wildlife areas for tourism **$0.5M** | | |
| Maintenance and infrastructure development | | Develop needed infrastructure (equipment, personnel) **$2.0M** | Maintain infrastructure **$2.5M** |
| More responsible ecotourism | Certification costs; other costs covered by industry **$0.1M** | | |
| **Subtotals** | **$50.7M** | **$197.8M** | **$246M** |
| **Totals** | **$63.6M** | **$246.9M** | **$304.2M** |
| **Ten-year total for Patagonian large marine ecosystem** | | | **$615M** |

feeding and breeding areas for major marine organisms. These GIS map layers should be verified by field investigations and local knowledge. Major data gaps should be identified and filled (Kramer & Kramer 2002).

- These governments should determine how their EEZs are being used. Major cities, pollution sources, shipping lanes and ports, fishing areas, aquaculture areas, mining areas, tourist attractions and marine protected areas (MPAs) should all be mapped into GIS formats. Assessment of these use patterns should engage the public and allow extensive input and comment.
- GIS analysis and assessment should also be done for current and potential future threats to marine ecosystems including pollution, habitat destruction, invasive species, offshore oil and gas operations and the potential impacts of global climate change.
- Extensive GIS analysis and public input should be followed by scoping meetings, draft zoning plans, comment periods and final implementation of a zoning plan that separates potentially conflicting human activities and builds sufficient conservation and management protection through MPAs. This zoning plan should be promulgated for a trial period of five years, followed by public review and reauthorization.
- Marine protected areas and no-take zones should be set up that adhere to scientific principles. Attention should be paid to physical and biological connectivity between the areas, ecosystem function and species protection.

Table 4
**Estimated costs for ocean-use zoning and management of large marine ecosystems (LMEs)**

| | |
|---|---|
| **Patagonian LME** | |
| Year 1 | US $64 million |
| Years 2–6 | US $247 million |
| Years 6–10 | US $304 million |
| 10-year total | US $615 million |
| **Fifteen highest-priority LMEs** | |
| 10-year total | US $9.2 billion |
| **All LMEs worldwide** | |
| 10-year total | US $39.4 billion |

- We propose the following guidelines be used to gain a better understanding of human exploitation of ocean resources, plan for better enforcement and support growing stewardship.
  - The "precautionary principle" should be used for any new fisheries or coastal development projects.
  - Over time, all commercial operations exploiting ocean resources should be required to use an onboard vessel monitoring system (VMS) tracked by satellite, as well as onboard observers or video systems to verify fishing techniques, catches and fishing areas.
  - At a minimum all licensed vessels should have VMS and be required to submit vessel tracks with regular catch reports.
  - The exploiting industry should be required to pay for these systems.
  - The licensing of commercial vessels should also require a VMS be onboard and functioning. Only when a vessel has a functioning VMS and a record of fishing location and/or travel route, should it be allowed to dock, or land cargo or catch.
  - Bycatch, gear effectiveness and other aspects of exploitation must be quantified before commercial activity is allowed. As a first step to gain adherence to these requirements, we recommend that catches be landed at the dock only from regulated vessels.
  - Tolerable levels of bycatch should be determined. The level of bycatch should be set for fisheries and areas. Governments should make information on bycatch levels—both limits set and actual catches—available to the public before and after landing.

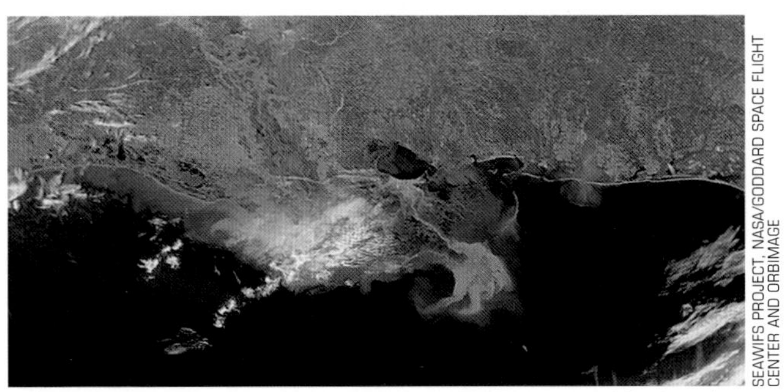
SEAWIFS PROJECT, NASA/GODDARD SPACE FLIGHT CENTER AND ORBIMAGE

**Figure 5** Sediment plumes from the Mississippi River extend far into the offshore waters of the Gulf of Mexico.

  - Mother ships and at-sea processing used in industrial-scale fishing require regulation.
  - Strengthened regional fishery organizations and additional multinational treaties are needed to manage the high seas open-access ocean areas.

## Conclusions

An unprecedented decline in the health of coastal ocean ecosystems over the past few decades has occurred in concert with a general global decline in ocean health. With the 2002 World Conference on Sustainable Development and various national commissions and other international meetings highlighting the importance of ocean resources and ecosystem services, we now have a window of opportunity to implement new ecosystem-based policies which recognize pervasive human disturbances. We must enter into a new era of energized, reorganized, ecosystem-based management, stewardship, sustainable use and science in the service of conservation.

People respond to big ideas. We should implement a decade-long effort to create ocean-use planning for the EEZs of coastal nations. We have outlined herein our vision of the major steps required to create an ocean-use plan in a case study of the Patagonian large marine ecosystem and have estimated its costs over a 10-year development period. There are other excellent case studies showing how ocean-use plans are structured and implemented including the Great Barrier Reef Marine Park and the Mesoamerican Coral Reef (MACR) project of the WWF and the World Bank. In each case, geographic information systems (GIS) and overlay maps provide the basis for visualization and decisions on balancing protection, development and sustainable use of marine resources.

## Acknowledgements

*We thank Conservation International for hosting the* Defying Ocean's End *Conference and Linda Glover, Conference organizer, for her patience and editorial help. A number of participants provided input. Ken Sherman of NOAA provided background on the long-standing effort to promote management and governance of large marine ecosystems. Ellen, Claudio and Dee are grateful for the support of the Wildlife Conservation Society and the Pew Charitable Trusts. Rodrigo Bustamante acknowledges the continuing support of CSIRO Marine Research and the Charles Darwin Foundation.*

## Literature Cited & Consulted

Briggs, J. C. 1987. *Biogeography and Plate Tectonics.* New York: Elsevier Science Publishing Co.

——. 1995. *Global Biogeography.* Amsterdam: Elsevier Science Publishing Co.

Duda, A. M. and K. Sherman. 2002. A new imperative for improving management of large marine ecosystems. *Ocean & Coastal Management* 45:797–833.

Esteves, J. L., N. F. Ciocco, J. C. Colombo, H. Freije, G. Harris, O. O. Iribarne, I. Isla, P. Nabel, M. S. Pascual, P. E. Penchaszadeh, A. L. Rivas, N. Santinelli. 2000. The Southeast South American Shelf Marine Ecosystem: The Argentine Sea. In *The Seas at the Millennium,* ed. C. Sheppard, 105–127. Amsterdam: Elsevier Science Publishing Co.

Kramer, P. A. and P. R. Kramer. 2002. *Ecoregional Conservation Planning for the Mesoamerican Caribbean Reef,* ed. M. McField. Washington, D. C.: WWF.

Longhurst, A. 1998. *Ecological Geography of the Sea.* New York: Academic Press.

Murray, S. N., R. F. Ambrose, J. A. Bohnsack, L. W. Botsford, M. H. Carr, G. E. Davis, P. K. Dayton, D. Gotshall, D. R. Gunderson, M. A. Hixon, J. Lubchenco, M. Mangel, A. MacCall and D. A. McArdle. 1999. No-take reserve networks: sustaining fishery populations and marine ecosystems. *Fisheries* 24 (11): 11–25.

Myers, R. A. and B. Worm. 2003. Rapid worldwide depletion of predatory fish communities. *Nature* 423:280–283.

NOAA. 2004. *NOAA Fisheries, Large Marine Ecosystem Program: Status Report.* NOAA Technical Memorandum NMFS-NE-183.

Norse, E. A. 1993. *Global Marine Biological Diversity. A Strategy for Building Conservation into Decision Making.* Washington, D.C.: Island Press.

——. Forthcoming. Ending the range wars on the last frontier: Zoning the sea. In *Marine Conservation Biology: The Science of Maintaining the Sea's Biodiversity,* ed. E. A. Norse and L. B. Crowder. Washington, D.C.: Island Press.

Olson, D. M. and E. Dinerstein. 1998. The global 200: A representation approach to conserving the Earth's most biologically valuable ecoregions. *Conservation Biology* 12:502–515.

Pikitch, E.K., C. Santora, E. A. Babcock, A. Bakun, R. Bonfil, D. O. Conover, P. Dayton, P. Doukakis, D. Fluharty, B. Heneman, E. D. Houde, J. Link, P. Livingston, M. Mangel, M. K. McAllister, J. Pope and K. J. Sainsbury. 2004. Ecosystem-based fishery management. *Science* 305:346–7.

Ray, G. C. and J. McCormick-Ray. 2004. *Coastal-Marine Conservation: Science and Policy.* Malden, Massachusetts: Blackwell Science Ltd.

Sherman, K., L. M. Alexander and B. D. Gold, eds. 1990. *Large Marine Ecosystems: Patterns, Processes and Yields.* Washington. D.C.: American Association for the Advancement of Science.

Sherman, K. 1994. Sustainability, biomass yields, and health of coastal ecosystems: An ecological perspective. *Marine Ecology Progress Series* 112:277–301.

Sobel, J and C. Dahlgren. 2004. *Marine Reserves: A Guide to Science, Design, and Use.* Washington, D.C.: Island Press.

Venter, J. C., K. Remington, H. F. Heidelberg, A. L. Halpern, D. Rusch, J. A. Eisen, D. Wu, I. Paulsen, K. E. Nelson, W. Nelson, D. E. Fouts, S. Levy, A. H. Knap, M. W. Lomas, K. Nealson, O. White, J. Peterson, J. Hoffman, R. Parsons, H. Baden-Tillson, C. Pfannkock, Y. H. Rogers and H. O. Smith. 2004. Environmental genome shotgun sequencing of the Sargasso Sea. *Science* 304 (5667): 66–74.

Watson, R., D. Pauly, V. Christensen, R. Froese, A. Longhurst, T. Platt, S. Sathyendranath, K. Sherman, J. O'Reilly and P. Celone. 2003. Mapping fisheries onto marine ecosystems for regional, oceanic and global integrations. In *Large Marine Ecosystems of the World: Trends in Exploitation, Protection, and Research,* ed. G. Hempel and K. Sherman, 375–395. Amsterdam: Elsevier Science Publishing Co.

CHAPTER 7

# *Rationality or Chaos? Global Fisheries at the Crossroads*

Rod Fujita, *Environmental Defense*
Kate Bonzon, *Environmental Defense*
James Wilen, *University of California, Davis*
Andrew Solow, *Woods Hole Oceanographic Institution*
Ragnar Arnason, *University of Iceland*
James Cannon, *Conservation International*
Steve Polasky, *University of Minnesota*

KIM WESTERSKOV

## Introduction

The depletion of commercially exploited fish populations is well documented in the scientific literature (Pauly *et al* 1998), with new studies and perspectives being published at a rapid rate. Economic problems in fisheries, such as extreme economic inefficiency and escalating government subsidies, are also fairly well documented (Arnason 1991). However, in spite of a number of fundamental theoretical contributions (Schrank, Arnason & Hannesson 2003), the literature has been less successful in identifying socially and politically practical solutions to these problems.

It is too often overlooked that the interests of commercial fisheries and environmental conservation overlap to a large extent. Fisheries are typically managed for maximum sustainable yield (or are depleted), rather than managed for longer-term maximum profits. Long-term profit maximization in commercial fisheries normally implies fish population sizes well in excess of the levels corresponding to the maximum sustainable yield. Compared to the current state of most ocean fisheries, this means an increase in the size of the fish populations and a corresponding reduction in fishing fleets and fishing effort. This, of course, is exactly what conservationists would like to see. At the same time, larger fish populations (Figure 1) maximize long-term profits in the commercial sector. Sensible fisheries management thus offers the opportunity for mutually beneficial cooperation between environmental and commercial sectors. Both groups have an interest in sustainable management of the ocean's living resources, and the alignment of these interests can produce ecological and economic benefits simultaneously.

In spite of this potential, attempts at fisheries management reform are often blocked by social, economic and technical factors. These include:

- Inadequate technical and administrative abilities to run a fisheries management system
- Lack of funds to finance the transition of a fishery to sustainability, which usually requires temporarily reduced harvests
- The risk perceived by fishermen and other members of the fishing industry of adopting a new and untried fisheries management regime
- Lack of understanding of the basic issues involved on the part of many fishermen and managers

These obstacles can be overcome by the appropriate provision of resources in the form of finance and expertise. In many commercial fisheries, management reform can generate substantial economic benefits, so support of this kind can be repaid with an appropriate rate of return in due course. Fisheries management reform often constitutes a potentially profitable investment opportunity. There is little doubt that under the right circumstances (e.g., reasonably secure resource-use privileges) private investment capital would be attracted to this opportunity.

Consensus around the need for fisheries reform is building and numerous reform processes are underway, but lack of funding hampers wide-scale change. To provide ongoing resources for fisheries reform, we propose the creation of a Global Fisheries Reform Fund (the Fund), a revolving loan and investment vehicle designed to improve both the economic and conservation performance of fisheries. The Fund is one example of a range of institutions that could potentially address the widespread lack of funding for fisheries reform by providing low-interest loans and investment

© RICHARD HERRMANN

**Figure 1** Large school of commercially fished yellowfin tuna.

capital aimed at specific fisheries with good potential for economic and environmental improvement. Importantly, the Fund will be structured to recover its outlays with the appropriate return on capital, rather than continuing the tradition of sinking resources into an industry through vessel buybacks and subsidies without receiving any return on the investment and without management reform. Therefore, with relatively modest initial capital, the Fund will be able to assist sequentially in fisheries management reform in a large number of fisheries as it expands.

## State of the World's Fisheries: Biological and Economic Problems are Linked

There have been numerous recent reports and studies documenting the status of the world's fisheries and the downfall of many current management schemes. The United Nations Food and Agriculture Organization (FAO) reports that 47% of the world's fisheries that have been assessed are fully exploited, 18% are over-exploited and 10% are significantly depleted or recovering (FAO Fisheries Department 2002). Furthermore, it has been shown that fisheries are generally shifting effort to smaller, lower trophic level (lower on the food chain) fish as populations of larger predatory fish decline (Pauly *et al* 1998).

Fisheries mismanagement—or often, lack of management—is central to the decline of many living ocean resources. From the standpoint of environmentalists, fisheries mismanagement results in a host of ecosystem impacts that extend beyond dangerously low fish populations. These include high levels of bycatch and serious damage to marine habitats. Bycatch is the accidental killing of species that are not targeted by the fishery—like seabirds, sea turtles and marine mammals—where the unintended catch is almost entirely thrown over the side, dead or dying. In fact, the report of the Pew Oceans Commission (2003) highlights overfishing, bycatch and habitat alteration due to fishing as among the greatest threats to ocean ecosystem health. Many of these ecological impacts are direct results of poor management. The ecological impacts are highly publicized and well known, but there are also severe adverse economic effects from poor management. Global fisheries are contributing much less to the economic welfare of both fishermen and the global economy than if fisheries were well managed. Nations and their fishermen, no less than environmentalists and marine ecosystems, stand to gain from improved management of fisheries.

The United Nations Food and Agriculture Organization estimates that global marine capture fisheries production is just above 80 million metric tons annually (FAO Fisheries Department 2002), generating first-hand sale revenues of about US $75 billion per year. This value is derived by multiplying the first-hand sale value of total capture fisheries production (US $81 billion) by 0.91, the marine proportion of total capture fisheries production (FAO Fisheries Department 2002). But numerous reports (e.g., FAO Fisheries Department 2002, Milazzo 1998, Virdin & Schorr 2001) show that fishing costs substantially exceed revenues, resulting in grossly inefficient fisheries. How can such a high level of economic waste occur? There are two major factors contributing to this deficit, each of which acts to reinforce the other: extensive subsidies to the fishing industry and the open-access nature of most fisheries.

Fisheries are usually managed as common property, open-access resources, making them vulnerable to the "tragedy of the commons" as described by Garrett Hardin (1968). Common property is property held collectively by a group of people. In the case of domestic fish populations, this group of common rights holders is typically an entire nation; when fish populations occur outside national exclusive economic zones, the property holders are the entire human population. The common property arrangement creates competition among fishermen for the available catch, which is limited by nature. Because individual shares of the catch are not specified, an uncontrolled number of fishermen compete to maximize their catch. Under this system, individual fishermen have incentives to catch as much as possible before others have decimated the populations. To do so, fishermen invest in excessive fishing capital, such as bigger and faster boats, and adjust their fishing practices in socially detrimental ways, for example, by fishing under dangerous ocean conditions. Powerful fishing fleets, advanced fish-finding technology and gear capable of catching large numbers of fish (Figure 2)—and damaging ocean habitats—are all logical outcomes of uncontrolled competition for fish in a commons. Additionally, new entrants will seek access to the fishery, and expansion of the fishery will occur as long as it is individually profitable (or at least perceived as such). These incentives remain until the populations of fish have been reduced to the level where all profits have disappeared and "economic rents"—the difference between the price and the production cost of a good—are completely dissipated. For a natural resource like fisheries, this is often referred to as "resource rent," or the productive value of the natural resource, with long-term profits earned when the "property owner" limits inputs to an economically efficient level.

The current situation described above is referred to as the "race to fish." While this approach often creates severe economic, biological and safety problems, traditional fisheries management has generally failed to address this problem. In fact, one of the most common management methods—the imposition of a total allowable catch (TAC)—only exacerbates the economic problem. Faced with an overall catch limit, the incentive is significantly increased for individual fishermen to fish as fast as possible before the TAC is reached. Therefore, the fishery proceeds even more quickly and more wastefully. The results are economically inefficient fisheries in which there is too much capital investment and effort, deployed over seasons that are artificially shortened, catching fish that are not as valuable, sometimes severely damaging habitats and all-too-often inadvertently taking seabirds, mammals, small fish and non-targeted species (known at "bycatch") as unintended victims of this race (Figure 3). Economic distress in turn tends to exacerbate conservation problems since conservation measures such as marine reserves, bycatch controls and catch reductions are perceived as threats to livelihood rather than as investments in the future. The race to fish is thus responsible for not only many of the environmental problems plaguing the world's ocean, but also for the fact that living ocean resources are not generating any net economic surplus and actually result in a net economic loss. This impoverishes fishermen, as well as the environment.

© WOLCOTT HENRY 2001

**Figure 2** Powerful fishing fleets and advanced gear, such as these massive tuna nets in American Samoa, are contributing to overfishing.

The other important factor contributing to the substantial global fishery deficit is the high level of subsidies to fisheries. Most available estimates place the global value of these subsidies between US $15–30 billion (Milazzo 1998, Virdin & Schorr 2001) or about 20–25% of first-hand sale revenues from world capture fisheries (Milazzo 1998). Some estimates are as high as US $54 billion (FAO Fisheries Department 1994). Subsidies add to the common property problem described above by artificially supporting the fishery. Subsidies artificially raise profits and reduce fishing costs, creating the perception that fishing is profitable even when it is not. This encourages even more entrants and greater over-capitalization and creates further economic inefficiency and environmental destruction, leading to the description of this global practice as "perverse subsidization."

Analysis shows just how economically inefficient world capture fisheries are and roughly what the benefits of ending the "race to fish" might be. Recall from above that global first-hand sale revenues are approximately US $75 billion (FAO Fisheries Department 2002). While some fisheries are already under good management and are profitable, the level of global subsidies suggest an overall net revenue shortfall (or economic loss) of up to US $15 billion. To this loss one must add the substantial global costs (Statistics Iceland 2003) of largely ineffectual fisheries management and research, which can easily be in the neighborhood of US $5 billion annually, generally also financed with public funds. Rough estimates suggest that, if better managed, world fisheries that are worth US $75 billion today could actually generate annual revenues of approxi-

© RICHARD HERRMANN

**Figure 3** Sealions—unintended bycatch—lie dead in a commercial fishing net.

mately US $100 billion (modified from Wilen 2004). After rationalization, there are typically changes in yield as well as in product quality, marketing, creation of new markets, etc., that lead to revenue increases. Wilen (2004) conservatively suggests a 30–35% revenue increase due to these changes; thus, global increases in fishing revenue would be expected to be more than US $25 billion (33% of US $75 billion). It has been estimated that the global "maximum sustainable yield" from current fisheries could be close to 100 million metric tons—instead of the current 80 million metric tons—if fisheries were better managed. Furthermore, eliminating the "race to fish" will not only increase revenues, but will also generate additional rents, or profits, since the capital investment necessary to catch the fish will fall. Based on case studies of the British Columbia halibut and Bering Sea pollock fisheries, Wilen calculates that following rationalization and elimination of excess inputs, profits are approximately 60% of revenues (Wilen 2004). A conservative estimate suggests that world marine capture fisheries should be generating US $60 billion in profits each year, rather than losing billions of dollars in subsidies and other excess costs (Wilen 2004).

This analysis is necessarily incomplete because it includes only the direct costs of over-capitalization and poor fish quality. It does not include indirect costs in the "race to fish" that are difficult to calculate but are most likely very high: the excess management costs incurred by supporting excessive idle fishing capacity, the ecosystem values being lost from excessive bycatch and the value of habitat lost or degraded by fishing.

Inadequate protection of ecosystem values and underproduction of potential economic surplus differs in extent from fishery to fishery and among management regimes. Most of the world's fisheries are conducted within the 200-nautical-mile Exclusive Economic Zones (EEZs) claimed by coastal nations that include both developed and developing countries. The remainder of global fisheries occur in the open ocean outside of the legal jurisdiction of any coastal nation. Each of these situations has spawned management structures unique in design, enforcement and performance, ranging from lack of regulation to area-specific and season-specific closures to complex international regimes. However, many countries have not yet managed to solve the common property problem with regard to their fisheries.

## Minimizing Impacts, Maximizing Profits

Given that the common property problem and the "race to fish" are the root causes of both environmental problems and poor income generation from the rich resources of the ocean, what are the possible solutions? The most effective solutions revolve around replacing the perverse incentives operating under "race to fish" conditions with incentives aimed at conservation and at generating value from the world's ocean resources. As it turns out, there are several tried and true ways to alter incentives. All of these methods closely emulate a private property rights system—but with a major difference. Instead of granting direct property rights to the resource itself, most of these systems are based on the granting of transferable privileges to harvest a certain fraction of the total allowable catch to individuals or groups. There is an important distinction between granting exclusive private property rights versus privileges to use a public resource. While there are benefits to granting full property rights, they raise the specter of governments having to compensate rights holders if allowable catches must be reduced for any reason. Since the ocean is common property held in the public trust, the government must retain authority to manage fish populations and fishing activities, free from the threat of a legal "compensable takings" claim. By granting use privileges instead of property rights in the strict sense, governments retain the ability to manage fisheries while still providing fishermen with strong incentives to maximize the value of their catch and improve their conservation performance.

The granting of "harvest" privileges can take many forms that can be tailored to specific fisheries, including allocations to groups of fishermen using similar gear types, to fishery cooperatives, to community-based co-management entities or to individual fishermen. When harvest privileges are held by individuals, they are termed Individual Transferable Quotas (ITQs)—also known as Individual Fishing Quotas (IFQs). ITQs represent individual percentage shares of scientifically determined allowable catches, thus removing the need to race to catch fish before one's competitors.

ITQs have been adopted in several hundred fisheries in over sixteen countries, and the results have generally been very positive for both conservation and income generation. We believe that having an ITQ changes the focus of fishermen from maximizing one's share of allowable catch (which is uncertain under open- or limited-access programs) to maximizing the value derived from a secure share of the total allowable catch. Following ITQ implementation, it has been suggested that fishermen in the British Columbia halibut fishery now time their halibut fishing trips according to expected ex-vessel prices—the prices paid to fishermen for seafood products right off the boat (Casey *et al* 1995). Freed of the need to compete with other fishermen to catch as many fish as fast as possible, a fisherman can tune his/her operation to maximize profits by fishing at times when market prices for the catch are high or when product quality is high. Fishermen can also reduce costs by trading their large boats for smaller vessels more closely aligned with their fish quota. This can result in more income with less investment and cleaner, less destructive fishing.

In the vast majority of ITQ fisheries that have been established around the world, the "race to fish" has ended, profits have increased and environmental performance has improved substantially. For example, ITQs transformed the Alaska and British Columbia halibut fishery from a five-day frenzied race, prosecuted by a bloated and over-capitalized fleet, into a much less intense fishery delivering fresh fish nearly continuously throughout the year. In British Columbia, ex-vessel halibut prices rose by 40% following ITQ introduction and total ex-vessel revenues increased by an average of Canadian $5.8 million per year (Environmental Defense 2003). ITQs introduced in stages from 1976 to 1990 in various Icelandic fisheries greatly reduced both fishing effort and vessel numbers and transformed a previously revenue-losing industry into a profitable one: in 1988, profits as a percentage of gross revenue for all Icelandic fisheries was –5%, compared to +12% in 2002 (Statistics Iceland 2003). Similar results apply to the New Zealand fisheries, where ITQs were introduced between 1982 and 1984, as well as other ITQ fisheries around the world. The value of New Zealand fisheries doubled in real terms between 1990 and 2000, much of which is attributable to having dedicated-use privileges in the form of ITQs and gains from trade (Sanchirico & Newell 2003, Newell, Sanchirico & Kerr forthcoming, J. Sanchirico personal communication). After its transition to ITQs, the multi-species groundfish fishery off British Columbia, Canada, showed improved compliance with catch limits and virtually eliminated bycatch and unaccounted fish mortality (Environmental Defense 2003). This fishery also substantially improved its economic performance. Ex-vessel prices increased by US $0.32 per pound for Pacific ocean perch and US $0.37 per pound for lingcod, and the landed value of the total catch increased by US $13 million (Environmental Defense 2003).

ITQs generally result in improved environmental performance, with respect to compliance with total allowable catch levels, bycatch reduction and other measures.

Rather than opposing conservation measures and complaining about the lack of adequate research, ITQ holders in some ITQ fisheries have actually invested in conservation, research and management (Statistics Iceland 2003). In New Zealand, for example, quota shareholders have asked for voluntary reductions in TAC to improve population status and have funded research related to population assessments, enhancement programs and bathymetric surveys. Since fisheries become profitable and are able to thrive, even without indirect government subsidies, increased profitability may help to reduce opposition to conservation measures. Under an ITQ regime, conservation measures may be perceived as investments in the future of the fishery, producing a flow of increased benefits to the ITQ holders. Also as a result of increased profitability, ITQ fisheries are often able to pay management costs formerly paid by the government. For example, New Zealand fisheries achieve almost full cost recovery and now pay annual fees that cover nearly all management and research costs (J. Sanchirico personal communication).

In a number of countries, approaches analogous to ITQs have evolved. These systems often involve the allocation of exclusive privileges—revocable by the government and without compensation—to use marine resources to defined groups rather than individuals (e.g., Mexican fishery cooperatives and European Union producer organizations). Some coastal communities, fishing cooperatives or fishermen's unions have also been granted management responsibilities—including the authority to set, enforce and implement rules to ensure sustainability. These innovations reduce the "race to fish" driven by both local unrestrained fishing effort and poaching by fishermen who are not members of local communities. Cooperatives, community-based management systems and other institutions promise some of the same benefits seen with ITQ systems, including decreased damage to valuable, rare or unique components of ecosystems, and fisheries that maximize economic value and contribute to economic development in settings where income generation opportunities are scarce. Traditional marine tenure systems may have operated in similar ways (Johannes 2002).

Despite strong evidence that secure access to a guaranteed share of sustainable catch cures many of the environmental and economic ills present in modern fisheries, progress toward adopting these measures has been disappointingly slow. In some countries, there is a deeply felt concern about the allocation of exclusive-use allowances for resources that have traditionally been open to everyone. These sentiments must be balanced against the economical, biological and environmental gains associated with dedicated-use privileges. Other more specific concerns focus on the potential difference between individual and collective economic gains or losses, the potential for economic windfalls resulting from the sale of fishing privileges, the possibility of excessive levels of industry consolidation and, related to all this, perceived inequitable distribution of privileges. However, both theory and experience show it is possible to design programs in such a way to adequately address these concerns. For example, Alaskan fishermen insisted on a requirement that ITQ holders be on board the vessel to prevent a fishery dominated by "absentee landlords." They also implemented a cap on the maximum amount of ITQ that an individual or firm can accumulate (1% of the allowable catch) to prevent a shift from a fleet dominated by small businesses to one dominated by large firms. In *Sharing the Fish*, the National Research Council recommended ways to allocate ITQs more fairly and to address other concerns (National Research Council 1999). But fear of change and deeply held concerns, coupled with a lack of funding for education, consensus building and implementation continue to block progress toward a more rational way of managing precious marine resources. This applies especially to less developed countries which, in addition to all of the concerns above, also face the difficulties of limited alternative employment opportunities, inadequate financial resources and, in many cases, the lack of administrative and fisheries management capacity and infrastructure.

## The Global Fisheries Reform Fund: Investing in Conservation and Sustainable Fisheries

If fisheries are to be transformed on a scale that will make a difference for the world's ocean, a catalyst for change is needed—one that can overcome the obstacles that currently hinder the designation of use privileges. We propose the establishment of the Global Fisheries Reform Fund, a revolving loan and investment vehicle focused on rationalizing world fisheries, as an example of the kind of institution that can address this need. The Fund would target candidate "win-win" fisheries that promise high financial returns, social benefits and environmental benefits from rationalization simultaneously (the transition to a system of designated-use privileges). The Fund would be a streamlined, independent institution that can take investments from a variety of sources—including bilateral, aid and governmental organizations, as well as private foundations and individuals—without being beholden to a partner with alternate goals and values.

In order to be successful, the Fund must have three core attributes: research expertise in fisheries; sensitivity to social, cultural and ecological impacts of implementing use privileges; and the ability to raise funds and remain financially solvent. In general, the Fund would work to connect available capital with appropriate candidate fisheries in the form of a loan or other instrument, and then monitor the investment to ensure that the fishery meets the loan requirements. The Fund could be organized in a variety of ways. A non-profit, non-governmental organization with a governing board of experts and a commitment to a triple bottom line—measuring success in economic, environmental and social benefits—may provide the right kind of institution to carry out the mission of the Fund. The organization could have two distinct divisions that would work together to achieve fisheries reform. A research and accountability team would identify candidate fisheries, provide the biological and political expertise necessary to carry out the reform and follow through with the fisheries to monitor and analyze success and ensure compliance with the loan conditions. An investment team would provide financial expertise, structure the loan and/or financial instruments, recruit and interface with potential investors, negotiate transactions and manage investments. The Board would be responsible for holding the Fund accountable to the triple bottom line.

With all divisions working together, the Fund would first identify candidate fisheries, analyze the ecological and economic potential of reform and create a process (with meaningful stakeholder consultations and participation) for carrying out a transition to a rights-based management system, including the structure of an appropriate financial instrument such as a loan or bond. The Fund would then approach investors, targeting those with a commitment to social and environmental investments, and raise money to fund the reform process. All of the investment would be directed at reforming the fishery, for example, as a direct application to reducing excess fishing capacity or funding the work of the experts employed or retained by the Fund.

The Fund could support education, analysis and deliberative processes to help each fishery realize its economic and biological potential through the dedication of use privileges. The relationship between the Fund and the fisheries management entity (e.g., a fishery cooperative or other management institution) could take the form of a contract setting forth the terms of a loan, bond or other instrument. Various conditions on the instrument could be negotiated, including conservation measures. Following ITQ implementation, the fishery would be responsible for paying back the loan, with interest, or providing the promised rate of return. The Fund would assess a fee as a percentage of the increased revenues enjoyed by fishermen following ITQ implementation, with the exact structure depending upon the specific cultural and political context of each fishery. Therefore, some of the enhanced revenue stream resulting from ITQ implementation would be retained by the fishermen, some would be paid to the Fund to cover operating expenses and some would be used to replenish the Fund and pay back investors, thereby achieving self-sustaining fisheries reform. Importantly, the financial instrument would be structured to achieve a desired rate of return for the investors, similar to a more traditional instrument. Following payback from a specific fishery, investors could choose to keep their money in the Fund, thereby reinvesting in the next fishery reform candidate, or they could exit the Fund with a return.

The Fund would seek to work with fisheries of all sorts, from both developed and developing countries. The specific attributes of loans made by the Fund would be malleable enough to accommodate differences in fishing history and sociopolitical structure in each fishery. The specific method of rationalization would depend on the fishery—taking into account each fishery's unique biological, social and political context—and could range from ITQs to fishing cooperatives to area-specific use privileges. For example, a subsistence fishery (Figure 4) would likely require a different approach than a commercial fishery to achieve environmental and economic benefits. In addition, the Fund could finance viable business plans to add value to fishery products such as the creation of cooperatives made up of fishers, processors and retailers who all abide by sustainable seafood criteria.

Fisheries that have serious economic and conservation problems, but show potential for economic recovery by implementing designated-use privileges, could be chosen for investment. Such fisheries could include those in fragile tropical systems with unique ecosystems and species of global interest from a conservation perspective. However, in many developing countries where the fisheries management infrastructure may lack the capacity to administer complex programs like ITQs or may prove incapable of generating sufficient returns to satisfy investors, the Fund could take different approaches. One approach might provide loans to fishery or area cooperatives (which would then assign use privileges) or grants to develop capacity for community-based management (CBM) in selected coastal zones. Community-based management has proven effective in various countries, including the Philippines, Samoa and Fiji (Johannes 2002, King & Faasili 1999, Russ & Alcala 1999). While opportunities to generate both economic and conservation benefits would be targeted, specific options for reforming fisheries would be generated by communities and supported by the Fund. In addition to reforming the fishery, the Fund could support alternative employment activities and sustained resource management by communities, including enforcement of fishery regulations, marine reserves and subsistence fishing.

The groundfish fishery off the west coast of the United States is characteristic of fisheries in many developed countries and can be considered a case study for rationalization by the Fund. This fishery has been declared a "federal fishery disaster"; we suggest that biological decline and economic collapse were due primarily to the lack of designated-use privileges leading to "race to fish" conditions and depletion of both the target population and non-target populations. Economically inefficient, the fishery lost 49% of its value from 1995 to 2001 (Pacific Fishery Management Council 2003). In addition, environmental problems include excessive bycatch of non-targeted fish species, associated accidental killing of marine birds and mammals and degraded habitat. Obstacles to rationalization have included fear of excessive fleet consolidation, fear of inequitable initial ITQ allocations and concern about windfall profits resulting from an inequitable allocation of public resources.

© SYLVIA A. EARLE

**Figure 4** Local subsistence fishing can be benign or destructive when dynamite or poisons are used.

External funding by philanthropic foundations to support education and research on ITQs has contributed to the development of a new process to rationalize the fishery, and the federal government has provided funds to buy out about half the productive capacity of the trawl fleet with a $10 million grant and $36 million loan. Even with these efforts, the fishery is still in danger due to the lack of dedicated-use privileges, and a major constraint to rationalization is a lack of funds to support and encourage change (U. S. General Accounting Office 2003). Investment by the Fund could finance ITQ program development, analysis, stakeholder processes, administration, observers and enforcement. It could also be conditioned on a commitment to improve environmental performance standards such as sustainable (lower) limits for bycatch, implementation of habitat damage performance standards for all gear types, and marine reserves to reduce the ecological damage imposed by the fishery on society. The Fund would leverage investment for the fishery from a variety of sources interested in improving both economic and conservation performance in fisheries, including local governments, private investors and conservation funders.

We anticipate that changes resulting from the application of monies from the Fund would not only improve the environmental performance of the fishery, but also enhance its revenues and profitability. The revitalized fishery could then pay back the loan and eventually share most or all of the cost of program administration, enforcement and scientific research. Substantial cost sharing is typical of many ITQ fisheries (National Research Council 1999). This collaboration of different interests would simultaneously tackle the four major problems in this fishery: any excess capitalization of fishing capacity remaining after the federal buyout, lack of secure use privileges, lack of incentives for stewardship under "race to fish" conditions and the need to protect habitats and reduce bycatch. Other promising candidates for reform (Figure 5) may include Gulf of Mexico fisheries, the Gulf of California shrimp fishery (Conservation International 2003) and the West Africa continental shelf fishery (Pauly 2002).

**Figure 5** The Gulf of California shrimp trawling fleet is a prime example of a fishery in which monies from a Global Fisheries Reform Fund could improve both environmental performance and profitability of the fishery.

## Summary

Many fisheries worldwide are economically inefficient and in biological decline. Numerous entities have recognized this and are independently attempting to address both the symptoms and causes of global depletion of the world's fisheries. Some countries—such as New Zealand, Iceland, Australia and Canada—have successfully reformed some of their fisheries through designating use privileges in the form of ITQs, and governments in other countries are cautiously experimenting with various types of rationalization. In many cases, fishermen themselves are designing and promoting innovative new methods for removing "race to fish" incentives (At-Sea Processors Association 2004). Participants in the West Coast Pacific whiting fishery, for example, formed the Pacific Whiting Conservation Cooperative (PWCC) to improve biological and economic conditions of this fishery. The PWCC functions similarly to ITQ fisheries and has had numerous benefits (At-Sea Processors Association 2004). In addition, conservation interests are seeking and developing innovative ways to halt destructive fishing methods and set aside unique and fragile marine ecosystems from extractive practices. We believe that large gains are possible by combining forces, experience and funding. A Global Fisheries Reform Fund offers a model to coordinate these various activities into a unified and practical effort that tackles the challenges of fisheries management. Given the current environment of uncoordinated actions and the lack of sustainable funding for fisheries reform, the Global Fisheries Reform Fund holds promise for helping to achieve the vision of a healthy ocean—articulated by the Pew Oceans Commission, the U.S. Commission on Ocean Policy, the World Summit on Sustainable Development and by many other bodies, laws and treaties—by providing sustained funding to improve both the economic vitality of fisheries and their environmental performance.

## Literature Cited & Consulted


Arnason, R. 1991. Efficient management of ocean fisheries. *European Economic Review* 35:408–417.

At-Sea Processors Association. 2004. *A case study of fish harvesting cooperatives: The Pacific Whiting Conservation Cooperative (PWCC). http://www.atsea.org/concerns/pwcc.html* (accessed August 10, 2004).

Casey, K., C. Dewees, B. Turris and J. Wilen. 1995. The effects of individual vessel quotas in the British Columbia Halibut Fishery. *Marine Resource Economics* 10 (3): 211–230.

Conservation International. 2003. *Resource Economics Program.* Conservation Programs: Policy and Economics. *http://www.conservation.org/xp/CIWEB/programs/policyeconomics/policyeconomics.xml#rep* (accessed August 11, 2004).

Environmental Defense. 2003. *IVQs and Multi-Species Groundfish Fisheries: The British Columbia Experience.* Environmental Defense Brochure available from Environmental Defense at 5655 College Ave., Suite 304, Oakland, CA 94618.

FAO Fisheries Department. 1994. Marine fisheries and the law of the sea: A decade of change. Special chapter (revised) of *The State of Food and Agriculture 1992.* Rome: Food and Agriculture Organization of the United Nations.

——. 2002. *The State of World Fisheries and Aquaculture.* Rome: Food and Agriculture Organization of the United Nations. *http://www.fao.org/docrep/005/y7300e/y7300e00.htm* (accessed August 10, 2004).

Hardin, G. 1968. The tragedy of the commons. *Science* 162 (3859): 1243–1248.

Johannes, R.E. 2002. The renaissance of community-based marine resource management in Oceania. *Annual Review of Ecology and Systematics* 33:317–340.

King, M. and U. Faasili. 1999. Community-based management of subsistence fisheries in Samoa. *Fishery Management and Ecology* 6 (2): 133–144.

Milazzo, M. 1998. *Subsidies in World Fisheries: A Re-Examination.* Washington D.C.: World Bank Technical Paper. 406, Fisheries Series.

National Research Council. 1999. *Sharing the Fish: Toward a National Policy for Individual Fishing Quotas.* Washington D.C.: National Academy Press.

Newell, R., J. N. Sanchirico and S. Kerr. Forthcoming. Fishing quota markets. *Journal of Environmental Economics and Management.*

Pacific Fisheries Management Council. 2003. Amendment 16-2: *Rebuilding plans for dark blotched rockfish, Pacific ocean perch, canary rockfish, and lingcod.* Environmental impact statement and regulatory analyses. *http://www.pcouncil.org/groundfish/gffmp/gfa16-2/am16-21203.pdf* (accessed August 10, 2004).

Pauly, D. 2002. A Symposium with Results. *The Sea Around Us Project Newsletter* 12, July/August. *http://saup.fisheries.ubc.ca/Newsletters/Issue12.pdf* (accessed August 11, 2004).

Pauly, D., V. Christensen, J. Dalsgaard, R. Froese and F. Torres, Jr. 1998. Fishing down marine food webs. *Science* 279 (5352): 860–863.

Pew Oceans Commission. 2003. *America's Living Oceans: Charting a Course for Sea Change.* Arlington, Virginia: Pew Oceans Commission. *http://www.pewoceans.org/oceans/downloads/oceans_report.pdf* (accessed August 10, 2004).

Russ, G. R. and A. C. Alcala. 1999. Management histories of Sumilon and Apo Marine Reserves, Philippines, and their influence on national marine resource policy. *Coral Reefs* 18 (4): 307–319.

Sanchirico, J. and R. Newell. 2003. *Catching Market Efficiencies: Quota-Based Fisheries Management.* Resources, No. 150, Spring. *http://www.rff.org/Documents/RFF-Resources-150-catchmarket.pdf* (accessed August 11, 2004).

Schrank, W., R. Arnason and R. Hannesson, eds. 2003. *The cost of fisheries management.* Aldershot, UK: Ashgate.

Statistics Iceland. 2003. *Afkoma Sjávarútvegs.* Reykjavik.

U. S. General Accounting Office. 2003. *Commercial Fisheries: Entry of Fishermen Limits Benefits of Buyback Programs.* Report to House Committee on Resources GAO/RCED-00-120. *http://www.gao.gov/archive/2000/rc00120.pdf* (accessed August 10, 2004).

Virdin, J. and D. Schorr. 2001. *Hard Facts, Hidden Problems: A Review of Current Data on Fishing Subsidies.* World Wildlife Fund Technical Paper, October 2001. *http://www.worldwildlife.org/oceans/pdfs/hard_facts.pdf* (accessed August 10, 2004).

Wilen, J. 2004. Property rights and the texture of rents in fisheries. In *Evolving Property Rights in Marine Fisheries,* ed. D. Lead. Lanham, Maryland: Rowan and Littlefield Publishers.

CHAPTER 8

# A Global Network for Sustained Governance of Coastal Ecosystems

Stephen B. Olsen, *Coastal Resources Center*
Richard Kenchington, *University of Wollongong, Australia*
Neil Davies, *Richard B. Gump South Pacific Research Station*
Guilherme F. Dutra, *Conservation International*
Lynne Zeitlin Hale, *The Nature Conservancy*
Alejandro Robles, *Conservation International*
Sue Wells

© SYLVIA A. EARLE

## Introduction

Seven-tenths of the planet's surface is covered by ocean. Yet the forces of change that increasingly threaten the quality and function of marine ecosystems all originate on the three-tenths that is land. These forces, in turn, are concentrated at the margins of the planet's ocean, seas and great lakes on lands that comprise less than 20% of the land-space inhabited by people. By 1995, nearly half of the world's human population was concentrated in these coastal lands, and twelve of the fifteen largest cities were coastal (Cohen *et al* 1997). Demographers predict that by mid-century, three-quarters of a much larger population will be living in these coastal regions. This concentrates energy production and consumption, manufacturing, waste production and built-up environments (Figure 1) on the lands that are directly adjacent to coastal marine waters having the highest concentrations of marine biological production and marine biodiversity. Any comprehensive set of actions designed to sustain and restore the qualities of the world's ocean cannot succeed without concerted efforts to address the root causes of ocean change that are being generated in the coastal ecosystems that lie at the boundary between the land and the sea.

A global set of coastal stewardship incubator projects, operating as nested efforts of different sizes culminating at the large marine ecosystem scale, could be the catalyst for the changes in societal values and behavior that are the precondition to reversing current negative trends. Only nested systems that address ecosystem processes at a range of scales will successfully counter the complex web of forces that is threatening the quality of coastal and marine ecosystems on planet Earth. This chapter proposes the structure for such an approach, describes the benefits it would produce and explains why this effort will not be realized without an independently funded and administered initiative. Our approach avoids the weaknesses in many coastal management programs that have emerged since the Rio Conference in 1992—the absence of common frameworks for assessing progress and disseminating innovations, the absence of funding to reward and sustain sound programs in

© WOLCOTT HENRY 2001

**Figure 1** Present-day Cancun is an example of an extremely over-built environment where near-wilderness existed fifty years ago.

low-income nations and the absence of global mechanisms to encourage collaborative learning.

## The Evolution of Integrating Approaches to Coastal Governance

Contemporary coastal management—variously termed coastal zone management (CZM), integrated coastal management (ICM) and integrated coastal zone management (ICZM)—has its roots in the seminal Stratton Commission Report released in 1969 that recommended a "new approach" to planning and decision-making along the coastlines of the United States. The Stratton Commission declared that the coastal zone was the nation's most valuable physical feature and found that local and state governments were not capable of the planning and decision-making necessary to halt and reverse the degradation of important coastal assets. Their recommendations were the basis of the Coastal Zone Management Act of 1972. This innovative legislation put a novel nested system of management in place that required identifying and accommodating national interests along the coastlines of each U.S. state, setting policies to guide future development and conservation and demonstrating that the state had the necessary authorities and resources to implement its CZM Program. In return, the federal government offered sustained funding for the implementation of approved state programs and pledged that federal actions within the designated coastal zone would be consistent with the state program.

A decade later, the EPA's Estuaries Management Program expanded upon these ideas through a program designed to address the many forces creating the loss of quality in the nation's estuaries. The U.S. experience, and early efforts in developing nations sponsored by the U.S. Agency for International Development (USAID), influenced Chapter 17 of Agenda 21—the complex, non-binding resolution that was a major product of the United Nations Conference on the Environment and Development held in Rio in 1992. The Conference found that integrated coastal management (ICM) is the most promising approach to address the complex web of pressures and actions in coastal regions. Unfortunately, the many ICM programs initiated in response to the Rio Declaration—largely in developing nations—did not have the features that have been central to progress in the U.S.: sustained funding for accredited programs and consistency of subsequent investments with the approved program. ICM programs have, in varying degrees, embraced and further developed the features of public involvement, application of the best available science, transparent decision-making and issue-driven management. By 2000, according to one estimate, 345 ICM initiatives had been undertaken in 95 sovereign and semi-sovereign states—of which 70 are classified as developing nations (Sorensen 2000). The majority have been designed and implemented as four- to eight-year "projects" by international donors and development banks. The total investment has been in the range of hundreds of millions of dollars (Olsen & Christie 2000). Ten years later in 2003, the World Summit on Sustainable Development in Johannesburg reemphasized the need for comprehensive approaches that integrate the needs of both conservation and development. The Johannesburg Conference placed a special emphasis on an ecosystem perspective that addresses issues linking coastal watersheds to the coastal ocean.

While the CZM practiced in the U.S. has evolved into a largely regulatory process directed at the strip of land adjacent to marine waters and the Great Lakes, ICM has evolved as a more inclusive approach that addresses the processes of coastal planning and decision-making. ICM can be considered a stepping-stone toward making the principles of ecosystem management an operational reality. In ecosystem management, both the biophysical components of the environment, the associated human population and its economic/social systems are all seen as integral parts of complex and interdependent systems. It is designed and executed as an adaptive, learning-based process that applies the principles of the scientific method to the processes of management. Ecosystem-based management signals the transition from traditional sector-by-sector planning and decision-making to a holistic approach based on the interactions within ecosystems (Table 1).

## Priority Coastal Issues during the Anthropocene

In the final decades of the 20th century, the Earth entered a new geological era in which the activities of one species —human beings—became the dominant force shaping the ecology of the planet. This era has been named the "Anthropocene" (Crutzen & Stoermer 2000). Attention to human-induced changes has been focused largely on climate change, since it is in the planet's atmosphere that substances mix and are transported from one region to another most quickly. While it is not possible to predict the outcomes of

Table 1

**Transition from sector-specific to integrated coastal management (ICM)**

| From | To |
|---|---|
| Individual species | Ecosystems |
| Small spatial scale | Multiple scales |
| Short-term perspective | Long-term perspective |
| Humans independent of ecosystems | Humans as integral parts of ecosystems |
| Management divorced from research | Adaptive management |

**Source:** Lubchenco 1994.

climate change region by region, it is now well established that current concentrations of carbon dioxide in the atmosphere released by the burning of fossil fuels exceed the maximum levels of any time in the past 400,000 years. This single expression of anthropogenic (human-induced) change has placed Earth in a "no analogue state" in which the study of past events is of limited use in forecasting the future (International Geosphere Biosphere Program 2001). Climate change is already altering well-established patterns of rainfall and temperature and having multiple impacts on coastal ecosystems. Estuaries that have evolved to adapt to seasonal pulses of freshwater inflow and to long-established temperature regimes are changing. In Narragansett Bay (Rhode Island, U.S.), for example, average winter temperatures have increased by 3°F since the 1890s (Nixon, Granger & Buckley 2004). This is believed to be a root cause for dramatic changes in the abundance and distribution of commercially important finfish, the persistence of predatory zooplankton (small drifting animals) and major shifts in what were formerly predictable spring and winter plankton blooms. In the tropics, the prevalence of unusually high water temperatures has caused massive stressing or deaths of corals in "bleaching" events.

Climate change is only one expression of global change. Coastal ecosystems are being reengineered on a massive scale. The accelerating competition for fresh water from the expansion of human populations, agriculture, industry and cities is bringing a sharp reduction in the flows of freshwater estuaries in many regions, particularly in the tropics. It is the quantity, quality and pulsing of freshwater inflows that give individual estuaries their unique identity and makes them among the most naturally productive ecosystems on the planet. The recent and ongoing changes in estuaries have enormous implications in temperate regions. Roughly 80% of all commercially sold finfish depend on these estuaries at some stage in their lifetime. Estuaries also play a critical role in nutrient and waste reprocessing and these important services to society are also under attack. In arctic regions where global warming is most dramatically apparent, the melting of icecaps and permafrost may be generating a massive pulse of fresh water to arctic oceans (Peterson *et al* 2002). This could disrupt the long-established planetary patterns of ocean circulation driven by differences in seawater temperature, salinity and density. One dramatic consequence could be the "switching off" of the Gulf Stream (Broecker 1995). This would have catastrophic consequences for the climate in Europe and massive implications for all forms of life—terrestrial, coastal and oceanic—in and around the North Atlantic.

A more familiar expression of change in coastal regions—and a major theme in this volume—is the degree to which the combined impacts of habitat destruction and intense fishing pressure are causing a fundamental change in marine food webs. A vast majority of the world's fisheries are taken from shallow coastal waters. Beginning in Europe in the 1920s, trawlers have massively increased our ability to chase and capture fish. The wastage from bycatch (unintended and non-commercial species of ocean life caught and killed through modern fishing methods) and the repeated scouring of seafloor habitats is now being documented. One of the many impacts of contemporary fishing technology is "fishing down the food chain," a process in which the higher trophic levels (larger predator species at the top of the food chain) are removed and ever greater threat is posed to the plankton-eating species at the base of the chain (Pauly *et al* 1998).

The combined impacts of climate change, the reengineering of coastal ecosystems and fundamental changes to marine habitats and trophic structure make coastal regions

© WOLCOTT HENRY 2001

**Figure 2** Merchant ships carrying water between ports in their ballast tanks can inadvertently introduce "exotic" or "invasive" species that often have a very destructive effect on local marine ecosystems.

increasingly susceptible to additional pressures. In coastal seas where the wastes of human society are concentrated, such as in the Baltic and the Gulf of Mexico adjacent to the mouth of the Mississippi River, we are seeing high concentrations of algae growth known as algal "blooms"—some of which are toxic—and expansion of areas called "dead zones" where depletion of oxygen in the water kills off most of the marine life (Raloff 2004, Malakoff 1998). Dead zones are being identified in coastal waters worldwide.

Another threat to coastal waters occurs when merchant ships fill their ballast tanks with seawater and unwittingly transport coastal waters and the life within them from one region to another (Figure 2). When this ballast water is released before entering the next port, "exotic" species can be introduced and the ecosystem of the coastal waters disrupted, sometimes disastrously. For example, when the carnivorous comb jelly *Miniopsis lydii* was accidentally introduced into the Black Sea, it multiplied to extraordinary densities and drastically reduced the eggs and larvae of the dominant finfish (Mee 1992, Zaitsev 1992).

Future scenarios are further complicated by new forms of human activity, not least of which is the rapid expansion of marine aquaculture, or "mariculture." As the volumes of wild stocks of fish and shellfish taken by fishing plateau and decline, and the demand for seafood simultaneously increases, sea farming is becoming increasing prevalent. The major form of this new activity has been mono-cultures (single-species cultivation) of such high-value species as salmon and shrimp. These have caused serious, localized negative impacts. Shrimp farming has been particularly destructive since it is conducted in ponds in the tropics that are frequently built by clearing coastal lagoons and mangrove wetlands that are critical to the productivity of many marine species (Figure 3). An alternative strategy known as ecological aquaculture (Costa-Pierce 2002) proposes mimicking the diversity of natural systems and, if energetically pursued, could permit the growth of this new industry in a more ecologically responsible manner.

Of potentially great long-term concern is the degree to which the ecosystem processes that give coastal waters their identity and significance are being altered. Human beings are now fixing nitrogen and making it biologically available (putting it into compounds that can be ingested by animals) at a rate greater than all natural processes combined. A primary example is the use of nitrogen in compounds, like fertilizers, that wash into coastal waters and create nutrients for small marine life, like algae. This can cause high-concentration algal "blooms" that use all the oxygen in the water (eutrophication), killing off other life forms and creating "dead zones." Nitrogen plays the primary role in productivity and eutrophication of estuaries. Recent forecasts are that nitrogen inflows to coastal waters will more than double by 2050 (Seitzinger *et al* 2002).

Strategies that address this complex set of issues cannot ignore their human dimensions and particularly the differences that separate the minority living in high-income societies from the majority living in low-income societies.

ARLO H. HEMPHILL

**Figure 3** Although shrimp farming avoids the bycatch problems of shrimp trawling, it can also be particularly destructive. For instance, these shrimp farming ponds near Bahía de Caráquez, Ecuador, were created by destroying coastal lagoons and mangrove habitats critical to the productivity of many marine species.

These inequities have been explored as "the digital divide" in a recent article in *Science* (Kates *et al* 2001). It is popular to think that the solution is for the low-income societies to adopt the consumption ethic and economic growth patterns of the world's wealthy. But such thinkers as E. O. Wilson (1998) point out that if the entire planet's current population were to live in the contemporary style of the "North," we would require seven more planets.

## What Are We Learning as Coastal Stewards?

Many results from various attempts to "see it whole" when trying to manage the complex interactions and interdependencies in human-dominated ecosystems are now unfolding along the world's coastlines. What can be learned from these efforts?

The most obvious, but often overlooked, conclusion is that modulating or reversing the negative trends outlined in the previous section requires changes in human behavior at the societal—rather than the individual—scale. Methods for analyzing changes in human behavior at the larger scales are being designed and applied. The "Orders of Outcomes" framework (Olsen 2003), for example, recognizes that coastal governance is the process by which societal goals are defined and the mechanisms for achieving them are selected and implemented (Figure 4).

The framework highlights the importance of changes in state (such as the abundance of fish or quality of life), but also recognizes that for each change in state correlated changes in the behavior of human groups and institutions within the ecosystem in question are required. "First order" outcomes are defined as actions required when a society commits to a plan to modify the course of events in a coastal ecosystem. At the national level, first order outcomes are expressed as a formal commitment to a governance program and the establishment of the "enabling conditions" that ensure policies, plans and actions can be successfully implemented. The setting of goals is an essential element required for successful implementation. Achievement of first order outcomes requires building the constituencies and the institutional capacity to undertake integrated coastal planning and decision-making and ensuring the authority, funding and other resources are in place to implement the selected policies and actions. "Second order" outcomes are evidence of successful implementation of an ICM program. This includes evidence of new forms of collaborative action among institutions, including State-public partnerships, and demonstrable

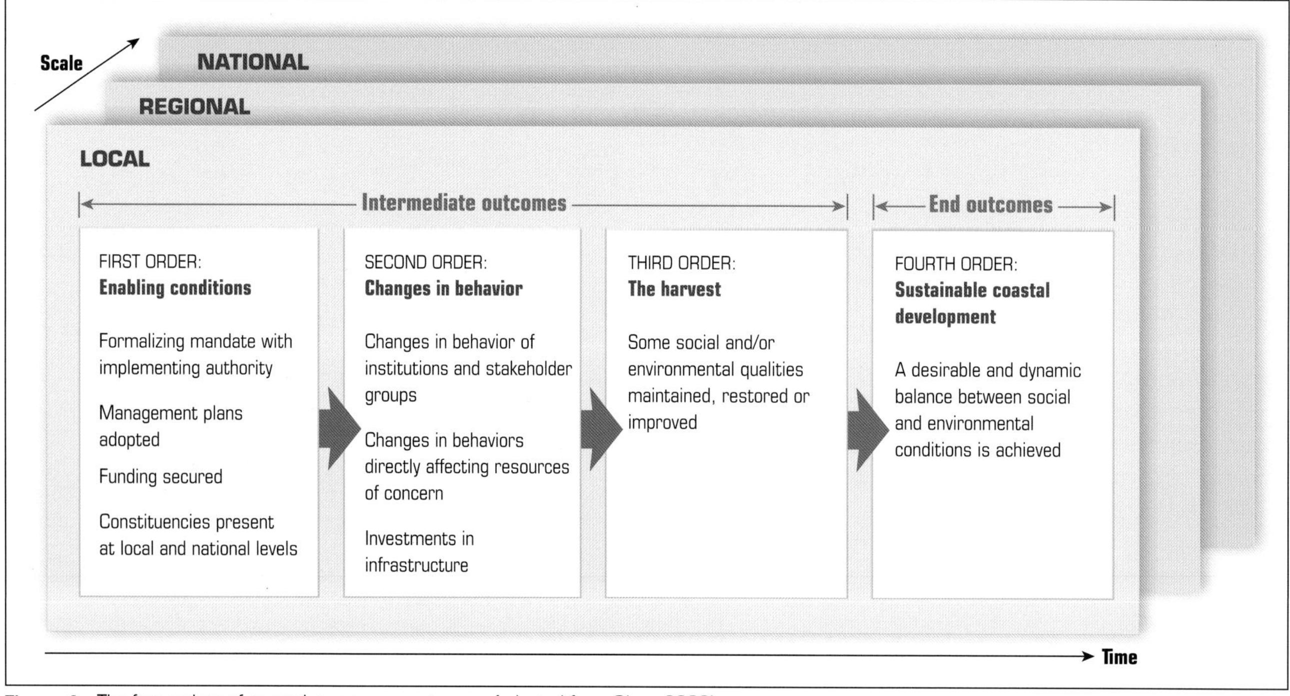


**Figure 4** The four orders of coastal governance outcomes (adapted from Olsen 2003).

behavioral changes of resource users. Second order changes in the behavior of organizations and user groups are the precursors to "third order" socioeconomic and environmental outcomes that embody physical evidence of progress towards desired environmental and social conditions.

In the Anthropocene, behavioral changes must occur simultaneously and at several scales. For example, increasingly widespread events of coral bleaching—death of the algae living in coral polyps that sometimes results in death of the corals—are the result of both local stresses and elevated water temperatures related to global climate change and behaviors that may be occurring on the other side of the planet. Cyanide fishing for the aquarium trade that is degrading coral reefs in the Philippines and Indonesia is being driven by the demands of aquarium hobbyists in Europe and the U.S. On the other hand, coral reef degradation caused by dynamite fishing—the result of local poverty and overfishing—may be addressed by a localized strategy to change this form of behavior.

Changing societal behavior at any scale is difficult and requires sustained effort. In democratic societies, people must understand that change is necessary and must support the efforts to bring it about. This is the core of constituency building and its importance is perhaps the major lesson to be drawn from coastal management efforts since Rio. In heavily populated coastal regions, where competition for a dwindling resource base is usually intense, building a foundation of popular support—a constituency—for a coastal stewardship program is the number one challenge. It is equally essential that efforts be sustained over the long term since the costs and benefits of implementing a plan of action are often widely separated in time. A degraded estuary will not recover from years of pollution and overfishing within the time frame of the typical "project." On the other hand, the benefits of stewardship from small, community-managed marine protected areas can be documented in restored coral cover and more abundant fish in as little as two years. Successful strategies artfully combine long-term investments with targeted activities that show quick results and generate hope.

Another major conclusion is that inadequate science is not the limiting factor to progress. We know enough about how coastal ecosystems function and respond to make far greater progress than has been achieved thus far in the vast majority of the world's coastal regions. Yet many projects invest heavily in descriptive science that is of marginal usefulness to supporting an effective course of action. The most critical limiting factor—equally clear but rarely addressed directly—is the lack of institutional capacity to support improved coastal management. There is an abundance of guidelines, but these rarely offer an approach that ensures the development of local and regional skills necessary to instigate behavioral change and appeal to the needs, values and heritage of the people who will be most directly affected by the program. Large short-term infusions of funds do not solve long-term problems in complex systems. Importing expensive external "specialists" to design and administer externally funded short-term projects continues to be the dominant approach of the many institutions investing in coastal management. It is hardly surprising that the outcomes are frequently disappointing.

While early efforts in coastal stewardship at the large ecosystem scale present many disappointments, there are also some remarkable successes that support some approaches that have been recommended and encourage us to press ahead with renewed vigor. Where coastal stewardship has been adequately funded for a sustained period—and has adhered to the principles of balancing the scope of the approach with local/regional institutional capacity for constituency-building and enforcement of adopted policies—the outcomes are remarkable. Three outstanding examples are programs that emerged in the early 1970s to address the concerns raised by human impacts on Australia's Great Barrier Reef, Europe's Wadden Sea and on Chesapeake Bay (Figure 5)—the largest and most commercially important estuary in the United States (Olsen & Nickerson 2003). These three programs evolved to address multiple issues simultaneously at the regional/large coastal ecosystem scale, and they share the following characteristics:

- Encompass areas of coastal waters and adjoining watersheds in the range of 10,000 to 100,000 square kilometers
- Have defined these boundaries by linkages among ecosystem features and processes rather than by political or administrative boundaries
- Contain both exceptional natural qualities and a diversity of human activities
- Have established functioning governance systems that set clear goals for desired social and environmental qualities, monitor change as it relates to those goals and successfully adapt to both their own experience and ongoing ecosystem change

PROVIDED BY THE SEAWIFS PROJECT, NASA/GODDARD SPACE FLIGHT CENTER AND ORBIMAGE

**Figure 5** The Chesapeake Bay—shown here in a Sea-viewing Wide Field-of-View (SeaWiFS) satellite image—is the largest and most commercially important estuary in the United States.

These programs demonstrate that success lies in achieving the full suite of first order outcomes. In particular, they all have:

- Built constituencies around unambiguous goals
- Promoted decentralized planning and decision-making in support of those goals
- Analyzed indictors associated with those goals
- Implemented strategically targeted public education programs

Together these efforts can result in changes in values and behavior at a geographic scale large enough to address forces and complex interactions in the coastal environment that cannot be successfully addressed at a smaller scale.

The time required to establish the necessary legal mandates and institutional capacity to sustain a governance initiative at these large scales, and then to successfully implement a plan of action, is measured in decades. Successful programs demonstrate that while education and constituency building are essential, the importance of enforcing the adopted rules is equally critical. A program's credibility rests on its ability to require that everyone be held accountable for their actions and that the sanctions on those who disregard the formally adopted rules are vigorously applied.

## Our Vision

By 2025, we imagine a global set of "incubator" programs for large coastal ecosystem stewardship that are demonstrating significant progress towards both societal and environmental quality goals simultaneously. While each incubator program would be tailored to its own societal needs and environmental conditions, the principles underlying the program design and strategies selected to meet their goals will be sufficiently similar to encourage exchange of knowledge across the network. The incubator network would therefore demonstrate how the principles of coastal ecosystem stewardship are being made operational in a wide diversity of settings. The program would feature competitive bidding for seed money to support innovative ideas. The central goal would be to construct nested systems of governance that address processes at the scale of large coastal ecosystems.

In the Anthropocene, the forces driving ecosystem change operate at large scales. Efforts to improve conditions and change human behavior at a small scale are too often overwhelmed by external forces they cannot control. Our concept, therefore, calls for constructing nested systems of governance that build to such scales as the Great Barrier Reef, the Chesapeake Bay, the Gulf of California or the Baltic Sea. Any newly developed programs for managing large coastal ecosystems must explicitly address the resources, processes and human activities in the coastal watersheds and adjoining estuarine and marine waters. The boundaries of such programs must be set in accordance with the scale of issues that must be addressed. The Chesapeake Bay program is an example of how the boundaries of the management area have expanded as the program has gathered strength and knowledge. In its first phase, only the immediate watershed was subject to management. In its most recent expression—following the discovery that atmospheric deposition is a major source of increased nutrients that are negatively altering the Bay's ecosystem—the Program's plan of action expanded to include a very large "airshed." The large marine ecosystem (LME) approach advocated by the World Bank's Global Environmental Facility and detailed by Sherman (Sherman & Duda 1999) is a major step in this

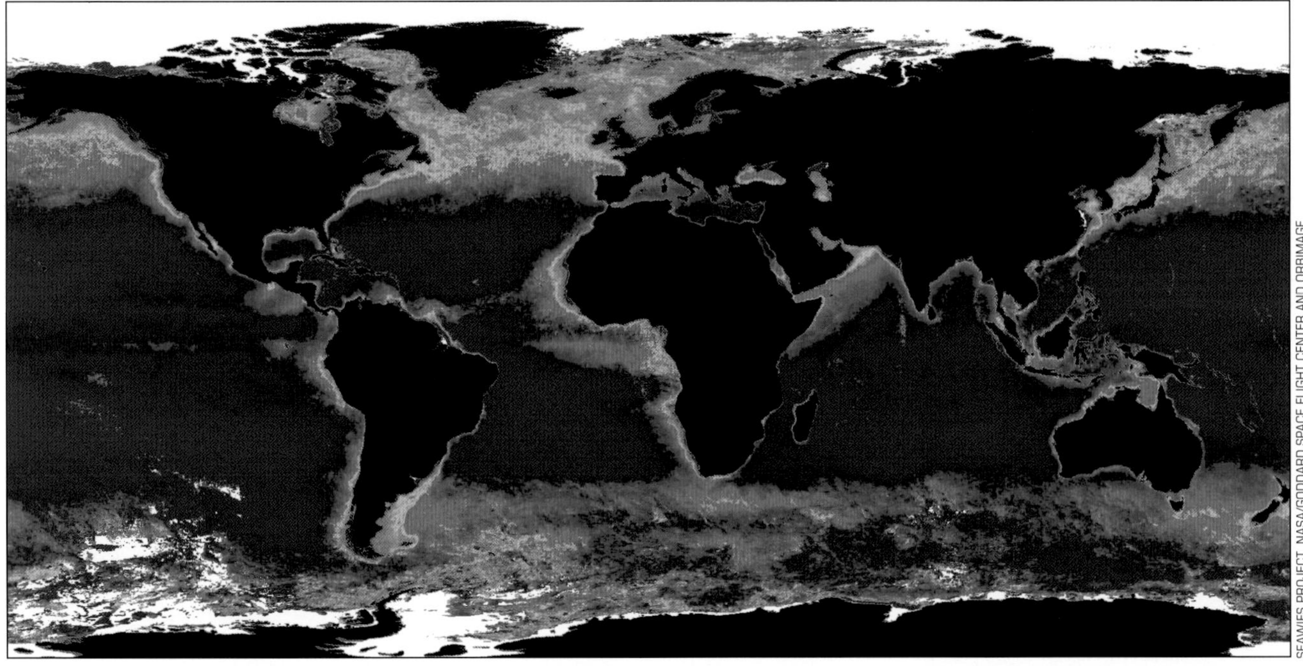


**Figure 6** This global composite of Coastal Zone Color Scanner (CZCS) data from the Nimbus 7 satellite collected from November 1978 to June 1986 highlights the high chlorophyll levels—indicating high productivity—in estuaries, coastal waters and areas of upwelling. White areas are ice; red/orange/yellow areas indicate highest chlorophyll concentrations (SeaWiFS Project 2004).

direction. LMEs are concerned primarily with the management of fisheries, however. The LME approach also does not currently embrace the concept of a nested governance system in which management systems in sub-areas are encouraged to operate at smaller scales with considerable independence so long as their approach contributes to the overall goals of the program.

Estuaries are the most biologically productive ecosystems on the planet (Figure 6) and a habitat critical to the life history of most commercially important fish species. Our proposal addresses the shortcomings amply illustrated by the first generation of attempts at coastal ecosystem governance. The first of these is the imperative of sustained effort. The outcomes of effective stewardship cannot be achieved through large but short-term infusions of money. We propose, as a first step, the identification of existing programs that are adhering to principles of sound ecosystem stewardship and assuring they have a minimal level of core funding. The strategy will not initiate a fresh set of initiatives but rather will nurture appropriate existing efforts and encourage them to adhere to an explicit set of good practices. The highest priority is to build well-documented examples of sustained efforts in low-income countries, particularly in the tropics.

Documentation of the experiences in the supported programs would provide a basis for public education and local capacity-building efforts. Documentation and analysis of change in the targeted ecosystems would also be the basis for an agile and accessible knowledge management system within the network, which would provide ready access to the most pertinent information, ideas and growing knowledge on how the principles of stewardship can be made an operational reality. Successful approaches to stewardship would be disseminated not only to all other programs in the network but also to the public. The message conveyed to the public cannot always be dreary statistics on all the threats and problems. The network must also demonstrate why we should have hope, and a primary focus for the network's education and outreach programs will be youth. Because behavioral change in specific local groups, in governmental agencies and industries and in societies as a whole will be the root of success, it is essential to design systems for monitoring and rewarding incremental levels of behavioral change.

## The Investment Strategy

We propose establishing a Global Fund for Coastal Stewardship with the following characteristics:

- Formulation of global standards of accountability for program goals, transparency and inclusion of appropriate stakeholders designed to produce both coastal ecosystem health and societal well-being
- Establishment of a Coastal Stewardship Incubator Network of approximately 20 programs operating at ecosystem-wide regional scales. (Where necessary, the Fund will provide core support to sustain coastal management programs in low-income nations if they meet the global coastal stewardship standards and are making measurable progress on high-priority land/ocean management issues through changed behavior.)
- Application of monitoring and trend analysis protocols that recognize outside influences, not just the program's actions alone, that assess progress toward unambiguous, time-limited goals and contribute to desired social and environmental outcomes
- Provision of a regionally grounded, but globally accessible, knowledge management system that promotes the dissemination of innovations and good practices
- Establishment of a regionally grounded, but globally overseen, program of small- to medium-sized grants to support quick response to emerging opportunities for good marine conservation practices
- Development of a regionally grounded, but globally coordinated, public education program that disseminates quality information on the societal behavior changes needed to mitigate the root causes of land-driven marine degradation and on the hope for the future implicit in improving stewardship practices

The annual costs of regional coastal management programs at the scale of the Great Barrier Reef and the Gulf of California are in the range of $15 million, translating to about $20 per square kilometer per year. The estimated cost for an effective Global Fund for Coastal Stewardship program is $150 million for the initial ten years. This assumes $5 million per year for global collaborative learning and knowledge management and $0.5 million per year in sustaining core funds for each of 20 regional integrated land/ocean management initiatives selected to participate in the Coastal Stewardship Incubator Network. In all cases the Fund will avoid supporting the creation of new organizations, unless through the cost-efficient fusion of two or more existing ones. We have not estimated the costs of an associated global education program since this was the subject of the Communications theme group at the *Defying Ocean's End* Conference (Chapter 10).

We recommend that *within three years,* the global standards—framed as a Code of Conduct for the Stewardship of Coastal Ecosystems—and the regionally grounded but global coordinating institution both be in place. Initial members of the global incubator network should be identified and their designation appropriately celebrated. At this point, it is our hope that the world community will recognize that the purpose of the Fund is not to initiate and finance stand-alone programs but to recognize and celebrate those that meet its standards. The Fund's second purpose will be to promote collaborative learning and dissemination of successful innovations that translate a stewardship ethic into an operational reality.

*Within ten years,* we expect the Fund will be fully operational as a well-recognized global network of incubator programs demonstrating sound stewardship practices in a range of settings. The good practices will be well documented, as will the outcomes of behavioral change and improved ecosystem condition. These achievements will be widely disseminated and will be influencing the investment practices of many institutions involved in development and conservation. The result will be greater efficiency and effectiveness in coastal ecosystem management worldwide.

## Expected Long-term Benefits Versus the Cost of No Action

We expect that *within ten years,* the outcomes in areas of the Fund's incubator programs will be:

- A well-documented and widely disseminated body of knowledge regarding successful application of stewardship principles applied in a range of contexts will be inspiring hope and catalyzing new initiatives
- Reduced fishing pressure and healthy fishery "nurseries" will have restored depleted fish and shellfish populations
- Major reductions in nutrient inflows from land-based activities will have slowed the over-fertilization of coastal waters and prevented the worldwide expansion of dead zones (Figure 7)
- Networks of marine protected areas will be flourishing as elements of integrated coastal management programs

- Nested ecosystem management regimes will be adapting to major threats like global change and to lessons from collaborative learning
- Coastal communities will be modifying the values and behaviors that lead to coastal and marine degradation and actively supporting a coastal stewardship ethic

The Global Fund for Coastal Stewardship will be a major counterpoint to the current pattern of investments in coastal governance. Governments and international development banks are focusing their investments on areas where coastal ecosystems have deteriorated so severely that restoration is seen as a necessary response to a crisis and thus economically justifiable. Restoration was the motivation for the hundreds of millions of dollars invested in the Great Lakes and Chesapeake Bay programs in the United States. In Europe, the Wadden Sea, the Baltic Sea, the Black Sea and the Mediterranean have all crossed a threshold where restoration is seen as a social and environmental necessity. Elsewhere, large sums are being poured into restoration and management strategies for scattered lagoons and bays in low-income nations that have become recognized as marine and estuarine "ecological disasters."

In contrast, the benefits of the proposed network of incubator programs for coastal ecosystem governance at the large coastal ecosystem scale are:

- Taking preemptive action in a variety of coastal ecosystems, in both high- and low-income settings, before there is a need for major restoration
- Creating conditions for sustained action required to achieve the desired environmental and social conditions
- Creating a knowledge management system that can inform and inspire adaptation both within the network and elsewhere
- Supporting functional incubator programs with similar goals, structures and processes across a range of contexts and disseminating knowledge on the practices of efficient and effective stewardship

Without the inspired leadership of a Global Coastal Stewardship Fund, the path to a desirable coastal future will remain unknown or opaque. As a result, overfishing will continue, critical coastal habitats will degrade and disappear and the eutrophication of estuaries and coastal waters with their attendant dead zones will proliferate. The majority of the marine protected areas that are established will

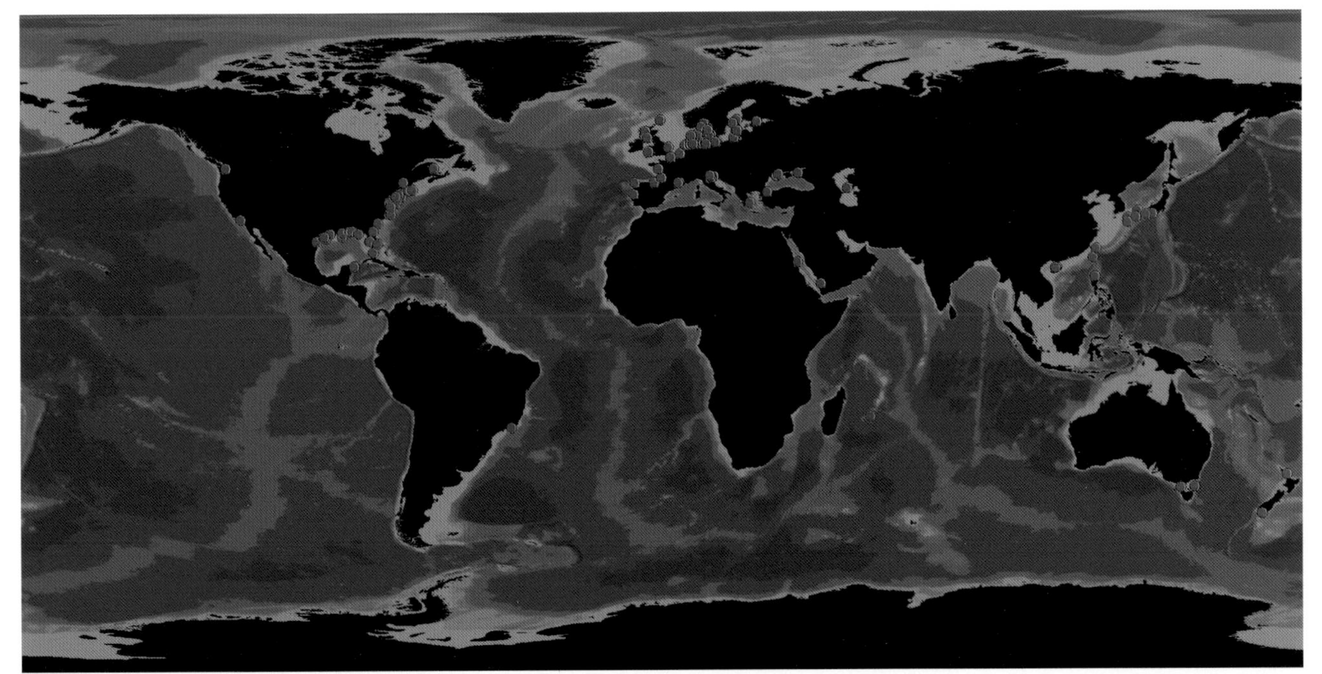

**Figure 7** "Dead zones," created largely by agricultural run-off, are now found in many areas of the world's coastal waters. Existence and areal extent of the zones vary greatly over time, so only generalized locations are shown (from Diaz & Rosenberg 1995).

be overwhelmed by external forces and will fail. Massive losses in the goods and services provided by healthy coastal ecosystems will accumulate, eroding the prospects for an acceptable quality of life for large portions of the planet's population. Efforts to address the complex forces driving the degradation and misuse of coastal and marine systems and their associated productivity will continue to evolve as unconnected initiatives, primarily in a scattering of high-income nations. As a result there will continue to be no accessible body of knowledge and experience on how the principles of equity and stewardship can be made operational and yield desired social and environmental outcomes. And there will be no well-documented and widely disseminated alternative strategy to counter the current pattern of large, but short-term, investments in responses to disasters. If we do not implement an approach like that proposed for the Global Fund for Coastal Stewardship, far greater investments will be required to restore degraded coastal ecosystems and mediate the effects of this degradation on the primary human habitat of this planet.

## Acknowledgements

*The authors gratefully acknowledge the stimulating conversations and the many contributions made by all participants during the sessions on this topic at the* Defying Ocean's End *Conference.*

## Literature Cited & Consulted

Broecker, W. S. 1995. Chaotic Climate: Global temperatures have been known to change substantially in only a decade or two. Could another jump be in the offing? *Scientific American* 273 (5): 62–68.

Cohen, J. E., C. Small, A. Mellinger, J. Gallup and J. Sachs. 1997. Letter: Estimates of coastal populations. *Science* 278:1209–1213.

Commission on Marine Science, Engineering and Resources (The Stratton Commission). 1969. *Our Nation and the Sea: A Plan for National Action.* Washington, D.C.: U.S. Government Printing Office.

Costa-Pierce, B. A., ed. 2002. *Ecological Aquaculture: The Evolution of the Blue Revolution.* Oxford: Blackwell Science.

Crutzen, P. J. and E. F. Stoermer. 2000. The "Anthropocene." *Global Change Newsletter* 41:12–13.

Diaz, R. J. and R. Rosenberg. 1995. Marine benthic hypoxia: A review of its ecological effects and the behavioral responses of benthic macrofauna. *Oceanography and Marine Biology: An Annual Review* 33:245–303.

Earl, S., F. Carden and T. Smutylo. 2001. *Outcome mapping: Building learning and reflection into development programs.* Ottawa, ON, Canada: International Development Research Centre.

Hershman, M. J., J. W. Good, T. Bernd-Cohen, R. F. Goodwin, V. Lee and P. Pogue. 1999. The effectiveness of coastal zone management in the United States. *Coastal Management* 27 (2): 113–38.

International Geosphere Biosphere Program. 2001. Global change and the Earth system: A planet under pressure. IGBP Science Series #4.

Kates, R. W., W. C. Clark, R. Corell, J. M. Hall, C. C. Jaeger, I. Lowe, J. J. McCarthy, H. J. Schellnhuber, B. Bolin, N. M. Dickson, S. Faucheux, G. C. Gallopin, A. Gruebler, B. Huntley, J. Jager, N. S. Jodha, R. E. Kasperson, A. Mabogunje, P. Matson, H. Mooney, B. Moore III, T. O'Riordan and U. Svendin. 2001. Environment and development: Sustainability science. *Science* 292 (5517): 641–642.

Lubchenco, J. 1994. The scientific basis of ecosystem management: framing the context, language and goals. In *Ecosystem Management: Status and Potential,* ed. J. Zinn and M. L. Corn, 33–39. 103 Congress, 2nd session, Committee Print. Washington D.C.: U.S. Government Printing Office.

Malakoff, D. 1998. Death by suffocation in the Gulf of Mexico. *Science* 281:190–92.

Mee, L. D. 1992. The Black Sea in crisis: A need for concerted international action. *Ambio* 21 (4): 278–287.

Nixon, S. W., S. Granger and B. Buckley. 2004. *The warning of Narragansett Bay.* 41 (12): 1 *http://seagrant.gso.uri.edu* (accessed August 22, 2004).

Olsen, S. B. 2003. Frameworks and indicators for assessing progress in integrated coastal management initiatives. *Ocean and Coastal Management* 46 (3–4): 347–361.

Olsen, S. B. and P. Christie. 2000. What are we learning from tropical coastal management experiences? *Coastal Management* 28:5–18.

Olsen, S. B. and D. Nickerson. 2003. *The governance ecosystems at the regional scale: An analysis of the strategies and outcomes of long-term programs.* Narragansett, R.I., USA: Coastal Resources Center.

Olsen, S. B., J. Tobey and L. Hale. 1998. A Learning-Based Approach to Coastal Management. *Ambio* 27 (8): 611–619.

Pauly, D., V. Christensen, J. Dalsgaard, R. Froese and F. Torres, Jr. 1998. Fishing down marine food webs. *Science* 279:860–863.

Peterson, B. J., R. M. Holmes, J. W. McClelland, C. J. Vorosmarty, R. B. Lammers, A. I. Shiklomanov, I. A. Shiklomanov and S. Rahmstorf. 2002. Increasing river discharge to the Arctic Ocean. *Science* 298:2171–2173.

Raloff, J. 2004. Dead waters. *Science Weekly* 165 (23): 360.

SeaWiFS Project. 2004. *Nimbus-7 Coastal Zone Color Scanner Data: GLOBAL–18 kilometer, Sea-viewing Wide Field-of-View Sensor (SeaWiFS) Project, NASA/Goddard Space Flight Center. http://seawifs.gsfc.nasa.gov/SEAWIFS/CZCS-DATA/global_full.html* (accessed October 12, 2004).

Seitzinger, S. P., C. Kroeze, A. F. Bouwman, N. Caraco, F. Dentener and R. V. Styles. 2002. Global patterns of dissolved inorganic

and particulate nitrogen inputs to coastal systems: Recent conditions and future projections. *Estuaries* 25:640–655.
Sherman, K. and A. M. Duda. 1999. An ecosystem approach to global assessment and management of coastal waters. *Marine Ecology Progress Series* 190: 271–287.
Sorenson, J. 2000. Baseline 2000. Background paper for Coastal Zone Canada 2000: *Coastal stewardship—lessons learned and the paths ahead.* September 17–22, 2000, New Brunswick, Canada, *http://www.sybertooth.ca/czczcc2000/* (accessed August 20, 2004).
Turner, E. and N. Rabalais. 1991. Changes in Mississippi River water quality this century. *Bioscience* 41:140–147.
Wilson, E. O. 1998. *Consilience: The Unity of Knowledge.* New York: Vintage Books.
Zaitsev, Y. P. 1992. Recent changes in the trophic structure of the Black Sea. *Fisheries Oceanography* (2) 180–189.

CHAPTER 9

# *Restoring and Maintaining Marine Ecosystem Function*

Les Kaufman, *Boston University*
Jeremy B. C. Jackson, *Scripps Institute of Oceanography*
Enric Sala, *University of California, San Diego*
Penny Chisolm, *Massachusetts Institute of Technology*
Edgardo D. Gomez, *University of the Philippines*
Charles Peterson, *University of North Carolina*
Rodney V. Salm, *The Nature Conservancy*
Ghislaine Llewellyn, *Ecosafe Consultants*

MATTHEW POTENSKI, MDP PRODUCTIONS

## Introduction

There is ample evidence that the world's marine ecosystems have degraded badly in the past few centuries, with accelerated declines since the Industrial Revolution (Kaufman & Dayton 1997, Pauly *et al* 1998, Auster 1998, Palmer *et al* 2000, Agardy 2000, Jackson 2001, Jackson *et al* 2001, Myers & Worm 2003). Perhaps the most striking bellweather for these changes is a very recent, precipitous decline in the health of coral reefs and coastal environments, particularly in the northwestern Atlantic (Hughes *et al* 1985, Kaufman 1993, Hughes 1994, Jackson 1997, Aronson *et al* 2002, Belliveau & Paul 2002, Boesch 2002, Fox *et al* 2003). The ecological and social processes involved in this great ecological unraveling are not simple. Our defiance of the global decline in ocean health must be based upon a firm understanding of these complex processes. The simplest and most reliable policy for ocean sustainability—though admittedly not the highest in economic yield from exploitable resources—would be to back off rapidly from all offending human impacts and activities, adopt a position of extreme precaution and curtail anthropogenic impacts to the point where they can be offset by natural rates of regeneration (Johannes 1998, Hilborn, Punt & Orensanz 2004). Social and economic forces will create even heavier pressures on ocean resources in the coming years, however. It is theoretically possible to relax the position of extreme caution through enhanced monitoring and a more sophisticated understanding of marine populations and communities as components of complex ecosystems. This ecosystem-based approach to impact management can in principle allow for some leniency in exploitation levels while minimizing risk of collapse or fundamental alteration of a marine ecosystem (Link 2002, Butterworth & Punt 2003, Kaufman *et al* 2004, Coleman & Walters 2004, Pikitch *et al* 2004).

Because of human nature, the challenges faced in marine environmental health are very similar and closely related to those in public health (Palmer *et al* 2004). Prevention and precaution are unpopular, and it is unlikely that the "precautionary principle" will be applied at the level necessary for it to work as the sole approach (Gerrodette *et al* 2002). Without precautionary approaches, scientists are pressured to calculate, and even to advocate, levels of exploitation and disturbance just at the edge of what the supporting systems *might* be able to bear, as if the natural ecosystems could be managed by human artifice (Larkin 1977, Holling 1978, Holling & Meffe 1996). Inordinate faith is placed in the use of technology to restore natural systems when their limits are overshot through failures in management or enforcement. These are highly imprudent courses. Getting society to adopt approaches more appropriate to the problems of marine conservation is largely a social challenge, requiring political action, public awareness and education, legislation and enforcement. Science must provide the foundation for action and accountability in these efforts, however. Our theme group on Restoring and Maintaining Marine Ecosystem Function at the *Defying Ocean's End* Conference created a package of proposed scientific initiatives that would provide planning guidance to marine environmental decision-makers, accountability and assessment for their actions, preservation of an adequate reserve capacity in natural systems and support for restoration and maintenance of marine ecosystems. The goal of our plan is to ensure that marine ecosystems remain productive, diverse and resilient in the face of human activities, and to restore them to such condition where past stewardship efforts have failed.

## Defining the Relationship between Humanity and the Sea

Although human impacts on marine ecosystems are very diverse, cumulatively their effects can be represented on a single graph (Figure 1). The abscissa (horizontal scale) in this graph is the cumulative anthropogenic (human-induced) impact on the system. The right-hand ordinate (vertical scale) is the ratio of productivity to biomass in an ecosystem;

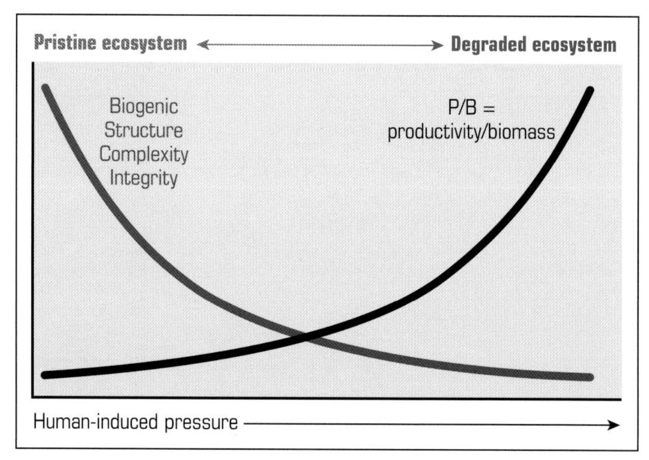


**Figure 1** As ecosystems degrade, the amount of biological productivity needed to preserve a certain amount of biomass rises (or the biomass for a given amount of biological production falls), and ecosystem complexity and integrity are undermined.

the ordinate on the left side is the biological integrity of the ecosystem and the prevalence of biogenic structures—physical structures made up of living organisms—like coral reefs. Toward the left side of the graph, the ratio of productivity to standing biomass is very low, but the levels of biomass, biogenic structure and species diversity are high. A great deal of the biomass is found in organisms that are very large or those that sequester biomass in the creation of biogenic habitat structures that define the marine communities of which they are a part—for example, corals in reefs, kelp in beds, seagrass in underwater meadows (Figure 2) and mangroves in mangrove forests. Human impacts on the ocean tend to throw systems from the left-hand to the right-hand side of the graph. Systems with low ratios of productivity to biomass (P/B) often have high value for their wilderness, conservation or ecotourism characteristics but may not be sustainable for human extraction of biomass. Our job as "clinical ecologists" is to guide systems toward a position on the graph that provides a balance between ecosystem services and ecosystem complexity and stability. One way to do this is through areal management. Most of this effort relates to diagnosing problems in the ecosystems and managing people for preventative behavior, but there is some room for science and technology to manipulate the systems toward natural regeneration and away from the unfavorable state caused by perturbation (Sutherland 1974).

There are three major domains in which human society interacts with the ocean: wilderness areas, humanized areas and urbanized landscapes. True wilderness can be defined mathematically as a region in which human influence is within "one standard deviation of the mean" of all other species in the system, meaning human influence is no greater than that of any other species. The urbanized domain is overwhelmingly dominated by human activities. The humanized domain is everything in between—essentially a buffer zone in which some level of human exploitation, agriculture and outdoor recreational activity takes place. These three domains interact dynamically. Any proposed conservation strategy can be framed in terms of the way it seeks to balance the three domains and to harmonize the manner in which they interact. The strategies must also consider that human impacts are imposed on natural systems across a hierarchy of spatial scales some of which—like global warming and changes in atmospheric chemistry—transcend the jurisdiction of any one habitat, community type or ecosystem. The most recent report of the Intergovernmental Panel on Climate Change (IPCC 2001) is regarded as a scientifically valid demonstration that global climate change and changes in atmospheric and ocean chemistry pose a major challenge for marine conservation. The future of life in the world ocean is, however, dependent on concerted regional action that can and must be taken in addition to global climate change mitigation initiatives.

GREG FOSTER

**Figure 2** Seagrass, such as this Caribbean meadow beneath a passing southern stingray, defines its marine community and is its major source of biomass.

## Initial Guiding Principles

The plan recommended by our *DOE* theme group is designed to provide sufficient scientific information to make it possible for all human activities in the world ocean to be conducted in accordance with the "precautionary principle." We endorsed the comprehensive statements of best management practices articulated in the Pew Oceans Commission report (2003) and the draft of the U.S. Commission on Ocean Policy report (2004). The past decade has witnessed an increased interest in the sea and the future of marine resources. This history of ideas and the development of key terms and concepts for marine conservation provide philosophical and scientific support for our theme group's recommendations:

### Ecosystem integrity

This is the sum of dynamic forces that allows an ecosystem to maintain its sustainability, complexity and characteristic structure in the face of disturbance. It is this character of marine ecosystems that ensures the continual provision of natural goods and services to human society. Ecological integrity is sometimes loosely referred to as ecosystem "health" but can be tied to quantitative criteria (Bremner, Frid & Rogers 2003, McGlade 2003). We endorse human behavior that nurtures ecosystem integrity.

**Shifting the burden of proof**

All human activities must be conducted in concert with the "precautionary principle," meaning in the face of uncertainty about the impacts, the parties proposing the activity must demonstrate that its impact will be minimal (Gerrodette *et al* 2002). We endorse adherence to the "precautionary principle."

**Adaptive management**

This refers to the process of "learning by doing" initially proposed by C. S. Holling (1978). In this approach, interventions designed to mitigate marine conservation problems are conducted in such a way that their consequences can be rigorously assessed and the resulting knowledge used to modify future management approaches (Holling 1978, Smith & Walters 1981, Kendall 1984, Collie & Walters 1991, Walters, Goruk & Radford 1993, Halbert 1993, Lee 1993, Imperial, Hennessey & Robadue 1993, Hennessey 1994, McLain & Lee 1996, Thom 1997, Walters 1997, Weinstein *et al* 1997, Zedler 1997, Smith *et al* 1998, Williams 1999, Castilla 2000, Dichmont, Butterworth & Cochrane 2000, Gray 2000, Walters *et al* 2000, Bearlin *et al* 2002, Clark 2002, Kaufman *et al* 2004). Adaptive management in its most rigorous (and truest) form is attained only when all conservation management activities are conducted as controlled experiments, allowing the consequences of management in one area to be distinguished from natural changes in the ecosystem in other areas. Our theme group endorsed rigorous, experimentally controlled adaptive management as the paradigm for marine conservation and restoration.

**Marine reserves**

A marine reserve, or marine protected area (MPA), is an area in the ocean where some to all harmful and extractive activities are prohibited. A reserve where all harmful activities are prohibited is called a Totally Protected Marine Reserve (TPMR). Marine reserves are an essential component of integrated ocean management (Carr & Raimondi 1999, Murray *et al* 1999, Castilla 2000). A properly designed network of functionally linked marine reserves including TPMRs within larger environmental management regimes—ecoregions, seascapes, marine corridors or large marine ecosystems—provides both a robust safety net for conservation, and the TPMRs provide undisturbed "control" areas against which the effectiveness of adaptive management experiments can be measured (Ballantine 1994, Roberts *et al* 2001, Kaufman *et al* 2004, Pikitch *et al* 2004).

**Ecosystem-based management**

This is the management of human impacts on ecosystems that takes into consideration both immediate and future effects of any conservation activities, including not only changes in the targeted species or habitat but also all those linked to it through ecological interactions (Christensen 1996, Walters, Christensen & Pauly 1997, Auster 1998, Crowder & Murawski 1998, Gentile *et al* 2001, Gunderson 2001, Olsson & Folke 2001, Bolker *et al* 2002, Day 2002, Keedwell, Maloney & Murray 2002, Link 2002, Butterworth & Punt 2003, Bundy 2004, Pikitch *et al* 2004).

We recommend a *large marine ecosystem* (LME) approach (Alexander 1993, Sherman 1994, Duda & Sherman 2002) for the creation of marine conservation plans. Within each region, attention must be given to the full range of marine environments and their dynamic interactions, including tropical nearshore, temperate nearshore, estuaries, open ocean, seamounts, other deep ocean areas and polar seas. And marine conservation approaches must be implemented at different scales, with totally protected marine reserves (TPMRs) nested within larger reserves, and these in biologically designed MPA networks or functional seascapes within large marine ecosystems.

## Four Action Items

After long deliberation, our *DOE* theme group distilled its recommendations into four action items:

- Create a global set of marine reserves
- Implement integrated ocean management
- Provide environmental conditions necessary for successful restoration of ecosystems
- Carry out interventions to accelerate ecosystem recovery

**Create more marine reserve networks worldwide**

Totally protected marine reserves—sometimes called "no-take zones"—offer a means of safeguarding areas deemed to be of great importance to the well-being of an ecosystem and the people dependent upon it. An increased number of marine reserves of all kinds, including totally protected marine reserves (TPMRs), and MPA networks of "functional seascapes" (Figure 3), are needed within each large marine ecosystem (LME). Collectively, marine reserves aid in buffering human impacts and in preserving the wilderness character of a marine ecosystem. Political realities dictate that a majority of marine reserves will arise from compromise between biological best practice and political inertia. Biological realities dictate that small marine reserves must be numerous and interconnected via dispersal "corridors" that allow natural movement among the reserves and avoid problems like in-breeding and extirpation (complete destruction) of a species due to confinement in small reserve areas. In other words, a reserve network must be designed to ensure high likelihood and short turn-around time for "rescue effects" among the reserves for every species in the system.

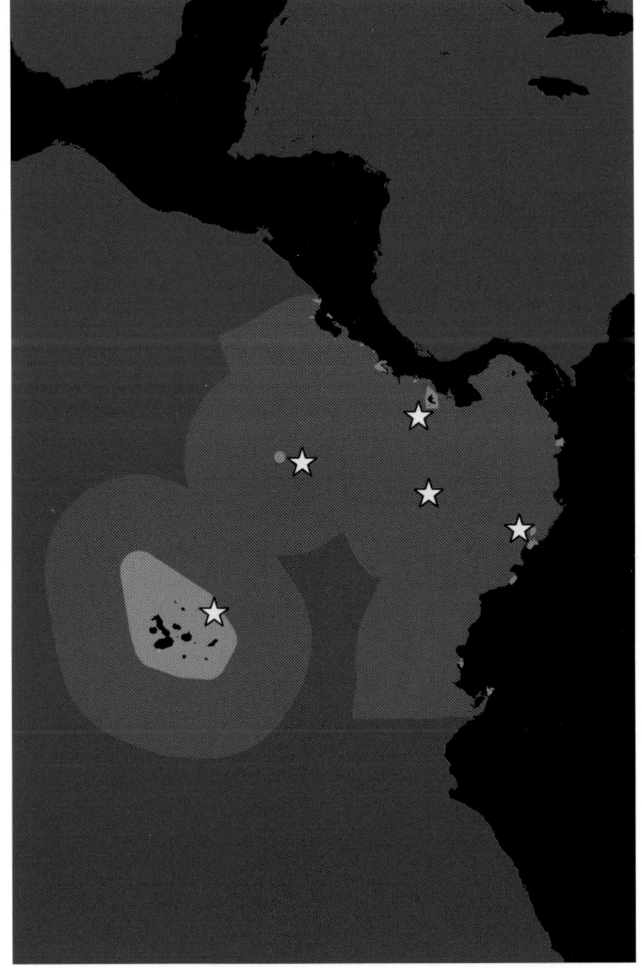

**Figure 3** In 2004, Costa Rica, Panama, Colombia and Ecuador joined with non-governmental partners to establish the Eastern Tropical Pacific Seascape (ETPS) shown in dark green. Centered around smaller marine protected areas and world heritage sites (light green), the ETPS will be a functional marine conservation corridor—or "seascape"—with a network of MPAs across the 211 million hectare expanse of ocean that falls within the four countries' EEZs (Rodriguez 2004). The gold stars denote initial management focal areas.

Seascapes, or MPA networks worldwide, should include a healthy proportion of TPMRs and other kinds of marine reserves to provide each LME with a foundation of functional wilderness areas sufficient to resist fundamental changes to the entire ecosystem; continually renew areas that suffer from management errors such as overexploitation or under-enforcement; and provide "control areas" with no direct human impacts, so local anthropogenic influences can be assessed against other factors contributing to ecosystem change, such as global climate change.

Increased sophistication in the selection of marine reserves is best achieved through actual practice in adaptive management. Additional basic research is not required before we take action in the field. Several issues related to the selection will, however, benefit greatly from additional research. First, the criteria currently used to identify candidate marine reserves are sound, but the disproportionate weight given to political considerations in making the final selection can invalidate or greatly weaken the process. Once established, the efficacy of marine reserves must be vigilantly assessed and success or failure made clearly attributable to specific management actions. We are still in the earliest stages of learning how to map and quantify ecological and demographic connectivity among different reserve areas within a protected area network. And new scientific effort is also needed to improve our design of appropriate and successful schemes of human exclusion that will permit the designated wilderness areas within an LME network to be self-sustaining.

Until very recently, efforts of the world's major conservation institutions have been focused on preventing species extinctions and preserving the last remaining pristine habitat, and these are high priorities. But the marine conservation paradigm *must* change to give equal weight to the overall ecological functioning of a site, not just its levels of species diversity or endemism (the degree to which species are unique to a particular site). In other words, *the biodiversity "hotspot" approach has been over-emphasized in marine conservation.* It is important, but has been given too much weight as opposed to working toward conserving broader ecosystem functionality or "health." New analyses must consider whether a network of protected areas designed specifically to prevent extinction of individual species can also be ecologically self-sustaining for a broader range of representative community types. A TPMR has the highest conservation value of any type of marine reserve, serving both as a safe haven for biodiversity and as part of a "life support system" for surrounding TPMRs in a well-designed network. The wilderness area required to ensure the safety of both endangered species and the full suite of marine communities, must be much larger than the distribution of any one species because local habitats change in quality and range as the environment changes over time. A TPMR can be functionally intact though it contains few species, and a TPMR with high biodiversity may be able to lose some species and still remain a valuable and self-sustaining component of a network's wilderness domain. Resolving terms like "ecological integrity" and "functional intactness" into measurable parameters will be a very important aspect of marine conservation in the coming decade and is critical to implementing the global marine conservation agenda that emerged from the *DOE* effort.

The importance of understanding connectivity among TPMRs is well recognized in a general sense, but the science needed to assess it effectively is proceeding much too slowly. The chief obstacle is one of sample size. Though marine reserve effects can be very large, it takes a great deal of time, energy, infrastructure and funding to measure them with statistical confidence. The best way to learn more quickly is to have more examples to study. There may be enough researchers and hardware available to do the work, but it is being done mostly by individual researchers within the "basic research" community doing fundamental science with no immediate applications to societal problems and is not being coordinated or managed as an urgent applied research problem. Marine protected area managers should address the establishment and monitoring of marine reserve networks in the same light as the eradication of polio: a discrete mission that must be accomplished without delay.

Population structure is not even well understood for the major commercial finfish species. Fishery impacts are still being managed on the basis of untested assumptions about the real spatial and temporal scales of fish population reproduction rates, response to environmental change and other population dynamics. There has been too little attention focused on the need to match the scale of exploitation to the scale of the biological processes that generate the so-called "surplus" on which current paradigms of marine resource management are based.

Our ability to make ecological integrity or ecosystem health a useful concept is hobbled by the lack of realistic, dynamic computational models to quantify and predict the interactions among wilderness, humanized and urban domains within an LME. This science will be the foundation for "best practices" in the establishment and operation of marine reserve networks. Without it, the designation of marine reserves and their associated protected area networks will remain much more a political than a scientific exercise, and the effectiveness of marine reserve networks will be diminished by their dubious origins.

### Implement integrated ecosystem-based ocean management

By "integrated ecosystem-based ocean management" our *DOE* theme group means a scheme to manage human impacts on the ocean that brings together four essential elements, summarized here and presented in more detail below:

- A monitoring system that provides rapid and detailed information flow about the state of an ecosystem
- An experimental approach to statistically discriminate between regional human impacts, global human impacts like climate change, and changes in the system not induced by human activities
- A policy and enforcement infrastructure that can rapidly alter management practices in response to new information
- A link to the user community that supports rapid user response to changed conditions and knowledge

The importance of ocean monitoring and prediction is widely appreciated. The ocean is probed daily by a swarm of earth observing satellites (Hock 1984) and oceanographic monitoring devices, and the grid is slowly expanding and maturing. Scattered around the globe, mostly in developed nations, are determined efforts to extend the reach and resolution of ocean observing systems measuring biological as well as physical dynamics. U.S. examples include (Figure 4) GoMOOS (the Gulf of Maine Ocean Observing System), CalCOFI (California Cooperative Oceanic Fisheries Investigations) for the California coastal ocean system and CRANE (Cooperative Research and Assessment for the Nearshore) for California's narrow nearshore shelf. The global ocean

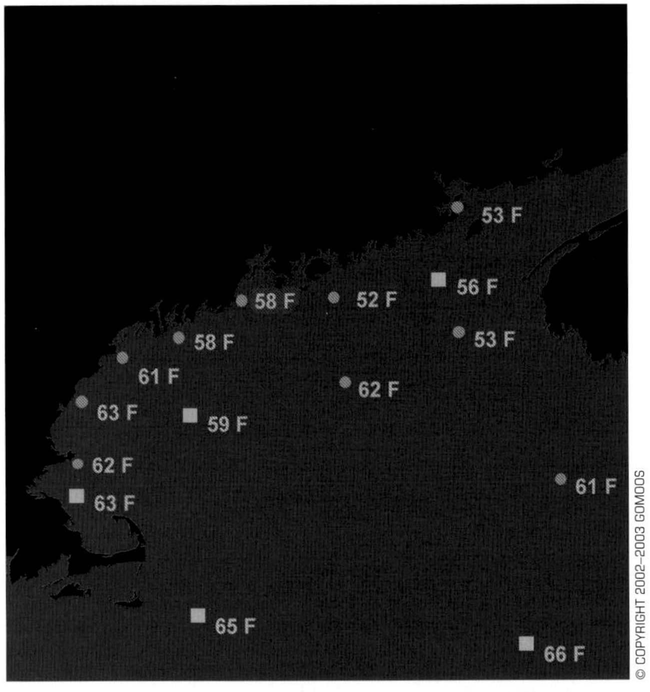


**Figure 4** The Gulf of Maine Ocean Observing System (GoMOOS) collects various types of marine oceanographic, atmospheric and biological data hourly from a stationary set of data buoys and displays the results freely on its website. Pictured here is the sea surface temperature (in °F) measured across the Gulf on September 9, 2004, at 3:00pm.

and atmosphere observation grid has saved thousands of lives and livelihoods by providing advance warning of severe weather phenomena and is now fueling a renaissance in our understanding of ocean/atmosphere and climate cycles. Oddly, little of this information has been incorporated into marine conservation. Long-term business plans for fishing fleets and limits set by fisheries managers should closely track fluctuating levels of productivity and species abundance patterns that result from large-scale climate cycles such as El Niño, the Pacific Decadal Oscillation (PDO) and the North Atlantic Oscillation (NAO). To the extent they do so now the behavior is mostly reactionary, not anticipatory, despite the fact that forecasting of such events has improved greatly.

A second and even more challenging deficit in ocean monitoring for marine conservation is the lack of a widely accepted standard set of ecosystem-based monitoring parameters to support measurement and regulation of human impacts. The need for ecosystem-based management (EBM) has finally begun to get the attention it deserves (Pikitch *et al* 2004) and resources have even been allocated to shift to this approach in the U.S. National Marine Fisheries Service. Unfortunately, almost none of these resources are allocated to the physical infrastructure and scientific activities required to make all the ongoing reorganization of personnel and departments a meaningful exercise. A segment of the research community has been addressing the biological problem of what to monitor with gusto, though with very limited resources (Goldberg 1975, Bonsdorff & Koivisto 1982, Eichbaum & Bernstein 1990, Ward & Jacoby 1992, Link 2002, Bremner, Frid & Rogers 2003, Kaufman *et al* 2004, Scientific Committee on Ocean Research 2004). Even within this group, however, attentions are divided between a perspective focused on managing commercial extraction rates, and those looking in a broader sense at the entire ecosystem and the impacts of human activities.

For the scientific needs of marine conservation to be met, we must accelerate the maturation of marine ecology as a predictive science. The creation of new institutes will be only as valuable as the useful science they produce. Funds must be directed predominantly to natural and social scientists and their science, and not to carpeting, furniture and administrators.

The use of adaptive management and the ecological equivalent of clinical experimental methods in resource management have long since become fundamental tenets in agriculture and forestry. In ocean impact management, however, the idea is being strongly opposed by powerful industry interests—even by many resource managers who fear the major change in the status quo that would accompany an end to the concept of the world ocean as a vast commons with many traditional freedoms of action allowed. A realization that this concept of an ocean commons, especially in areas such as the freedom of unregulated fishing on the high seas, has outlived its legitimacy is the central concept of *DOE*'s Ocean Governance theme group (see Chapter 11). Both the Pew Oceans Commission (2003) and the U.S. Commission on Ocean Policy (2004) strongly endorse areal management as a precept of ocean resource conservation. The transition from current management regimes to one that is truly ecosystem-based and adaptive is going to take time (Kaufman *et al* 2004, Pikitch *et al* 2004). Information is expensive, and decisions must be made about where investments in marine conservation management will rest along the cost-benefit spectrum for information richness (Coleman & Walters 2004). The relative costs and benefits of a given level of responsiveness to management failure must also be assessed. Is it more economical to invest heavily in monitoring to keep levels of exploitation right on the edge of sustainable, or is it a better investment in the long run to simply adopt an approach so precautionary that close monitoring for management purposes is unnecessary (Johannes 1998)? Or what is the optimal investment level between these two extremes?

Perhaps the weakest and most mysterious link in integrated ocean management is the role to be played by the parties whose behaviors are being managed—the immediate stakeholders (Wilson 2002). Abolishment of the global commons as a management paradigm will not relieve extractive industries from social culpability for activities that indirectly threaten others' quality of life. The only possible solution is to better integrate all of the social dimensions of society's interface with the sea. Conservation managers and scientists have long existed in separate worlds from each other, and both have been living and working in a manner very much removed from the much larger circle of humanity that is impacted by their decisions. In this information age, some of the best and most current information on status and trends in the ocean is available only to those who work daily on the sea—the fishermen and sea captains themselves. Nonetheless, these stakeholders have long been excluded from both the information-gathering and the decision-making processes. This has been changing recently, at least in the United States, Canada, Europe and Australia. The collapse of many of the world's fisheries has launched a barrage of lawsuits, and the fishermen, regulators and scientists have begun working together to avoid further litigation. Cooperative research programs, in which ocean exploiters and their vessels become an important link in the information-gathering for adaptive management, are now blossoming in New England and California. When this information is held in common trust and made transparent to all of the participants, the playing field of knowledge is leveled. When stakeholders can place greater faith in the information on which management decisions are based, compliance is more likely and enforcement may become less expensive.

Our *DOE* theme group therefore created the concept of "environmental town halls" to serve the important functions of information flow and social integration among all of the participants and beneficiaries in the ocean estate. Each LME should have at least one public facility, preferably under a single roof, to serve as this marine "environmental town hall." The town halls will be centers for promoting a comprehensive, multi-generational conservation plan and as the centers for public outreach, data management, training, research and development, restoration, policy, long-term planning and administration. Here scientists, stakeholders and decision-makers would periodically gather to observe the results of the adaptive management experiments and to produce new policy based on what has been learned. Existing institutions that sometimes offer the kinds of facilities and public exhibition capability required for a marine environmental town hall include public aquariums, some of which are already involved in aquatic conservation efforts. Another excellent model can be found among regional research and conservation consortia, such as the National Center for Caribbean Coral Reef Research in Miami, Florida.

**Provide global conditions for the successful restoration of marine ecosystems**

Fisheries and other forms of resource extraction are the human impacts most amenable to localized adaptive management. But they are not the only impacts and may not always be the most important ones. Marine ecosystems are exhibiting clear impacts from human alteration of the land masses, coastal waters, the atmosphere and the open ocean. Most marine ecosystem restoration will, naturally, be focused on nearshore waters and favored fishing grounds. While local efforts to restore marine ecosystem functionality will pay dividends, they will be much greater once human impacts from the land are brought under control. One example is the effects of poor land-use practices in a watershed leading to suspended sediment plumes and eutrophication (lack of oxygen in the water that causes most marine life to die) in coastal waters. Among the most egregious cases are the Gulf of Mexico, Lake Victoria, the Black Sea and the Baltic Sea. The contribution of watershed dysfunction to degradation in nearshore marine communities is not yet fully understood. The correlations may be affected both by complexity in the physical oceanography and hydrology of excess nutrient delivery and by variation in the capacity of reefs and other marine systems to deal with the stress created by nutrient over-loading.

Equally complex are the challenges to local or regional conservation efforts posed by changes in atmospheric chemistry. Submerged beneath the rubric of "global warming" are three marine impacts attributable to anthropogenic (human-induced) elevation of the levels of carbon dioxide in the atmosphere and ocean. First, a rise in sea surface temperatures has led to widespread coral bleaching (Figure 5).

**Figure 5** Widespread coral bleaching, as seen in the white areas of this colony from the Bahamas, is partially attributable to rising sea surface temperatures as a result of global warming.

**Figure 6** Ships such as the *M/V Tracy*, sunk in 1998 as an artificial reef off Florida's Broward County, are often used to provide new habitats for coral reef–associated marine life like this school of grunts.

Second, the rapid increase in carbon dioxide in the Earth's atmosphere—and its subsequent decrease in marine environments—has made it more difficult for marine organisms, especially corals and pelagic primary producers (open-ocean plankton that turn the sun's energy into food through photosynthesis) to create their calcium carbonate skeletons. Third, global climate change has resulted in thermal expansion of the ocean (warm water takes up more space than cold water) and a consequent rise in sealevel, which will be compounded by the effects of Earth's rapidly melting glaciers and polar icecaps. This sealevel rise further compounds the problems facing reefs and is a direct threat to coastal communities—both human and wildlife—worldwide.

Despite these obvious and formidable constraints on regional marine conservation programs, the common "think global, act local" admonishment remains a wise investment strategy for conservation science. For example, the local or regional restoration of oligotrophic conditions (the lowering of excessive nutrient levels and restoration of oxygenated waters) and high densities of grazing fishes and invertebrates (animals lacking a backbone) figure prominently among actions likely to help corals in dealing with the global threats. It thus makes great sense to mitigate local impacts as a way of combating the effects of global change. Whatever the level of effort dedicated to the root problem of elevated carbon dioxide in the atmosphere, the ability of marine communities to weather the decades or centuries required to rectify the global situation will rest on the extent to which locally manageable impacts can be reduced. This argument will be much more convincing when more is known about the cost/benefit of local restoration efforts. Meanwhile, optimal networks of TPMRs—with maximum environmental quality for coral growth within each reserve—are the best strategy for maximizing the natural regenerative and self-sustaining properties of marine communities.

### Carry out interventions to accelerate ecosystem recovery

For restoration and maintenance of ecosystem function, the rate of recovery must exceed the rate of decline. In many cases this will require that the decline be slowed *and* the recovery be accelerated artificially. Artificial enhancement of recruitment (reproduction or other introduction of new individuals), species assembly, growth and survival of key species in a marine community has already been implemented in restoration projects for seagrass beds, oyster bars, coral reefs and for populations of fishes, sea turtles and marine mammals. In the case of coral reefs, for example, there are practical and low-cost techniques for refurbishing reef frameworks with living corals and other framework-builders. Artificial structures are built or pre-existing man-made structures such as ships and oil rigs are sunk to provide new habitats for fishes and other marine life (Figure 6). These artificial reef structures are often covered with coral fragments tied or cemented into place in order to bypass the sometimes lengthy wait for corals and other reef-builders to arrive and settle naturally. Finally, conditions may be altered to enhance the growth and survival of the artificially placed corals through high-effort, low-cost, community-based methods. These may include the manual removal of predators and competitors and the addition of herbivorous (plant-eating) organisms to alter competitive interactions in a way that favors desired species. Practical restoration methods are being developed for several types of marine communities, but the field and its supporting framework of experimental ecology are still in an early stage of development (Yap & Gomez 1988, Clark & Edwards 1999, Coen & Luckenbach 2000, Hackney 2000, Yap 2000, McCook, Jompa & Diaz-Pulido 2001, Zarull & Hartig 2001, Szmant 2002). Many other kinds of intervention can be imagined to guide the process of growth and succession in marine communities that are pushed artificially toward naturally self-sustaining states (Sutherland 1974).

## A Proposed Program for the Tropical West Atlantic

Our *DOE* theme group illustrated the detailed intent of our recommendations by outlining a proposed ten-year work plan for one very large marine ecosystem (LME), the tropical west Atlantic. The region covered in our discussions encompassed both the northwestern and southwestern tropical Atlantic and extended into subtropical regions where typically tropical communities are known to exist. For comparison with discussions in Chapters 1 and 6 of this volume, our "large marine ecosystem" encompasses the Gulf of Mexico, Caribbean Sea and the north and east Brazilian Shelf, which are broken down into four separate LMEs in such schemes as Sherman, Alexander and Gold (1990). A detailed agenda for action was drafted for this region and organized according to each of the four action items discussed above, but implementation of all four is mutually supportive.

### Outcome one: Create networks of marine reserves within the region

Two objectives were recommended for the first year of effort toward creating functional networks of marine reserves in the tropical west Atlantic LME. The first is to create a composite map of ecosystem status based on existing information sources. Following this initial assessment, it is important to identify those ecosystems whose present integrity and vulnerability to human impacts justifies their immediate and complete protection from harm. These should become the first TPMRs in the system. Based on the experience of our theme group members, the following sites (Figure 7) were identified as being of the highest initial priority:

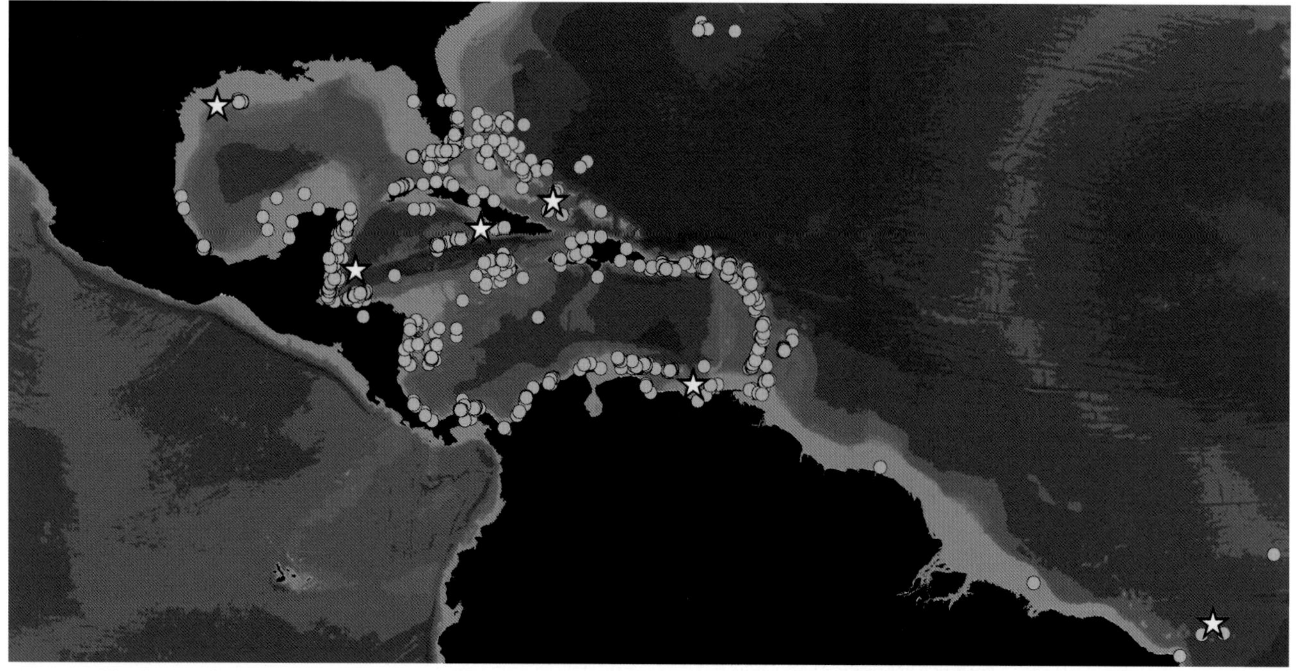

**Figure 7** Distribution of coral reefs in much of the tropical west Atlantic. Stars indicate our recommended areas for high-priority conservation action and establishment of "environmental town halls." (The high-priority Abrolhos shelf area of southeastern Brazil is not within this map.)

- Los Roques (Venezuela)
- Gardens of the Queen (Cuba)
- Turks and Caicos and other southern Bahamas areas
- Mesoamerican atolls: Glovers, Lighthouse and Turneffe (Belize)
- Brazilian Islands, especially Atoll Das Rocas and the Abrolhos shelf
- Flower Gardens (U.S. Gulf of Mexico)

All of these areas are already recognized for their biological significance, and efforts are well underway to designate portions of each of them as protected and/or wilderness areas. But these efforts are relatively modest in scope and are not knitted together into a coordinated regional plan. Our goal is to establish each of these priority areas as functional reserves *within three years,* chiefly by implementing the social and legal structures required for action and continuity, and by setting up the monitoring facilities and advisors needed to initiate adaptive management. Toward the end of the third year it will be possible to identify those areas that are ready to become part of a larger marine reserve network. *Within ten years* this first wave of marine reserve networks would be operational, and the lessons learned from this initial iteration of the process could be applied to the expansion and creation of new networks elsewhere in the tropical west Atlantic LME.

**Outcome two: Implement integrated ocean management**

*Within the first year,* a system for ocean monitoring in three of the priority areas should be established. This can be a GOOS-like (Global Ocean Observing System) approach of measuring physical and biogeochemical parameters at an appropriate scale to indicate the overall state of the ecosystem. The approach should include large-scale automated monitoring and smaller-scale, data-rich direct observations and experiments. Examples of the latter may include analysis of fish recruitment, invertebrate and algal growth on "settling" tiles placed to encourage new growth, reproductive growth and healing rates in corals, algae or seagrass, and measurements of grazing rates of fishes using video or "bite" recorders. These activities should be expanded into a standardized set of biological response parameters and systems for monitoring them.

It is already apparent that many of the impacts humans have on the ocean are completely unsustainable or should at least be sharply restricted to designated areas. Our *DOE* theme group recommended that Outcome Two activities *within the first year* target damaging forms of exploitation—such as benthic trawling (dragging trawls across the seafloor) and certain forms of gill-netting—and work toward a complete phasing-out of these activities in the designated areas *within ten years.* Such recommendations are dealt with by other *DOE* theme groups, but that policy change will require supporting science, and we recommend that some portion of the research resources be directed toward anticipatory science for policy decision-making in these areas.

*Within three years,* the results of monitoring should be passed on to conservation managers to support the first recommendations for regulatory revision. This process should be repeated annually thereafter. With initial field work underway, we also recommend *within three years* the establishment of six state-of-the-art marine research and monitoring centers to support expanded efforts in the tropical west Atlantic region in the high-priority areas listed above. These centers should be more than research facilities. They are also envisioned as centers for training and public outreach—with a central resource management library, broad-band Internet and conferencing facilities, exhibits explaining the vision for wilderness areas, humanized landscapes and urban development in the region, and exhibits on the regional history of management decisions, including the logic and justification for the current management regime. These centers should also provide facilities for storage and repair of monitoring gear and access to nearby waters for testing field methods. The term "environmental town halls" for these centers has taken hold since the *DOE* Conference.

With the environmental town halls, monitoring protocols and adaptive management in place for seven years, the efficacy of the entire enterprise should be rigorously assessed *within ten years* to determine whether environmental health, sustainability and quality of life have improved during the initial decade-long experiment.

**Outcome three: Providing optimal global conditions**

*Within three years,* existing information should be consolidated from the new environmental town halls on the sources and composition of all pollutants entering regional waters from land, atmosphere and ocean and on regional responses to global problems such as warming and elevated carbon dioxide. These town halls should join together to battle common threats to the tropical west Atlantic and launch an unprecedented and effective program for abatement of both physical/chemical and biological pollution (exotic species) throughout the entire tropical west Atlantic region. The shared information could be used to pursue prosecution of offending parties, beginning with those inside the region. It will take the better part of the first ten years to complete such a campaign, at which point the efficacy of the program should be assessed and a new ten-year plan created.

### Outcome four: Interventions to accelerate ecosystem recovery

Do we sit and wait for nature to restore herself, or do we try to support and speed up the process of recovery? Despite substantial investment in the restoration of marine communities, many projects have proceeded without having first answered the most basic questions about their purpose and likelihood of success. Our *DOE* theme group recommended that the network of environmental town halls conduct a series of restoration experiments to determine the methods and conditions under which direct intervention in habitat restoration is most useful. A few problems are common to most of the tropical west Atlantic region, like the need to speed up reestablishment of formerly extensive beds of staghorn and elkhorn coral (Figure 8).

DR. EUGENE SHINN, USGS

DR. EUGENE SHINN, USGS

**Figure 8** Since the 1960s, the tropical west Atlantic has seen dramatic losses in elkhorn and staghorn reef-building corals. A picture from Grecian Rocks in the Florida Keys taken in 1965 [above] depicts a star coral boulder surrounded by staghorn coral growth. A photograph of the same location taken in 1988 [below] shows the staghorn coral completely decimated.

There are four goals we recommend for accomplishment *within the first year:*

- Synthesize available information on tropical west Atlantic ecosystem dynamics and plan a series of experimental restoration projects
- Begin research and development targeted at innovative technologies to monitor and reduce impacts, such as changes in fishing gear, "biocapping" sediments filled with hazardous materials (introducing hardy animals and plants to grow over them), pollutant extraction and enforcement surveillance
- Evaluate previously started activities for their success in reestablishing marine communities
- Design a package of restoration tools for use through a series of carefully designed experiments

The experiments to assess methods for accelerating system recovery should begin *within three years* and a final assessment should be done at the *end of ten years.*

### Rapid incorporation of new approaches and technologies

Progress on all four actions can be accelerated by rapidly incorporating the most advanced technologies available, with particular attention to aspects of marine biology and marine ecosystem dynamics that we know are deficient in information. For instance, we know with confidence that marine biological processes—particularly in the open ocean—are all strongly driven by two things we know relatively little about: climatic forcing and microbial dynamics. A special session at *DOE* was devoted to identifying technologies needed for marine conservation (Table 1). Although much of the required technology already exists, a major effort must be dedicated to technology transfer from maritime industries and the military into marine environmental applications that are both effective and cost-effective. Microsensors and telemetry, for example, should

Table 1
**Goals for technology development**

**Within 1 year**

*Information management*
- User-friendly global databases for marine resources and environmental parameters

**Within 3 years**

*Fishing gear*
- Improve technologies to reduce bycatch and reduce the impact on the habitat
- Develop devices to facilitate the live release of bycatch
- Provide better information to fishermen on protected species and sizes

*Non-living natural resource extraction*
- Improve technologies to minimize impact of these activities
- Improve methods for geophysical exploration
- Develop methods to choose sites where there will be minimal ecosystem damage

*Monitoring the environment*
- Develop a standardized and affordable environmental measurement technology package
- Develop specialized sensors for oil spills, red tides, sedimentary loads, etc.
- Develop affordable new technologies for studying microbial genomics
- Standardize/calibrate sensors on observation platforms to support global databases

*Food technology*
- Develop techniques for making unthreatened species palatable and redirect fishing effort

**Within 10 years**

*Fishery independent surveys*
- Use autonomous and moored vehicles to measure abundance of non-commercial species
- Use chemical sensors for biomarkers
- Use genomics for analyzing population dynamics and genetics

*Monitoring organism activities*
- Invest in miniaturization of sensors
- Improve sensor reporting systems
- Expand use of satellite monitoring with sensors that see through clouds
- Focus more effort on data management and mapping

*Monitoring human activities*
- Develop enhanced Vessel Monitoring Systems to include data on fishing effort
- Develop realtime monitoring systems for catches and bycatch

*Mitigation technologies to improve habitat*
- Improve techniques for capping toxic areas
- Develop affordable techniques for oxygenating waters in dead zones
- Focus on technologies that prevent or mitigate oil spill damage quickly

*Treatment technologies*
- Develop technology to sterilize ballast water to prevent transporting of invasive species
- Improve treatment of sewage and chemical outfall materials
- Develop techniques to prevent eutrophication (severe oxygen depletion in the water that can kill plants and animals) on reefs

comprise the very backbone of marine environmental monitoring but their deployment is now severely limited by the high cost of proprietary technologies.

## Summary and Conclusions

Marine ecosystems worldwide have been severely degraded by the combined effects of pollution, land-based development, overfishing and global climate change. Marine communities tend toward the same response to these insults—shifting to a decreased abundance of large animals, reduced biological diversity and the degradation of complex seafloor habitats such as coral reefs. Associated with these losses may be a weakening of the self-restorative capacity of marine ecosystems, triggering a cycle of further decline.

Our *DOE* theme group on Restoring and Maintaining Marine Ecosystem Function has proposed an action plan to reverse this decline in the vitality and functionality of marine ecosystems and to maintain those systems that either are intact or can be restored in the seven marine domains—tropical nearshore, temperate nearshore, estuaries, polar seas, open ocean, seamounts and deep ocean. We prescribe four seminal actions, to be executed simultaneously in every large marine ecosystem worldwide: create functional seascapes or networks of marine reserves, implement integrated ocean management, provide global conditions for successful restoration of ecosystems and accelerate ecosystem recovery. It should be noted that the program outlined here is the result of deliberations that were aimed at identifying current best practices for immediate implementation. Each of the four principal actions are already in motion, at least in a nascent form, by pioneering projects around the world today.

We have presented the science required to restore and maintain ecosystem function under each of the four actions. Existing methods provide a strong basis for ecosystem mapping and assessment to support marine reserve planning. Marine reserve networks must include sufficient areas under total protection to serve as "control" areas for scientific comparison with less protected areas. Many new technologies can increase the power of marine ecological monitoring, but they must be assessed, adapted and made economical. Survey and monitoring data can guide efforts to reduce chronic anthropogenic (human-induced) impacts and highlight opportunities for natural regeneration. Direct intervention to speed restoration is appropriate under limited circumstances, but appropriate methods must be perfected. Each large marine ecosystem should have at least one central public facility to serve as a marine "environmental town hall" to promote a comprehensive, multi-generational conservation plan and support public outreach, data management, training, research and development, restoration, policy, long-term planning and administration. Here scientists, stakeholders and decision-makers can periodically gather to review the results of adaptive management experiments and develop new policy based upon lessons learned.

The maintenance and restoration of marine ecosystem function should be approached as one would a series of promising clinical trials for the treatment of a fatal disease—that is, in a fast-moving, aggressive and adaptive manner. Ecosystem-based management and experimental adaptive management for marine systems are being incorporated into standard practice directives by government agencies in the United States, Australia, New Zealand, Canada, Great Britain and other developed nations that can afford it. Therein lies the good news and the bad news. We know what we have to do, we know how to do it and we know how to do it better. And the overall cost/benefit ratio for society is highly favorable. The maintenance and restoration of marine ecosystem function is ultimately a problem of insufficient resource allocation.

## Literature Cited & Consulted


Agardy, T. 2000. Effects of fisheries on marine ecosystems: A conservationist's perspective. *ICES Journal of Marine Science* 57 (3): 761–765.

Alcock, G. and L. Rickards. 2002. OceanNET: Improving accessibility to marine data and information. *Hydro International* 6 (8): 28–29.

Alexander, L. M. 1993. Large marine ecosystems: A new focus for marine resources management. *Marine Policy* 17 (3): 186–198.

Annala, J. H., K. J. Sullivan, N. W. McL. Smith, M. H. Griffiths, P. R. Todd, P. M. Mace, A. Connell, compilers. 2004. *Report from the Fishery Assessment Plenary, April 2004: Stock Assessment and Yield Estimates.* Unpublished report. New Zealand Ministry of Fisheries.

Aronson, R. B., W. F. Precht, M. A. Toscano and K. H. Koltes. 2002. The 1998 bleaching event and its aftermath on a coral reef in Belize. *Marine Biology* 141 (3): 435–447.

Auster, P. J. 1998. A conceptual model of the impacts of fishing gear on the integrity of fish habitats. *Conservation Biology* 12 (6): 1198–1203.

Ballantine, R. 1994. The practicality and benefits of a marine reserve network. In *Limiting Access to Marine Fisheries: Keeping the Focus on Conservation,* ed. K. L. Gimel, 205–223. Washington, D.C.: Center for Marine Conservation and World Wildlife Fund.

Bearlin, A. R., E. S. G. Schreiber, S. J. Nicol, A. M. Starfield and C. R. Todd. 2002. Identifying the weakest link: Simulating adaptive management of the reintroduction of a threatened fish. *Canadian Journal of Fisheries and Aquatic Sciences* 59 (11): 1709–1716.

Belliveau, S. A. and V. J. Paul. 2002. Effects of herbivory and nutrients on the early colonization of crustose coralline and fleshy algae. *Marine Ecology Progress Series* 232:105–114.

Bernstein, B. B. and J. Dorsey. 1991. Marine monitoring in heterogeneous environments. *Journal of Environmental Management* 32 (3): 227–240.

Bilyard, G. R. 1987. The value of benthic infauna in marine pollution monitoring studies. *Marine Pollution Bulletin* 18 (11): 581–585.

Boesch, D. F. 2002. Challenges and opportunities for science in reducing nutrient over-enrichment of coastal ecosystems. *Estuaries* 25 (4B): 886–900.

Bolker, B. M., C. M. St. Mary, C. W. Osenberg, R. J. Schmitt and S. J. Holbrook. 2002. Management at a different scale: Marine ornamentals and local processes. *Bulletin of Marine Science* 70 (2): 733–748.

Bonsdorff, E. and V. Koivisto. 1982. Use of the log-normal distribution of individuals among species in monitoring zoobenthos in the northern Baltic archipelago. *Marine Pollution Bulletin* 13 (9): 324–327.

Bremner, J., C., L. J. Frid and S. I. Rogers. 2003. Assessing marine ecosystem health: The long-term effects of fishing on the functional biodiversity in North Sea benthos. *Aquatic Ecosystem Health Management* 6 (2): 131–137.

Bundy, A. 2004. The ecological effects of fishing and implications for coastal management in San Miguel Bay, the Philippines. *Coastal Management* 32 (1): 25–38.

Butterworth, D. S. and A. E. Punt. 2003. The role of harvest control laws, risk and uncertainty and the precautionary approach in ecosystem-based management. In *Responsible Fisheries in the Marine Ecosystem,* ed. M. Sinclair and G. Valdimarsson, 311–319. Rome and New York: FAO and CABI Publishing.

Carr, M. H. and P. T. Raimondi. 1999. Marine protected areas as a precautionary approach to management. *Reports of California Cooperative Oceanic Fisheries Investigations* 40:71–76.

Castilla, J. C. 2000. Roles of experimental marine ecology in coastal management and conservation. *Journal of Experimental Marine Biology and Ecology* 250 (1–2): 3–21.

Charles, A. T. 1994. Towards sustainability: The fishery experience. *Ecological Economics* 11 (3): 201–211.

Christensen, V. 1996. Managing fisheries involving predator and prey species. *Reviews in Fish Biology and Fisheries* 6 (4): 417–442.

Clark, M. J. 2002. Dealing with uncertainty: Adaptive approaches to sustainable river management. *Aquatic Conservation: Marine and Freshwater Ecosystems* 12 (4): 347–363.

Clark, S. and A. J. Edwards. 1999. An evaluation of artificial reef structures as tools for marine habitat rehabilitation in the Maldives. *Aquatic Conservation: Marine and Freshwater Ecosystems* 9 (1): 5–21.

Coen, L. D. and M. W. Luckenbach. 2000. Developing success criteria and goals for evaluating oyster reef restoration: Ecological function or resource exploitation? *Ecological Engineering* 15 (3–4): 323–343.

Coleman, F. and C. Walters, eds. 2004. Proceedings of the fourth William R. and Lenore Mote international symposium in fisheries ecology: Confronting trade-offs in the ecosystem approach to fisheries management. *Bulletin of Marine Science* 74 (3): 760.

Collie, J. S. and C. J. Walters. 1991. Adaptive management of spatially replicated groundfish populations. *Canadian Journal of Fisheries and Aquatic Sciences* 48 (7): 1273–1284.

Costanza, R., F. Andrade, P. Antunes, M. Van Den Belt, D. Boesch and D. Boersma, *et al.* 1999. Ecological economics and sustainable governance of the oceans. *Ecological Economics* 31 (2): 171–187.

Crowder, L. B. and S. A. Murawski. 1998. Fisheries bycatch: Implications for management. *Fisheries* 23 (6): 8–17.

Day, J. C. 2002. Zoning—lessons from the Great Barrier Reef marine park. *Ocean & Coastal Management* 45 (2–3): 139–156.

Dichmont, C. M., D. S. Butterworth and K. L. Cochrane. 2000. Towards adaptive approaches to management of the South African abalone *Haliotis midae* fishery. *South African Journal of Marine Science/Suid-Afrikaanse Tydskrif vir Seewetenskap* 22:33–42.

Duda, A. M. and K. Sherman. 2002. A new imperative for improving management of large marine ecosystems. *Ocean & Coastal Management* 45 (11–12): 797–833.

Eichbaum, W. M. and B. B. Bernstein. 1990. Current issues in environmental management: A case study of southern California's marine monitoring system. *Coastal Management* 18 (4): 433–445.

Ewel, K. C., C. Cressa, R. T. Kneib, P. S. Lake, L. A. Levin and M. A. Palmer, *et al.* 2001. Managing critical transition zones. *Ecosystems* 4 (5): 452–460.

Fox, H. E., J. S. Pet, R. Dahuri and R. L. Caldwell. 2003. Recovery in rubble fields: Long-term impacts of blast fishing. *Marine Pollution Bulletin* 46 (8): 1024–1031.

Gentile, J. H., M. A. Harwell, W. Cropper, Jr., C. C. Harwell, D. DeAngelis and S. Davis, *et al.* 2001. Ecological conceptual models: A framework and case study on ecosystem management for south Florida sustainability. *Science of the Total Environment* 274 (1–3): 231–253.

Gerrodette, T., P. K. Dayton, S. Mcinko and M. J. Fogarty. 2002. Precautionary management of marine fisheries: Moving beyond burden of proof. *Bulletin of Marine Science* 70 (2): 657–688.

Goldberg, E. D. 1975. The mussel watch—a first step in global marine monitoring. *Marine Pollution Bulletin* 6 (7): 111.

Gray, A. N. 2000. Adaptive ecosystem management in the Pacific Northwest: A case study from coastal Oregon. *Conservation Ecology* 4 (2): 6.

Gunderson, L. H. 2001. Managing surprising ecosystems in southern Florida. *Ecological Economics* 37 (3): 371–378.

Hackney, C. T. 2000. Restoration of coastal habitats: Expectation and reality. *Ecological Engineering* 15 (3–4): 165–170.

Halbert, C. L. 1993. How adaptive is adaptive management? Implementing adaptive management in Washington State and British Columbia. *Reviews in Fisheries Science* 1:261–283.

Hennessey, T. M. 1994. Governance and adaptive management for estuarine ecosystems: The case of Chesapeake Bay. *Coastal Management* 22 (2): 119–145.

Heyward, A. J., L. D. Smith, M. Rees and S. N. Field. 2002. Enhancement of coral recruitment by in situ mass culture of coral larvae. *Marine Ecology Progress Series* 230:113–118.

Hilborn, R. and C. Walters. 1981. Pitfalls of environmental baseline and process studies. *Environmental Impact Assessment Review* 2:265–278.

Hilborn, R., A. E. Punt and J. Orensanz. 2004. Beyond band-aids in fisheries management: Fixing world fisheries. *Bulletin of Marine Science* 74 (3): 493–507.

Hock, J. C. 1984. Monitoring environmental resources through NOAA's polar orbiting satellites. *ITC Journal* 4:263–268.
Holling, C. S. 1978. *Adaptive Environmental Assessment and Management.* London: John Wiley and Sons.
Holling, C. S. and G. Meffe. 1996. Command and control and the pathology of natural resource management. *Conservation Biology* 10:328–337.
Houghton, J. T., Y. Ding, D. J. Griggs, M. Noguer, P. J. van der Linden, X. Dai, K. Maskell and C.A. Johnson, eds. 2001. Climate change 2001: The scientific basis. Contribution of Working Group I to the *Third Assessment Report of the Intergovernmental Panel on Climate Change.* Cambridge, UK and New York, USA: Cambridge University Press. *http://www.grida.no/climate/ipcc_tar/wg1/index.htm* (accessed September 4, 2004).
Hughes, T. P. 1994. Catastrophes, phase shifts, and large-scale degradation of a Caribbean coral reef. *Science* 265:1547–1551.
Hughes, T. P., B. D. Keller, J. B. C. Jackson and M. J. Boyle. 1985. Mass mortality of the echinoid *Diadema antillarum philippi* in Jamaica. *Bulletin of Marine Science* 36 (2): 377–384.
Imperial, M. T., T. Hennessey and D. Robadue, Jr. 1993. The evolution of adaptive management for estuarine ecosystems: The national estuary program and its precursors. *Ocean & Coastal Management* 20 (2): 147–180.
IPCC. 2001. *Climate Change 2001: Synthesis Report. A Contribution of Working Groups I, II, and III to the Third Assessment Report of the Intergovernmental Panel on Climate Change,* ed. R. T. Watson and the Core Writing Team. Cambridge, UK and New York, NY: Cambridge University Press. *http://www.grida.no/climate/ipcc_tar/vol4/english/index.htm* (accessed October 10, 2004)
Jackson, J. B. C. 1997. Reefs since Columbus. *Coral Reefs* 16 (Suppl): S23–S32.
——. 2001. What was natural in the coastal oceans? *Proceedings of the National Academy of Sciences* 98:5411–5418.
Jackson, J. B. C., M. X. Kirby, W. H. Berger, K. A. Bjorndal, L. W. Botsford, B. J. Bourque, R. H. Bradbury, R. Cooke, J. Erlandson, J. A. Estes, T. P. Hughes, S. Kidwell, C. B. Lange, H. S. Henihan, J. M. Pandolfi, C. H. Peterson, R. S. Steneck, M. J. Tegner and R. R. Warner. 2001. Historical overfishing and the recent collapse of coastal ecosystems. *Science* 293:629–638.
Johannes, R. E. 1998. The case for data-less marine resource management: Examples from tropical nearshore finfisheries. *Trends in Ecology & Evolution* 13:243–246.
Kaufman, L. S. 1993. Why the ark is sinking. Chapter 1 in *The Last Extinction* (2nd ed.). Cambridge, MA: MIT Press.
Kaufman, L. S. and P. J. Dayton. 1997. Impacts of marine resource extraction on ecosystem services and sustainability. In *Nature's Services,* ed. G. Daily, 275–293. Washington, D.C.: Island Press.
Kaufman, L. S., B. Heneman, J. T. Barnes and R. Fujita. 2004. Transition from low to high data richness: An experiment in ecosystem-based fishery management from California. *Bulletin of Marine Science* 74 (3): 693–708.
Keedwell, R. J., R. F. Maloney and D. P. Murray. 2002. Predator control for protecting kaki *(Himantopus novaezelandiae)*—lessons from 20 years of management. *Biological Conservation* 105 (3): 369–374.
Kendall, S. 1984. Fisheries management planning: An operational approach to adaptive management of a renewable resource for multiple objectives using qualitative information. *Canadian Journal of Regional Science* 7 (2): 153–180.
Kershner, J. L. 1997. Monitoring and adaptive management. In *Watershed Restoration: Principles & Practices,* ed. J. E. Williams, *et al,* 116–131. Bethesda, MD: American Fisheries Society.
Larkin, P. 1977. An epitaph for the concept of maximum sustainable yield. *Transactions of the American Fishery Society* 106:1–11.
Lee, K.N. 1993. *Compass and gyroscope: Integrating science and politics for the environment.* Washington, D.C.: Island Press.
Link, J. S. 2002. What does ecosystem-based fisheries management mean? *Fisheries* 27 (4): 18–21.
McConnaha, W. E. and P. J. Paquet. 1996. Adaptive strategies for the management of ecosystems: The Columbia River experience. In *Multidimensional Approaches to Reservoir Fisheries Management,* ed. L. E. Miranda and D. R. DeVries, 410–421. Bethesda, MD: American Fisheries Society.
McCook, L. J., J. Jompa and G. Diaz-Pulido. 2001. Competition between corals and algae on coral reefs: A review of evidence and mechanisms. *Coral Reefs* 19 (4): 400–417.
McGlade, J. M. 2003. A diversity-based fuzzy systems approach to ecosystem health assessment. *Aquatic Ecosystem Health Management* 6 (2): 205–216.
McLain, R. J. and R. G. Lee. 1996. Adaptive management: Promises and pitfalls. *Environmental Management* 20 (4): 437–448.
Murray, S. N., R. F. Ambrose, J. A. Bohnsack, L. W. Botsford, M. H. Carr and G. E. Davis. 1999. No-take reserve networks: Sustaining fishery populations and marine ecosystems. *Fisheries* 24 (11): 11–25.
Musick, J. A., S. A. Berkeley, G. M. Cailliet, M. Camhi, G. Huntsman and M. Nammack. 2000. Protection of marine fish stocks at risk of extinction. *Fisheries* 25 (3): 6–8.
Myers, R. A. and B. Worm. 2003. Rapid worldwide depletion of predatory fish communities. *Nature* 423:280–283.
NCORE. 2004. National Center for Caribbean Coral Reef Research. *http://www.ncoremiami.org/* (accessed October 10, 2004 ).
Olsson, P. and C. Folke. 2001. Local ecological knowledge and institutional dynamics for ecosystem management: A study of Lake Racken watershed, Sweden. *Ecosystems* 4 (2): 85–104.
Pajak, P. 2000. Sustainability, ecosystem management, and indicators: Thinking globally and acting locally in the 21st century. *Fisheries* 25 (12): 16–30.
Palmer, M., E. Bernhardt, E. Chornesky, S. Collins, A. Dobson, C. Duke, B. Gold, R. Jacobson, S. Kingsland, R. Kranz, M. Mappin, M. L. Martinez, F. Micheli, J. Morse, M. Pace, M. Pascual, S. Palumbi, O. J. Reichman, A. Simons, A. Townsend and M. Turner. 2004. Ecology for a crowded planet. *Science* 304 (5675): 1251–1252.
Pauly, D., V. Christensen, J. Dalsgaard, R. Froese and F. Torres. 1998. Fishing down marine food webs. *Science* 279:860–863.

Pew Oceans Commission. 2003. *America's Living Oceans: Charting a Course for Sea Change*. May 2003. Arlington, Virginia: Pew Oceans Commission. *http://www.pewoceans.org/oceans/downloads/oceans_summary.pdf* (accessed September 4, 2004).

Pikitch, E. K., C. Santora, E. A. Babcock, A. Bakun, R. Bonfil, D. O. Conover, P. Dayton, P. Doukakis, D. Fluharty, B. Heneman, E. D. Houde, J. Link, P. A. Livingston, M. Mangel, M. K. McAllister, J. Pope and K. J. Sainsbury. 2004. Ecosystem-based fishery management. *Science* 305 (5682): 346–347.

Rinne, J. N. 1999. Fish and grazing relationships: The facts and some pleas. *Fisheries* 24 (8): 12–21.

Roberts, C. M., B. Halpern, S. R. Palumbi and R. R. Warner. 2001. Designing marine reserve networks. *Conservation Biology in Practice* 2 (3): 11–17.

Rodriguez, C. M. 2004. An ocean corridor. In *Our Planet,* the Magazine of the United Nations Environment Program 15 (1): 15.

Scientific Committee on Ocean Research-Intergovernmental Oceanographic Commission Working Group 119. 2004. *Home page for SCOR-IOC Working Group 119, Quantitative Ecosystem Indicators for Fisheries Management. http://www.ecosystemindicators.org/wg/wg.htm* (accessed October 10, 2004).

Sherman, K. 1994. Sustainability, biomass yields, and health of coastal ecosystems: An ecological perspective. *Marine Ecology Progress Series* 112 (3): 277–301.

Sherman, K., L. M. Alexander and B. D. Gold, eds. 1990. *Large Marine Ecosystems: Patterns, Processes and Yields.* Washington. D.C.: American Association for the Advancement of Science.

Smith, A. D. M. and C. J. Walters. 1981. Adaptive management of stock-recruitment systems. *Canadian Journal of Fisheries and Aquatic Sciences* 38 (6): 690–703.

Smith, C. L., J. Gilden, B. S. Steel and K. Mrakovcich. 1998. Sailing the shoals of adaptive management: the case of salmon in the Pacific northwest. *Environmental Management* 22 (5): 671–681.

Sutherland, J. 1974. Multiple stable points in natural communities. *American Naturalist* 108:859–873.

Szmant, A. M. 2002. Nutrient enrichment on coral reefs: Is it a major cause of coral reef decline? *Estuaries* 25 (4B): 743–766.

Thom, R. M. 1997. System-development matrix for adaptive management of coastal ecosystem restoration projects. *Ecological Engineering* 8 (3): 219–232.

U.S. Commission on Ocean Policy. 2004. *Preliminary Report of the U.S. Commission on Ocean Policy: Governor's Draft. http://www.oceancommission.gov* (accessed September 4, 2004).

Walters, C. 1997. Challenges in adaptive management of riparian and coastal ecosystems. *Conservation Ecology* [online] 1(2): 1. *http://www.consecol.org/vol1/iss2/art1* (accessed September 4, 2004).

Walters, C., V. Christensen and D. Pauly. 1997. Structuring dynamic models of exploited ecosystems from trophic mass-balance assessments. *Reviews in Fish Biology and Fisheries* 7 (2): 139–172.

Walters, C., R. D. Goruk and D. Radford. 1993. Rivers inlet sockeye salmon: An experiment in adaptive management. *North American Journal of Fisheries Management* 13 (2): 253–262.

Walters, C., J. Korman, L. E. Stevens and B. Gold. 2000. Ecosystem modeling for evaluation of adaptive management policies in the Grand Canyon. *Conservation Ecology* 4 (2): 1.

Ward, T. J. and C. A. Jacoby. 1992. A strategy for assessment and management of marine ecosystems: Baseline and monitoring studies in Jervis Bay, a temperate Australian embayment. *Marine Pollution Bulletin* 25 (5–8): 163–171.

Weinstein, M. P., J. H. Balletto, J. M. Teal and D. F. Ludwig. 1997. Success criteria and adaptive management for a large-scale wetland restoration project. *Wetlands Ecology and Management* 4 (2): 111–127.

Williams, J. G. 1999. Stock dynamics and adaptive management of habitat: An evaluation based on simulations. *North American Journal of Fisheries Management* 19 (2): 329–341.

Wilson, J. 2002. Scientific uncertainty, complex systems and the design of common pool institutions, in the drama of the commons. In *Report of the Committee on Human Dimensions of Global Climate Change,* ed. P. Stern, E. Ostrom, T. Dietz and N. Dolsak. Washington D.C.: National Research Council.

Yap, H. T. 2000. The case for restoration of tropical coastal ecosystems. *Ocean & Coastal Management* 43 (8–9): 841–851.

Yap, H. and E. Gomez. 1988. Aspects of benthic recruitment on a northern Philippine reef. In *Proceedings of the Sixth International Coral Reef Symposium, Townsville, Australia,* ed. P. Choat *et al.,* 279–283. Townsville, Australia: 6th International Coral Reef Symposium Executive Committee.

Zarull, M. A. and J. H. Hartig. 2001. Aquatic ecosystem rehabilitation: Targets, actions, responses. *Aquatic Ecosystem Health & Management* 4 (1): 115–122.

Zedler, J. B. 1997. Adaptive management of coastal ecosystems designed to support endangered species. *Ecology Law Quarterly* 24 (4): 735–743.

CHAPTER 10

# *Defying Ocean's End through the Power of Communications*

Robin Abadia, *Conservation International*
Brian Day, *International Institute for Environmental Communication*
Nancy Knowlton, *Scripps Institution of Oceanography*
John E. McCosker, *California Academy of Sciences*
Nancy Baron, *SeaWeb*
Aristides Katoppo, *Sinar Harapan Daily*
Hugh Hough, *Green Team Advertising, Inc.*

ROGER STEENE

## Overview

To accomplish any portion of the *Defying Ocean's End (DOE)* global marine conservation strategy, we must first understand that the problems to be addressed are caused by human behavior. The goal of this chapter from the *DOE* Communications theme group is to discuss the types of communications efforts necessary to bring about changes in these behaviors. We lay out some of the tools, ideas and approaches required to address the issues developed in more detail elsewhere in this volume, and we review the state of public awareness and understanding based on survey research. Each of the efforts described in other chapters will require development of its own tailored communications strategy to be successful. We in no way suggest that communications alone can accomplish the tasks, but we conclude they will be a critical element if we are to defy ocean's end.

## Introduction

Conservation begins with people—people cause the problems and people can solve them. Therefore, the goal of communications has to be to get beyond awareness and address behavior. For example, consumers drive demand for fisheries but can also push for making these food sources sustainable. Constituencies drive policymakers to develop laws and regulations to protect ocean ecosystems. Journalists can inspire public debate, create a sense of urgency around certain issues and inform the public about the issues in marine conservation. Social marketers can mobilize community support for new marine protected areas (MPAs), and educators can help fishermen adopt non-destructive fishing methods (Figure 1). These are among the many ways in which the power of communications may be harnessed to bring about change and help defy the practices and policies that are threatening an end to ocean life as we know it.

ROGER STEENE

**Figure 1** It is important to mobilize community support for new marine protected areas or the adoption of non-destructive fishing methods in places such as here in Milne Bay, Papua New Guinea.

An assessment of existing public opinion polling data on ocean issues (which is scarce) reveals that people know very little and are often misinformed about ocean issues; however, they care about the health of the ocean and express positive attitudes toward taking action to protect it. Many people around the world worry about the ocean's vulnerability in general but are unaware of the current degree of threat. Communicators must take advantage of the existing positive attitude, build awareness of the threats and instill a sense of urgency in target audiences before further widespread negative changes can occur. Each issue will require specific communications tools to address the behavior changes needed to bring about the necessary conservation goals.

## Communications Tools

Using communications to change behaviors and conservation outcomes is the goal. It is critical to identify tools and match them with desired outcomes. To accomplish the enormous array of tasks that must be completed to bring about the desired conservation outcomes, a number of tools must be considered. If we are going to change local or regional policy, then communications directed toward increasing the political will to act is critical. However, if you are trying to change general awareness of deterioration in fish populations, with the aim of getting consumers to change consumption patterns, then a whole different set of skills and communications tools are required. Each communications objective will need its own highly tuned strategy; a few generally applicable tools are discussed below.

### A World Advertising Council

There has been a non-profit organization in the U.S. since 1942 called the Advertising Council, better known as the Ad Council. During World War II, the Ad Council was formed by putting three very powerful interests together: the media, the advertising agencies and the largest advertisers. The Ad Council has become a powerhouse for public service advertising, a concept common in the United States and other developed countries, but random or non-existent in the majority of the world. In the 60-plus years since the formation of the U.S. Ad Council, the talent, the media and the corporations that do the advertising have all become multinational. Our strategy is to develop the first campaign for a World Ad Council by raising the attention of the world's populations that directly impact the ocean. This would be a global agenda-setting tool that would raise the saliency of ocean issues and set the stage for other communications tools to bring about specific changes in behaviors. It could be the key to creating a global ocean ethic. Because ocean issues are, by definition, in the public interest and global, this initial campaign would meet the needs of protecting the world's ocean and could open the world's media markets to public service advertising. This could rally a broader set of interests behind the effort. Creating a World Ad Council would take enormous effort but would reach global audiences at a tiny fraction of the commercial cost.

**Communicating with governments and donors**
**Example: A global fisheries reform fund**

Getting issue-specific media coverage can enhance the likelihood of attracting new donors to fund required efforts. Creating a set of news and feature articles on the need for fisheries management reform in major media outlets can help with the agenda and priority setting of funders who always have too many demands and not enough funding to address them. The actual application for funds is a fundraising/proposal writing process, but positioning the issue for high visibility is very much a communications task. Given that the need for fisheries reform has not been a major media topic, a significant effort to position the need for a reform fund would be required to raise the level of funds required to have significant impact. A media effort on the impact of reform would have to feature the long-term positive impacts for fisherfolk, communities and countries dependent on fishing, as well as the conservation message.

**A communications model for policy change**
**Example: Tanzania campaign**

In the East African coastal nation of Tanzania, there was a need to create a new coastal policy based on the results of many donors working with local people. A proposed model with three critical elements was designed to leverage political will. The first environmental communications element used was a "video newsletter" for policymakers about coastal issues to generate knowledge and a willingness to act. The second element was designed to create a visible constituency for the policy by mobilizing local people through a decentralized coastal awards scheme. The third was to motivate policymakers to act by helping reporters see the political dimension of the problems and the immediate need to address them. This was initiated by a workshop to make the press coverage of these issues easier. Collectively, these three steps formed a new model for the use of environmental communications for enacting policy change. By utilizing a set of environmental communications tools to educate policymakers, build constituency and enable reporters to cover complex environmental issues, the behavior of policymakers can be changed to implement a new policy.

**Public service advertising (PSA) campaign**
**Example: "What fish to eat?"**

In developed countries where public service advertising is already present, a PSA campaign can be critical in helping consumers realize that their seafood choices can make a world of difference to biodiversity, conservation on seamounts and sustaining fisheries. Because PSAs are by nature short, the goal must be to grab attention and then drive people to a website or an 800 number to get more information—like a "guide to fish eating" card.

**Generating outrage**
**Example: Understanding bottom trawling**

Few commercial activities in the world are as appalling in their impact as bottom-trawl fishing. It has been compared to harvesting vegetables with a bulldozer or clear-cutting the rainforest. In almost no place other than the ocean would such an activity be condoned, but it is invisible to the public because all of the resultant damage is under water (Figure 2). Harvesting tuna "on top of dolphins" was made public with a video of the real activities. A similar campaign will be required if we are to put outrage into those needed to change policy. It is in reality a very complicated issue, with governance, practice, equipment, financial and regulatory/enforcement issues. However, a well-designed communications campaign can reduce that complexity for the public into a reaction of simple outrage, which can provide the momentum to drive the other necessary changes.

## Types of Campaigns Necessary to Support the *Defying Ocean's End* Efforts

**Southern Ocean efforts**

*DOE* recommended that the management and governance structure of the Convention on Conservation of Antarctic

© 2004 GREENPEACE/ROGER GRACE

**Figure 2** Bottom trawling, seen here in the Tasman Sea, is very destructive to the underwater habitat, including ancient deepwater corals.

Marine Living Resources (CCAMLR) be more effective in addressing three specific fisheries: krill, Patagonian toothfish and icefish. Trying to rally the northern hemisphere nations around these issues in remote Antarctic areas would be expensive and likely ineffective. Segmenting the audience and addressing primarily the policymakers behind CCAMLR is a much more focused and effective approach. The campaign should identify the needs of the countries represented at the table, determine their specific incentives to change and then aim targeted communications campaign efforts at bringing about those changes. Policymakers in one country may need only some specific media attention in order to change their position, while in other countries significant changes in the position of their entire fishing industry may be needed to foster policy change. Campaigns aimed at changing the votes on the International Whaling Commission have led to many surprises, including vote purchasing and a total lack of impact of domestic and international pressure on the positions of some whaling countries. The task is not easy. It requires segmenting audiences, identifying the information needs of the decision-makers through focused research and then specifically targeting those needs through communications (and often financial and policy objectives) to bring about the required changes. Communications can bring about changes in how policymakers act; the key is knowing their constituents' needs and focusing on those needs.

**Caribbean region**

An ocean communications campaign for the Caribbean does not need to address global issues like overfishing but should specifically target regional land-based activities like agriculture, resource extraction, development and tourism. Communications strategies must address regulation and enforcement of land-based activities to solve the ocean pollution problems caused by those activities. Research on these islands may find that the best way to address ocean problems is to concentrate primarily on land-based issues such as decreasing the use of agricultural herbicides or improving human health and drinking water quality by the regulation of mineral extraction. In a region like the Caribbean, the approach could vary significantly from one island to another. The key is finding what it takes to get people to change their behaviors or policies to bring about the desired results and to provide the necessary enforcement. The *DOE* goal is preserving the ocean, but the actual reason for people behaving in the desired way may be quite a different story. Human behavior is complicated and must be sorted out using a host of research tools to identify its determinants and the opportunities to change them. Especially in developing countries where policy often exists only on paper due to weak enforcement, it is important to motivate individual, community or organizational behavior. Addressing the real needs of local people is required for effective communications and the desired conservation outcomes.

**Minimizing impact, maximizing benefits**

Changing people's perspectives on things they already care about can be very difficult. When working with those who want to conserve a resource and those who make a living from that same resource, getting all of them to understand that they likely have the same goals can be the major task. Both groups will probably need to change their perspectives—about the facts and about the other group—and that may take a series of communications tools. Key tools are those that support conflict resolution and consensus building. Helping to change the "race to fish" mentality to one of sustainable taking requires changing priorities from short-term to long-term goals. This is easier with policymakers than with individuals active in the industry at a particular point in time, who are often trying to make a living with an attitude of "get it while you can."

Policy change alone is often not adequate to bring about

the required changes. Educating fisherfolk about Individual Fishing Quotas or other conservation mechanisms can be difficult. Effective communication of these new policies requires someone who has enormous credibility with the targeted audience. Working with very targeted messages to effect major changes often requires multi-stage communications efforts, allowing small behavioral commitments to lead to larger ones after a comfort level is reached at each stage. Large changes can take long periods of time for acceptance and implementation.

## Current Public Opinion

The research already done provides some assistance in addressing three critical issues:

- An understanding of the current status of public opinion on the ocean around the world
- A baseline set of data with which to compare and gauge the success of resulting campaigns over time
- Information for formulation of appropriate messaging and tactics

An analysis of existing research revealed that relatively few research studies on ocean issues have been done, and those studies have been primarily focused on U.S. audiences, single species (primarily marine mammals) or single issues (e.g., coral reefs or coastal pollution). This constrains our ability to paint a full picture of global public opinion. Based on existing data summarized below, however, the primary findings across all the research were:

- People know very little about ocean issues
- But, they care about the health of the ocean and have a positive attitude about the ocean
- Many worry about the ocean's vulnerability
- But they are not at all aware of the current degree or sources of threat

Based on the polling data and experience of the *DOE* Communications working group, the following recommendations for communications initiatives emerged. Communicators should:

- Combine emotion with information
- Start with a values-based message, not an economic one
- Connect values to messages about recreation and healthy ocean environments for future generations
- Appeal to individual responsibility
- Provide more information about destructive fishing practices
- Position communications at points where people interact with the ocean—cruises, aquariums, vacations, diving, coastal communities
- Talk about the ocean as a *system with life* versus simply water
- Provide solutions along with threats and problems
- Use "social math" to communicate numbers in accessible and memorable ways
- Use scientists, not politicians, as primary spokespersons

Despite the scarcity of research, there are a few good studies worth noting. The Ocean Project, working with Belden, Russonello and Stewart (1999), conducted the study "Communicating about Ocean Health and Protection." They reviewed research from 1992 to 1999 as well as original research conducted in the U.S. using phone surveys with six focus groups. Their major findings conclude that Americans:

- Have little knowledge of ocean functions
- Believe that the health of the ocean is essential to human survival
- Don't think their individual actions affect the ocean
- Believe industry is responsible for the problems
- Do have a broad awareness of ocean vulnerability
- Do not, however, perceive the ocean to be in immediate danger
- View the ocean as a second-tier environmental problem, less important than air and water pollution, climate change and land-based species extinction

This study provided a set of recommendations for practitioners working on ocean-related communications projects in the U.S.:

- Combine emotion with information
- Start with values
- Appeal to individual responsibility
- Connect values to messages of recreation and healthy futures
- Emphasize impacts of destructive fishing practices and coastal development
- Specifically target women, African Americans, Hispanics and those who live near the ocean in communications messages because these groups expressed the most emotional connection to the ocean

In October 2002 Edge Research, together with SeaWeb and the Coral Reef Foundation, conducted a U.S. public opinion poll focusing on coral reef issues titled *Public Knowledge and Attitudes about Coral Reefs* (Edge Research 2002).

The survey revealed that Americans:

- Are concerned and believe that coral is being damaged and destroyed
- Think that an average of 27% of reefs worldwide are fully protected from the activities that can damage them (the real percentage is much lower)
- Express strong support for designating coral reefs as protected areas
- Believe the U.S. must take a leadership role on the issue, and regulations must be put into place
- Consider coral reefs and rainforests of equal importance

Based on this study, recommendations for ocean conservation communications projects will position communications where people interact with reefs (cruises, aquariums, vacations, diving); focus more on emotional connections to reefs versus their economic importance; and take advantage of an expressed public willingness to support reef conservation financially through vacation surcharges.

In 2002 the Frameworks Institute, Natural Resources Defense Council (NRDC), The Ocean Conservancy and the David and Lucile Packard Foundation conducted focus group surveys in the United States for the *Communicating Oceans to the Public* project (Frameworks Institute *et al* 2002). This study found:

- Most people have very little understanding of the ocean and lack a familiar "frame" sufficiently developed to connect and organize new information about the ocean
- Only among those strongly connected to the ocean is the environmental protection frame apparent
- There are four default "frames," or frames of reference, that people call upon in their minds when presented with new information about the ocean:
    - *Vast Unknown Ocean*—the ocean is vast and we know very little about it, so it must be able to sustain itself, and we need to do much more exploring before making any recommendations.
    - *Economic Model*—the ocean is a vast source of resources for human consumption.
    - *Emotional Frame*—people are emotionally connected to the ocean, despite very little knowledge.
    - *Pollution*—pollution is the greatest threat and evokes a sense of the *ocean as water* versus the *ocean as a system of life.*

There are a handful of non-U.S. surveys that also provide information on public response to the ocean. Two of these, one from Indonesia and one from Latin America, are worth noting. In Indonesia in 2000, Project Pesisir conducted a National Benchmark Study on *Exploration and Measurement of Public Attitudes to Conservation and Use of Marine Resources* (Farnell & Suhirman 2000). This public opinion poll reveals interesting information about public attitudes in Indonesia related to the ocean, including the following:

- Indonesians have a personal connection to the ocean because it is a major source of food
- They have many misconceptions, such as associating the word "coral" with stone
- Half of those who say they've seen a coral reef say they saw it on TV
- Many think that Japan has more marine life than Indonesia (in reality, Indonesia has far greater levels of marine biodiversity and endemism)
- Half of them consider the ocean condition to be getting worse
- Many consider trash as the most important problem related to the ocean
- They think that the responsibility for ocean conservation lies with the larger authorities, rather than with communities or groups at the grass roots level

The study includes a media analysis and recommendations for use of television and messaging that combines elements of comedy, simplicity and bluntness.

In Latin America, the Mellman Group (1998) together with SeaWeb, conducted the *Latin American Ocean Awareness Project* in Argentina, Chile, Mexico and Peru. This group conducted in-depth interviews with "elites" (government officials, NGO leaders, journalists, academics) and public opinion surveys among adults in the four countries. The survey investigates people's perceptions toward aquaculture, artisanal (local) fisheries, international pressures/autonomy, trustworthiness of various sectors, consumption practices and the media. The study reveals several interesting findings, including that people in those countries:

- Perceive a conflict between jobs and environment, but they are divided on which is more important
- Have an "emerging environmental consciousness"
- Care about the ocean, although political elites think the general public doesn't care
- Are more likely to say the condition of the ocean is personally important to them than is said in the U.S. public
- Have a high concern for commercial overfishing (seen as the fault of foreign fleets)

- See aquaculture as the answer to the major problems
- Generally maintain a positive and emotional perception about the ocean
- Think pollution in the ocean is the most important problem facing the marine environment

The *DOE* Communications theme group strongly recommended that more research be conducted concerning the knowledge, attitudes and behaviors of key audiences for ocean issues. This research should include international surveys focusing on critical countries and regions, focus groups with certain audiences, in-depth interviews with decision-makers and a global Internet survey. Despite its lack of statistical significance due to sampling issues on the international level, such a survey may reveal some interesting qualitative information.

## Strategic Recommendations: Global, Regional and Local Campaigns

The key to the overall communications effort is to communicate globally and advocate regionally and locally. To do this requires a multi-scaled approach to marine conservation communications: international, national and local campaigns. The reality is that communications must be tailored to reach target audiences in priority regions of the world in both urban and rural areas, must address behavior changes that support real conservation outcomes and must be done as soon as possible.

The benefits of a global campaign include large-scale, consistent outreach to broad audiences and a platform for major themes and issues. Local and regional communications can be highly specific to a particular set of social, political and economic issues, cultural factors and local scientific priorities.

*Global-scale outreach* will stimulate a universal ocean ethic, provide context and an introduction to the problem and address international policy and consumption issues. One major campaign per year should focus on thematic "hooks" such as overfishing and habitat destruction, seamounts, land/sea interactions and ocean governance. Global outreach would primarily use mass-media outlets in a combination of earned media, paid media, electronic media, an issue-based website and public service announcements distributed via the establishment of a World Ad Council.

*Regional-scale outreach* will be a more culturally appropriate communications approach to link with local conservation efforts. Regional outreach will determine its own communications "mix" of earned media, paid media, events, education and/or social marketing. To ensure long-term sustainability of the regional outreach, a network of regional communications nodes staffed by local people in high-priority countries should be established. These nodes will not only ensure that strategies are planned with socio-cultural characteristics in mind, but they will also involve and educate local partners to ensure that capacity is built for continued ocean communications into the future.

*Local-scale outreach* will focus on capacity building and supporting local efforts specifically designed to change threat behaviors, build local support for marine protected areas, change fisheries practices, provide education for local managers and establish enforcement tools. It will also acknowledge the enormous body of work done by local activists over time and provide them with the tools and resources they have historically lacked.

The campaigns on all levels can benefit from some common approaches that:

- Celebrate the importance of the ocean
- Communicate the urgency of the situation and convey that the time to act is now
- Stimulate an emotional connection to the ocean and convey the concept of an *ocean public trust*
- Dispel the myths that the ocean is vast and its resources limitless
- Describe key threats (overfishing, unsustainable consumption, coral reef degradation, lack of awareness) and their consequences
- Convey the personal benefits that individuals receive from healthy ocean ecosystems
- Highlight that there are solutions to the problems—they are not insurmountable
- Discuss alternative behaviors that should be exhibited
- Highlight specific case studies that are particularly relevant or poignant to target audiences

### Campaign goals

Looking at the major recommendations arising out of the overall *DOE* global agenda for action in marine conservation, the following goals should guide the communications campaigns:

*Garner support for conservation of key marine species*

Target consumption, fishing practices and policy measures that lead to long-term survival of threatened and endan-

Table 1
**Priority species**

Communications campaigns will focus on critically threatened or endangered species, either directly to save individual species or as "flagships" to draw public attention to broader ecoregion concerns. Campaigns will support IUCN Red List marine species, via input from the Species Survival Commission (IUCN 2003). It is important to note, however, that the Red List hosts only a small subset of the marine species that should be named due to a deficiency in data and scientific assessment. Therefore, campaigns may support species outside this listing. High-priority species to be used as the focus in campaigns include:

**Marine mammals**
Mediterranean monk seal
Vaquita (Gulf of California)
Irawaddi dolphin (Indonesia and Philippines)
Manatees and dugongs

**Sharks**
Deep-water sharks
Whale shark
Brazilian guitarfish
Sawfishes
Skates and rays
Basking shark

**Sea turtles**
Leatherback sea turtle
Hawksbill

**Commercially exploited fish**
Sturgeon (Beluga, Baltic)
Orange roughy
Boccacio rockfish—important fishery species
Totoaba—Gulf of California
Grouper (goliath, Warsaw, speckled hind, Nassau)
Bluefin tuna (especially critically endangered southern species)
Humphead wrasse
Atlantic halibut
Seahorses
Salmonids

**Invertebrates**
Abalone (several species)
Giant clams
Corals (especially reef-building stony corals)

**Seabirds**
Albatross (several species)
Andrews Frigate bird

gered marine species, including commercial fish species. These species are often keystones—meaning they are critical to preserving a whole ecosystem. Sometimes they are "flagships"—meaning they bring attention to the ecosystem failure due to their wide recognition or because they are on the brink of extinction and urgent action must be taken to save them (Table 1).
EXAMPLE OUTPUTS: Greatly reduced or no further consumption of orange roughy; new international policy to protect the Mediterranean monk seal; implementation of new fishing practices that reduce bycatch of the leatherback sea turtle (Figure 3).

*Support establishment of marine protected areas (MPAs) and MPA networks in critical regions*
Foster local stakeholder buy-in and policymaker support for new MPA networks.
EXAMPLE OUTPUTS: The establishment of a new network of MPAs in Indonesia; establishment of a multinational marine reserve system in the Caribbean.

*Influence key policy measures affecting critical issues, such as high seas bottom-trawl fishing*
Respond to emerging threats and use new science to influence policy changes. In the U.S., this might focus in part on recommendations of the National Ocean Commission and the Pew Oceans Commission reports. Internationally, this might focus on *DOE* or other recommendation sets.
EXAMPLE OUTPUTS: An international moratorium on high seas bottom-trawl fishing; a prototype high seas marine protected area for seamounts.

*Change the way people think about the ocean*
Condition the climate such that people and policymakers are more apt to support ocean conservation.
EXAMPLE OUTPUTS: Creation of a World Ocean Public Trust; formation of a World Ad Council; polling data that indicate a new level of public concern and understanding for ocean life; establishment of new management regimes that discontinue decision-making based on long-held myths; education that supports protection of seamounts and significantly changed attitudes about bottom-trawl fishing.

**Target audiences**
Understanding the makeup, demographics and motivations of the priority audiences is critical for successful outcomes. Target audiences will vary by region, by issue and in their level of priority. Special consideration and attention will be

focused on audiences outside of the U.S. as they have been under-represented in past campaigns. This presents a significant opportunity to influence the public in high-priority international sites who have not been exposed to these kinds of messages. Marine communications campaigns globally and regionally will target audiences such as:

- National decision-makers
- Local stakeholders
- Educators
- Journalists (multipliers of campaign messages)
- Conservation organizations (potential partners in campaigns)
- Specialized publics, such as tourism, fishing or other industry sectors
- Communities living near marine resources
- Consumers
- Donor organizations in the U.S. (a secondary target as a natural consequence of an effective overseas campaign)

### Geographic scope

The *DOE* Communications theme group developed a preliminary prioritization of geographic regions (Table 2) for communications outreach on ocean issues (Figure 4). We recognized, however, that this list would be subject to change as new science, issues, crises and opportunities emerged.

### Methodologies

Each campaign is unique to the local and sociocultural realities of the target audiences and the specific campaign objectives. However, campaigns are expected to follow a framework created by strategic communications practitioners. The following are some elements that may be incorporated into a campaign framework:

- *Formative research and baseline study* to determine knowledge and attitudes of target audiences to create campaign strategies, tactics and messages, as well as for later comparison with campaign results to gauge success.
- *Planning workshops* to ensure that partner organizations and stakeholders are integrated into the process. Strategies will be vetted through participatory gatherings to define a common work plan and delegate responsibilities.
- *Campaign identity* to ensure that campaigns have a specific identity that goes beyond any one organization, via a campaign logo, slogan and graphic identity.
- *Production of campaign materials* to ensure each campaign is armed with the appropriate portfolio of communications tools. These may include a combination of print elements (posters, brochures, stickers, print public service announcements), video tools (documentary, television PSAs), radio programs (PSAs, feature length programs), outdoor elements (banners, billboards, exhibits), informational websites and interpersonal communications (theater, meetings, workshops and face-to-face meetings). Materials must be locally appropriate for message, language and culture.
- *Campaign launch and materials distribution* to disseminate the message and reach an initial media saturation, campaign events will "kick off" in priority sites. The events should gather local celebrities and governmental

RODERIC MAST, CONSERVATION INTERNATIONAL

**Figure 3** Pacific leatherback sea turtles, such as these newly hatched juveniles, are "charismatic" animals that can function as potential "flagship" species in raising public awareness for marine conservation.

Table 2

**Regions considered high priority for marine conservation communications efforts by the *DOE* Communications theme group at the Conference**

| Within three years |
|---|
| Indonesia, Taiwan, China, Philippines, South Africa, Brazil, Jamaica, Ecuador |
| **Within ten years** |
| Japan, Gulf of Guinea, Mexico, India, Thailand, Argentina, Cuba, far eastern Russia, East Africa/Kenya, Madagascar, the Red Sea |

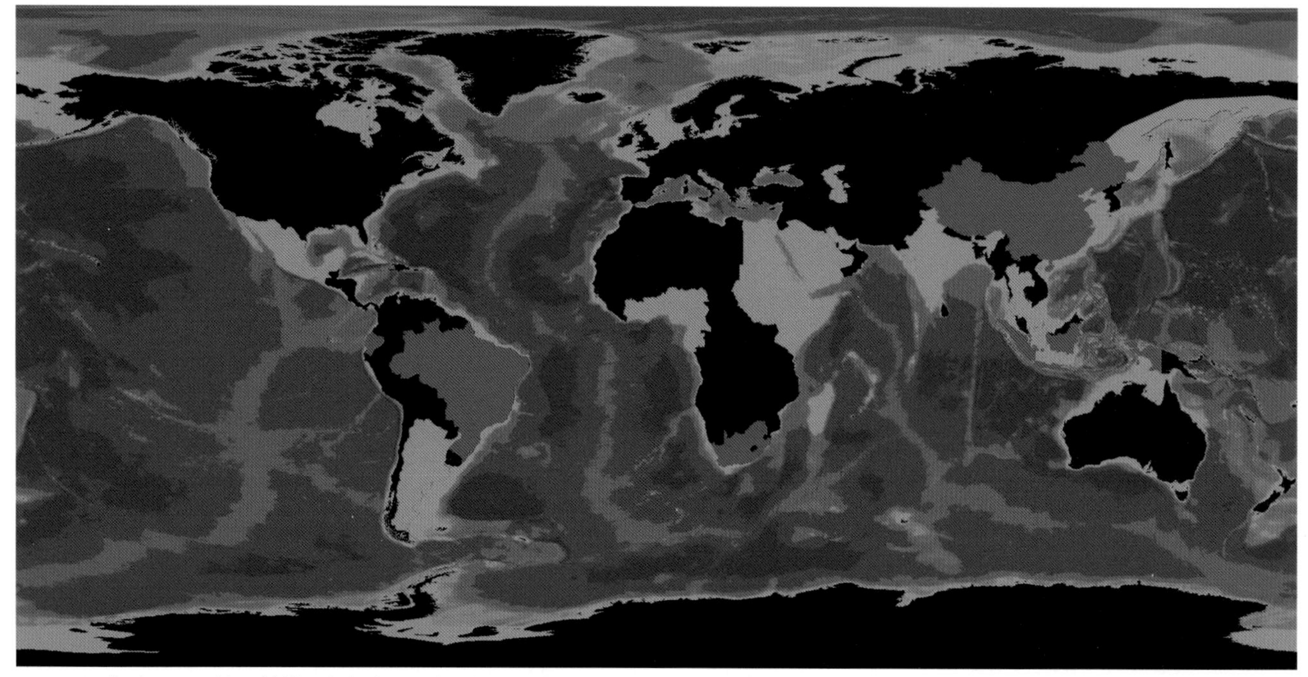

**Figure 4** Regions considered high priority for marine conservation campaigns and education by the Communications theme group at the *DOE* Conference, with areas in light green to be targeted *within three years* and dark green areas *within ten years.*

officials and may use "hook" dates of importance to the marine world.

- *Media materials, events and tours* (where appropriate) to enable journalists to cover campaign issues through the provision of standard press kits with changing press releases throughout the campaign and video news releases for use with the broadcast media. Journalists may also be invited to key sites important for marine conservation to create feature articles or coverage, or they may receive training that helps them better understand and communicate ocean issues.
- *Educational resources* (where appropriate) to help teachers educate youth about marine conservation concepts by developing resources and materials like educators' guides, by guiding the efforts of curriculum developers in key countries and by training educators in methodologies for teaching ocean issues.
- *Community outreach/social marketing* (where appropriate) to engage specific communities that may fall outside the reach of traditional media, via traveling exhibits, speakers and activities carefully crafted to introduce the issues to key local communities near critical marine sites around the world.
- *Evaluation* to assess campaign impact and provide a final report.

To implement this methodology, many local organizations will need support in the form of training, guidance in how to use new tools and financial assistance.

**Outcomes sought**

*Within ten years,* the campaigns will have contributed significantly to the accomplishment of three primary ocean conservation outcomes at a global level:

*Increased populations of critically threatened species*

Campaigns will generate increased levels of protection, change local behaviors and garner new local support and funding for conservation initiatives that support species conservation. Campaigns will focus on threatened or endangered marine species, such as the Mediterranean monk seal, the leatherback sea turtle, the whale shark and commercial fish species such as orange roughy that are threatened by overfishing.

*Creation of new marine protected areas (MPAs) and expansion and improved management of existing protected areas*

Campaigns will build the necessary level of local awareness

and support to facilitate creation of new protected areas, expansion of existing protected areas and improved management of marine protected areas in critical regions.

*Increased conservation measures at the ecoregional level*
Campaigns will promote improved, sustainable management practices and decision-making for larger marine ecosystems beyond specific species or protected areas. Changes to legal frameworks, national and international policy and management regimes will help ensure connectivity and viability of vital ecosystems and promote long-term, sustainable management.

Each global, regional and local campaign effort must be aimed at specific conservation outcomes if we are to conserve and sustainably manage our ocean. The progress toward outcomes must be tracked via measurable objectives.

The widely held assumption that increasing information and awareness leads to changed attitudes, which in turn changes behavior, has repeatedly been shown to be false. Sometimes these early steps are extremely helpful but not necessarily causal. The largest gap in human behavior is often between what we know and what we do.

Small behavior commitments can lead people to taking larger steps. Using communications to generate initial commitment—through education, appeals to need, avoidance of embarrassment, opportunities to improve one's own lot in life—is often critical. However, communications must also deal with the natural response of others to second guess new behaviors and must provide positive reinforcement for those initial behavioral steps.
EXAMPLE OUTPUTS: Decision-makers implement new policies; new MPAs are established; people make informed choices about sustainable seafood consumption in hotels, restaurants and markets; increased international news coverage of marine conservation issues and proposed solutions during the campaign timeframe.

## Monitoring and Evaluation

All campaigns will include a monitoring component to assess effectiveness. This will occur in four phases:

**Baseline study**
Prior to any communications outreach, a baseline study will determine the current status of target audience attitudes, knowledge and behavior, as well as collect existing information on the status of threatened species, the policy climate or any other outcome-level indicator that may be useful for later comparison.

**Formative research**
An effective campaign must be tailored for its specific target audience in order to be culturally appropriate. Formative research usually includes qualitative study via focus groups, one-on-one interviews or message testing.

**Mid-term evaluation**
All campaigns will be expected to submit a mid-term evaluation that reviews progress toward its outcomes and allows for adaptive management to reorient the effort if needed.

**Final evaluation**
Campaigns will include the preparation of a final evaluation that compares changes since the baseline study, campaign evaluation and methodology in a final report.

## A New Mechanism: The Ocean Communications Fund

After the *DOE* Conference, the need arose to find ways to move the communications agenda forward without the initial capital investment needed to open office nodes in regions around the world. While this infrastructure would certainly ensure a long-term commitment to sustained outreach in the regions, the costs—estimated at *DOE* to be approximately US $440 million—are large.

Thus, a new approach emerged to forward the recommendations of *DOE* through a new funding mechanism. Such a mechanism will leverage the presence of partners in the same priority regions, providing much needed funding and capacity-building opportunities.

As recommended by the *DOE* Communications theme group, the Ocean Communications Fund would catalyze communications campaigns in critical regions, focusing on threatened species, responding to key threats, building upon sound science and securing measurable marine conservation outcomes. The Fund would harness the collective enthusiasm of many partner organizations and leverage efforts toward marine communications strategies driven by measurable conservation outcomes.

An initial eight-year fund of $36 million would create a unified coalition of organizations and individuals using communications methodologies under one strategic umbrella for ocean conservation—a common voice for a sea change. The

Ocean Communications Fund would allow for multiple partners to launch more than 40 campaigns in various regions around the world. The Fund would use a set of geographic priorities that has been updated since *DOE* (Table 3), yearly focal themes (such as fisheries, coral reefs or seamounts) and a selection of communications methods (mass communications, films, advocacy, social marketing, community outreach, etc.).

Once established, the Fund would be an effective mechanism for additional donors to contribute resources for specific campaign themes and regions. They will be assured high-quality campaigns linked to conservation outcomes.

Grant-making mechanisms are nothing new to the conservation community. The Critical Ecosystem Partnership Fund (CEPF)—a partnership of Conservation International, the World Bank, the Global Environment Facility, the government of Japan and the John D. and Catherine T. MacArthur Foundation—is an excellent example of the effectiveness of pooling resources from many partners and then distributing them to a wide variety of local and international partners under a framework that clearly identifies priority needs (Conservation International 2004).

Table 3

**Regions currently considered high priority for focused ocean communications efforts**

**Asia**

- Coral Triangle: Indonesia, Philippines, Melanesia, Malaysia
- Major policy/consumption centers: China, Japan, Taiwan
- Australia

**Americas**

- Eastern Tropical Pacific: Ecuador (Galápagos), Peru, Colombia, Panama, Costa Rica
- Brazil
- Mexico (Gulf of California)
- Argentina
- Caribbean/Bahamas

**Africa**

- Sub-Saharan Indian Ocean: Kenya, Tanzania
- Southern Africa (eastern shore)
- Madagascar

While fund mechanisms have proven effective, there is no fund that specifically supports marine communications efforts in the places where they are needed most. Too often, campaigns or products, like films, are very creative and compelling, yet are done *"ad hoc,"* in the wrong places and without a clear connection to conservation outcomes. This new idea will harness lessons learned from funds such as CEPF and apply them to a new arena of partners that urgently need financing, technical support and capacity building to achieve marine conservation outcomes. Strengths of this approach include:

- Allows for geographic priority setting, based upon regional marine conservation needs
- Addresses emerging threats globally
- Empowers partners and local organizations
- Builds lasting capacity in marine communications
- Ensures quality control of science-based messages through technical oversight by Fund personnel
- Allows for quick dissemination of thematic priorities, with resources allocated to address these priorities
- Builds a network of global, regional and local marine communicators
- Pools resources for marine communications issues for greater impact

Recipients of grants would be local and international organizations with a proven track record of success in marine conservation and in using communications methodologies. Grantees will be required to provide matching funds, ensuring commitment and leveraging of additional resources. Additionally, implementing organizations would submit a strategic plan demonstrating how the campaign contributes directly to conservation outcomes and how results would be measured. Grantees would be given funds to implement over two years, with the second year funds doubling that of the first. Funds would be available for the three tiers of campaigns described earlier, at international, national and local/site scales.

A small central coordinating unit would ensure quality implementation, build the capacity of implementing organizations as needed, administer the grant mechanism, seek new funds and leverage the efforts of partners. Additionally, the Fund would initiate communications research efforts, monitor and evaluate the Fund's effectiveness, create training programs for grantees as needed and host a visual resource center making images for outreach accessible to global and regional campaigns. Establishing such a resource center for use in campaigns around the world will not only be cost-effective, but will also encourage more photographers and filmmakers to capture underwater and other marine-oriented images for the purpose of conservation. These resources historically have been lacking in the marine community yet are vital for getting effective messages into the mass media.

The Fund's management unit will receive guidance from an advisory council of key communications industry players and scientific experts knowledgeable in the marine conservation arena. This council would alert the Fund to emerging issues and threats in marine conservation as well as key opportunities in the communications industry. The council would also provide feedback on the funding portfolio in terms of selection and prioritization of issues, organizations and methodologies proposed. The communications arena offers a plethora of talented and innovative thinkers that could lend guidance to this Fund. They include individuals from international public relations and advertising firms, broadcasters, social marketers and influential spokespersons and entertainers.

## Conclusion

The ocean has always been treated as "worldwide commons," in which individuals and governments can plunder and pollute and yet remain unwilling to recognize the gravity of the problem. We must understand our shared responsibility for stewardship of the ocean and recognize that the high sea areas should be held in public trust. In order to meet these challenges, we must communicate with people in ways that are smart, effective, measurable and strategic—and most importantly, this must happen now and on a large scale.

The world environment is by definition a very dynamic system. Some changes occur due to natural cycles or evolution but many are human-induced. To address the latter, we must change human behavior. The *DOE* communications plan is an appropriate vehicle to use. Sometimes we can change policy and practice and thereby efficiently influence human behavior on a larger scale. In many parts of the world, however, there is no enforcement mechanism for new policies requiring a smaller-scale focus on the individual. Although this is an enormous task, focusing on behavior change is the key to long-term conservation and sustainable management of our ocean.

Scientists are realizing that we have a fairly narrow window of opportunity of less than a generation to protect ocean life and health—a window of opportunity we will likely never have again. The Communications theme group worked under this assumption, and took on the challenge of using this open window to save species, create new protected areas, establish a World Ocean Public Trust and fundamentally change the way people think about the ocean.

*DOE* came at a time when changing the way the world thinks about and conserves the ocean is imperative. The Conference organizers recognized that we must shatter pervasive myths that ocean resources are unlimited and invulnerable, and the international group of experts gathered at *DOE* established a clear need to conduct communications outreach at multiple levels in key regions, focusing on priority species and following sound communications methodology. Moving this agenda forward will require a solid investment of resources, which prompted the formulation of the Ocean Communications Fund proposal.

We invite everyone who reads this chapter to ask themselves what role they can play in helping to make the *DOE* recommendations a reality. We have institutions to build, funds to raise and manage, communications strategies to plan, local capacity to build, creative messages to develop and data to gather and analyze. Our communications efforts must have measurable results, be sustainable and flexible and be targeted yet inclusive. If we don't focus on human behavior, we will not succeed in supporting the global agenda for action in marine conservation that was developed during and since the *Defying Ocean's End* Conference.

## Acknowledgements

*We wish to acknowledge the valuable ideas contributed by several outside experts that joined our Communications theme group discussions during the* Defying Ocean's End *Conference: Jean Michel Cousteau, John Francis of the National Geographic Society, Julie Packard of the Monterey Bay Aquarium, Christina Mittermeier, Rod Mast of the IUCN marine turtle specialist group, Peter Knights of WildAid and Kraig Butrum and Robin Murphy from Conservation International.*

## Literature Cited & Consulted

Belden Russonello and Stewart Research and Communications. 1999. *The Ocean Project: Results of national survey executive summary.* Washington, D.C.: Belden, Russonello and Stewart Research and Communications in collaboration with the American Viewpoint.

Conservation International. 2004. *Critical Ecosystem Partnership Fund.* Conservation International, Global Environment Facility, Government of Japan, the John D. and Catherine T. MacArthur Foundation and the World Bank. *http://www.cepf.net/xp/cepf/* (accessed August 23, 2004).

Edge Research. 2002. *Public Knowledge and Attitudes about Coral Reefs.* Political briefing on the results of a national survey of U.S. adults. Conducted by Edge Research on behalf of Coral Reef Foundation, Munson and SeaWeb.

Farnell, J. and A. Suhirman. 2000. *National Benchmark Study: Exploration and Measurement of Public Attitude to Conservation and Use of Marine Resources.* Market research summary for national attitude survey. Indonesia: Consensus MBL for Proyek Pessir, Coastal Resources Management Project.

Frameworks Institute, Natural Resources Defense Council, The Ocean Conservancy and the David and Lucile Packard Foundation. 2002. *Communicating Oceans to the Public.* Internal Report.

IUCN. 2003. *2003 IUCN Red List of Threatened Species. http://www.redlist.org* (accessed August 23, 2004).

The Mellman Group. 1998. *Latin American Ocean Awareness Project.* Presentation and analysis of findings from elite interviews and public opinion surveys in Argentina, Chile, Ecuador, Mexico, Peru and Venezuela. Produced for SeaWeb.

CHAPTER 11

# *Ocean Governance: A New Ethos through a World Ocean Public Trust*

Montserrat Gorina-Ysern, *American University*
Kristina Gjerde, *IUCN–The World Conservation Union*
Michael Orbach, *Duke University*

GETTY IMAGES

## Introduction

Since the United Nations adopted the Stockholm Declaration on Human Environment as a non-binding instrument in 1972, the environmental and conservation movements have made significant strides. Designed to be a wake-up call for governments, organizations and public opinion, Principle 2 of the Stockholm Declaration provides that the

> *natural resources of the earth including the air, water, land, flora and fauna and especially representative samples of natural ecosystems, must be safeguarded for the benefit of present and future generations through careful planning or management, as appropriate.*

In Principle 4, we find humans exhorted to embrace

> *a special responsibility to safeguard and wisely manage the heritage of wildlife and its habitat which are now gravely imperiled by a combination of adverse factors. Nature conservation including wildlife must therefore receive importance in planning for economic development.*

The balance between environmental conservation and economic development lurks uneasily under the emerging principles outlined.

The Stockholm Declaration has served as the environmental policy yardstick over the last thirty years and has inspired the principles adopted under the 1992 United Nations Conference on Environment and Development (UNCED) and the call for action by governments gathered at the 2002 World Summit on Sustainable Development (WSSD) to protect the world's marine life and to ensure sustainable fishing. Again, the fragile balance between environmental conservation and sustainable development permeates the recommendations of UNCED and WSSD, as both seek to develop international law principles that will develop into binding rules for the governance of ocean space, marine wildlife and other living and non-living resources.

Led by Professor Michael Orbach of Duke University, the Ocean Governance theme group at the *Defying Ocean's End (DOE)* Conference was grounded in the principles of the Stockholm Declaration, UNCED and the WSSD. However, in its recommendations, *DOE* went beyond these and took issue with two prevailing principles that have undermined the protection of ocean space and wildlife beyond the limits of national jurisdiction. These principles are discussed below and include the high seas freedom of fishing and the principle of open access to high seas marine life and other non-living resources. By virtue of the high seas freedom of fishing and the open-access nature of the high seas, the flag-State of a fishing vessel or other commercial vessel retains exclusive jurisdiction to implement or enforce international law rules relating to the conservation of marine living and non-living resources and the protection of the marine environment on the high seas. Although restrictions on high seas uses exist and conservation principles have been developed, any vessel is entitled to use the high seas as an economic resource because access to living and biogenetic resources is either unimpeded or insufficiently regulated under international law.

In its final recommendations, the *DOE* Conference regarded current ocean governance as under-developed, sector-specific, inconsistent, conflicting or non-existent, especially for ocean space forming the high seas beyond the limits of national jurisdiction. The concept of a World Ocean Public Trust (WOPT) was therefore embraced as a means to limit access, or better manage access, to the high seas—the 60% of ocean space that lies beyond the limits of national jurisdiction (Figure 1). The WOPT concept is expected to lead to the effective implementation of governance principles currently developed under international law and to the widespread adoption of a new ocean ethos for the protection of ocean space and marine life. The *DOE* Conference placed its initial focus on the protection of rare and fragile ecosystems associated with seamounts under the waters of the high seas.

## Overcoming the Precarious Systems of Ocean Governance

The exercise of control over ocean space, over marine flora and fauna, and over the exploration and exploitation of natural resources (living and non-living) is complicated by many factors and is legally, technically and economically problematic. This state of affairs has allowed the erosion of marine ecosystems worldwide and will permit the continuing degradation of ocean habitats unless ocean governance structures are rapidly adopted and effectively implemented to combat pollution, overfishing and fishing practices destructive to ocean habitats.

The term "ocean governance" covers a set of rules—some legally binding and some not—adopted by the international community of States (in international law and treaties, nations are often referred to as "States") for the structured regulation, management and control of ocean uses. It also includes the persons, bodies and institutions entrusted with administering the rules that govern ocean space. One of the purposes of ocean governance is the conservation and protection of ocean habitat and marine life. Ocean governance also refers to the need to establish decision-making processes within global institutions to ensure that the persons charged with management within these institutions administer their mandates with predictability, efficiency, transparency and accountability. Efforts to adopt universal ocean governance agreements and to enforce the agreed upon rules tend to proceed very slowly because the international community includes nearly 200 nations with potentially conflicting interests that respond to many factors or "externalities." Two of the most important externalities are the will of States to take action and their level of scientific and economic capacity to support long-term and effective protection and preservation of the marine environment.

The *DOE* Conference recommended the initiation of international discussions leading to the "policy enclosure" of the world ocean through a "framework agreement addressing all human activities that affect the ocean and providing for ecosystem-based, integrated, precautionary management of high seas as a World Ocean Public Trust." The specific elements of this proposal include:

- Integrating governance regimes across use sectors, levels of government and the land-sea boundary
- Building scientific and policy capacity, especially in the less developed nations
- Applying principles in a comprehensive manner, including the best scientific evidence available, the precautionary approach, "polluter pays" practices and public trust compensation (cost recovery and economic rent)

The framework agreement advocated would require:

- Designating ocean living resources as "wildlife" using legal and policy frameworks analogous to those governing terrestrial and avian wildlife

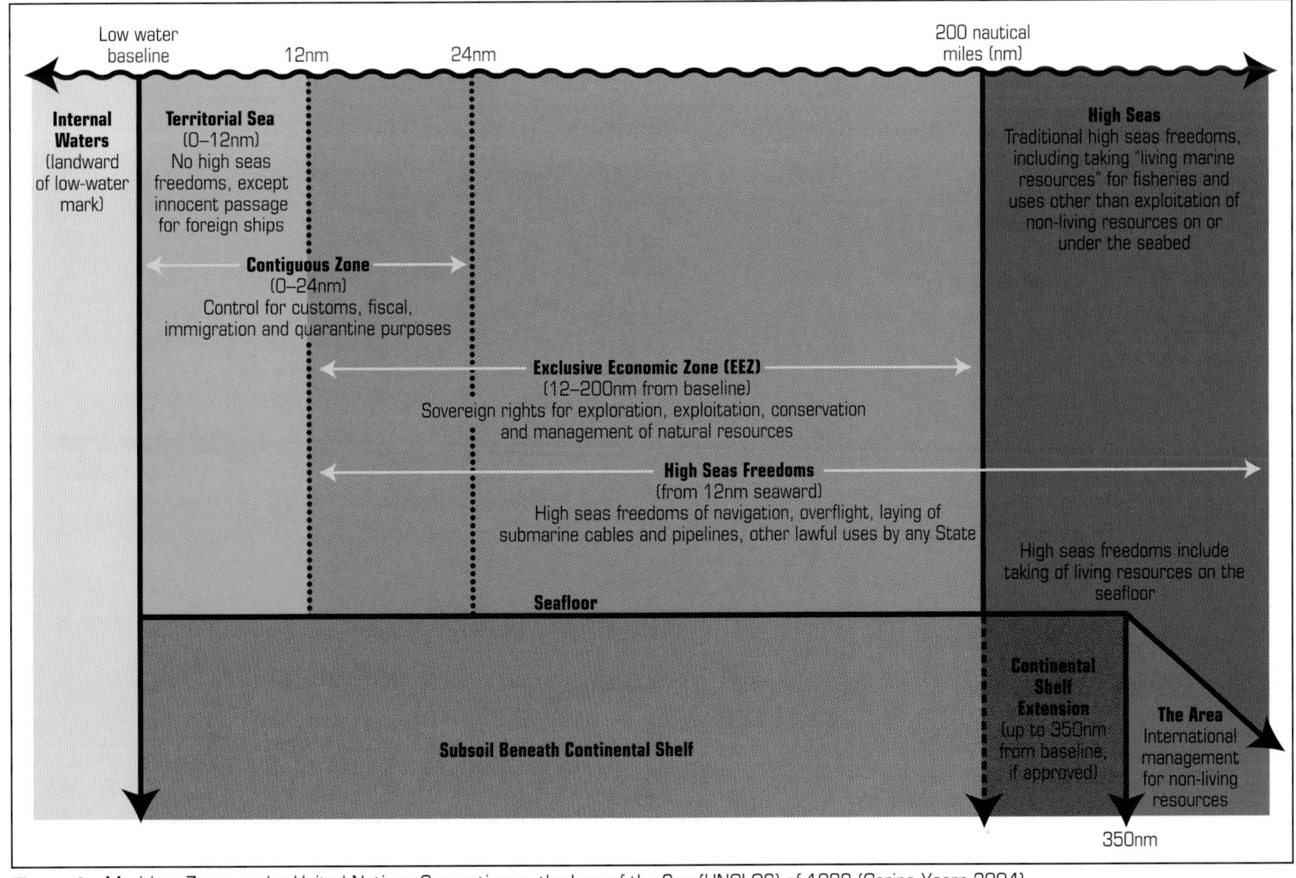


**Figure 1** Maritime Zones under United Nations Convention on the Law of the Sea (UNCLOS) of 1982 (Gorina-Ysern 2004).

- Shifting responsibilities for implementation and enforcement of ocean laws from flag- and port-States to an independent, verifiable international process
- Developing a decision-making process based on majority or super-majority voting, rather than consensus, for prompt international decisions

Finally, *DOE* recommended seeking a U.N. General Assembly Resolution establishing an immediate moratorium on bottom trawling of seamount resources in high seas areas until an effective management regime can be established. Seamounts are underwater mountains that rise at least 1,000 meters above the ocean floor. The long-lived biodiversity of seamounts is targeted and destroyed by newly developed bottom-trawl fisheries for deepsea species. Also recommended was the longer term goal of implementing an operational global network of high seas seamount marine protected areas (MPAs) by 2013 (*Defying Ocean's End: An Agenda for Action* 2003).

The success of the *DOE* recommendations depends in great measure on how effectively the conservation movement can garner institutional and international support for its vision in light of the many limitations that international law places on effective, integrated and coordinated ocean governance. One of these limitations is that States can pick and choose the rules by which they intend to be bound.

## Ocean Governance and the Consensual Nature of International Law

Public international law is the body of rules designed to regulate ocean, air and space realms outside of national jurisdiction. For the high seas ocean areas it includes marine environmental law, law of the sea rules, and hundreds of related bilateral, regional, sub-regional and multilateral treaties, codes of conduct and recommendations by expert bodies. In general, international law rules cannot be applied as effectively as a nation's domestic legal system because of the deficiencies within the international legal system itself. International law lacks a centralized legislative body; the United Nations does not have the power to enact laws that are universally binding. The consensual rules of conduct in force under international law permit an individual nation to consent or object to the creation of new rules and to withdraw its delegations from participating in treaty negotiations, to ensure that particular rules cannot become binding for that State. Even after multilateral treaties have been negotiated and adopted by a majority of States concerned with ocean uses, individual States or groups of States can opt out of binding treaties and can refuse to enforce treaty obligations despite originally agreeing to being bound by such obligations. Public international law is a primitive system of law indeed.

Moreover, the international legal system lacks an executive branch with powers to implement rules, as the Security Council of the U.N. is subject to the veto power of the five permanent members. Unlike domestic legal systems, international law also lacks effective law enforcement mechanisms to monitor infractions and apprehend offenders. There is an international judicial body, but States are not legally obligated to resolve their disputes before the judges of the international courts and tribunals (unless by express agreement). The inadequacy of this situation can be best illustrated through the analogy of a car driver who cannot be stopped for speeding on the highway by anyone, and even if stopped by some law enforcement official, could object to paying a fine or being brought before judicial authorities by simply saying (s)he has not consented to being bound by the rules or to submitting to the authority of the judge. This type of argument is perfectly plausible under the rules of public international law that regulate the exercise by States of jurisdiction and control over ocean space, and it is particularly deleterious to the protection of endangered high seas fish, biogenetic resources and other marine life.

## Maritime Zones and Their Freedoms and Governance Criteria

The World Ocean Public Trust proposed by the *DOE* Conference would further curtail the ability of States to justify unwise ocean uses on the basis of a legitimate exercise of high seas freedoms.

Under the international law of the sea, whose principles are codified in the 1982 United Nations Convention on the Law of the Sea (UNCLOS), maritime space is divided into zones of sovereignty, areas of sovereign rights with jurisdiction over resources only, and high seas with many freedoms and few restrictions on uses. A coastal nation's power over maritime space tends to decrease with growing distance from the land territory. Unless other special rules apply, a nation's claims over maritime space are measured from the low water line (at low tide) along the coastline as marked on the nation's official large-scale charts and known in international law as the "baseline." Waters inland of the low water line are internal waters and are completely under the sovereignty of the nation.

The *territorial sea* is an adjacent belt of sea extending seaward to a maximum of twelve nautical miles (nm) from the low water baseline. Within its territorial sea, a coastal nation exercises sovereignty over the water column, the airspace above it and the seafloor and subsoil beneath. Ships flying the flag of foreign nations can exercise a right of *innocent passage* through the territorial sea of another nation—provided that the foreign ship's passage is incidental to ordinary navigation, or is rendered necessary by *force majeure* (storms or other distress) and is not prejudicial to the peace, good order and security of the coastal nation. Nations may also declare a *contiguous zone* over the water column—extending to a maximum of 24nm offshore from the baseline—in which the coastal nation can implement measures relating to customs, fiscal, immigration or sanitary laws and regulations.

UNCLOS 1982 also recognizes the right of a coastal nation to proclaim an *Exclusive Economic Zone* (EEZ) consisting of the waters between 12 and 200nm seaward of the baseline. In the EEZ, coastal nations may exercise sovereign rights for the purpose of exploring, exploiting, conserving and managing the natural resources therein. It is estimated that 90% of all fish currently caught in the world's ocean are taken within EEZ boundaries.

A coastal nation also has sovereign rights to manage the exploration and exploitation of resources on or under the seabed of its *continental shelf* from 12 to 200nm seaward of the baseline. A nation may also claim an extension of the continental shelf under its sovereign rights up to 350nm from the baseline, where it can be shown to be a natural geologic prolongation of its land territory and evidence of the claim is internationally approved.

The concept of *high seas freedoms* has created some confusion for the layperson. It is therefore necessary to clarify which freedoms can be legitimately exercised within 12nm (the traditional territorial sea for most nations), between 12 and 200nm (the traditional EEZ for most nations) and on the high seas (outside the EEZ of any nation). For resource management, exploration and exploitation purposes, the high seas consist of the mass of waters—about 60% of the ocean's surface (Figure 2)—that are *not* included in the EEZs of any nation. The universal establishment of EEZs, and the increase in claims to extended continental shelves, have restricted the exercise of some traditional high seas

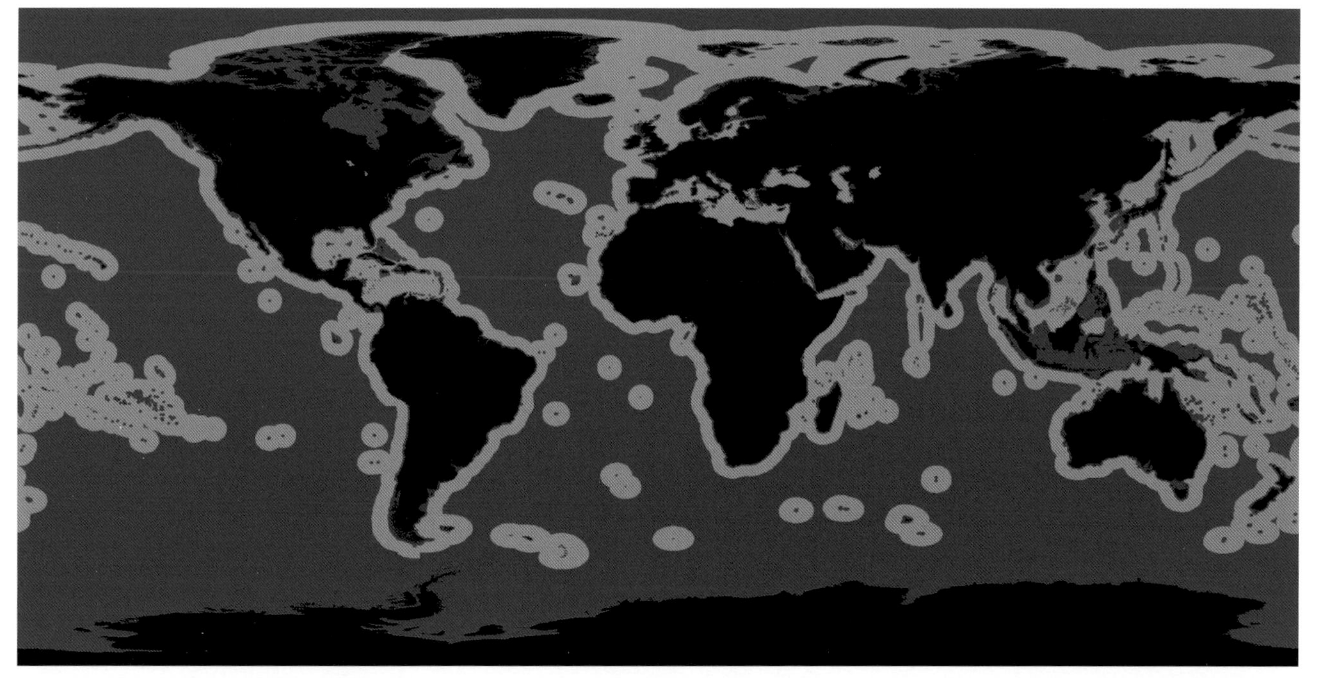

**Figure 2** Exclusive Economic Zones (EEZs) claimed by most nations cover roughly 40% of the world's ocean surface. The remaining 60% is the high seas, where governance mechanisms are very limited.

freedoms. In their EEZ, coastal States may exercise exclusive control over the construction of certain types of installations or structures (those erected for economic purposes or those that interfere with the coastal nation's economic interests), the conduct of marine scientific research activities and fishing by foreign vessels.

The traditional "high seas freedoms" of navigation, overflight, the laying of submarine cables and pipelines and other international lawful uses associated with the operation of ships and aircraft remain open to all nations outside the 12nm territorial sea—even within EEZs and over extended continental shelves. Lawful uses of the waters beyond 12nm include not only commercial navigation, overflight, cables and pipelines, but "complete freedom of movement and operation" for the following military activities (Department of the Navy 1995):

- Warship and military aircraft force maneuvering
- Flight operations
- Military exercises
- Surveillance and intelligence gathering activities
- Ordinance testing and firing
- Observing the naval exercises of other nations

The UNCLOS of 1982 codifies the high seas freedoms recognized earlier in Article 2(1) of the 1958 Geneva Convention on the high seas. With regard to the high seas beyond 200nm, UNCLOS Article 87 lists the freedoms of navigation, overflight, laying of submarine cables and pipelines, the freedom to construct artificial islands, installations and structures, the freedom of fishing and the freedom of scientific research. It explicitly states that these freedoms "Shall be exercised by all States with due regard for the interests of other States in their exercise of the freedom of the high seas, and also with due regard for the rights under this Convention with respect to activities in the Area." *The Area* beyond the limits of national jurisdiction regulates the exploration and exploitation of non-living resources of the seabed and subsoil under the high seas water column. However, the regime for marine life, biogenetic and other living resources in high seas waters is very limited or non-existent (Table 1).

The Ocean Governance theme group at the *DOE* Conference took issue with the precarious status of international

Table 1

**Governance principles in UNCLOS for high seas living resources**

**Straddling and highly migratory stocks**
UNCLOS provides that States "shall seek . . . to agree" on measures necessary to coordinate the conservation and development of these stocks (Article 63)

**Control over nationals for conservation of high seas living resources**
Establishes obligations for States to cooperate in the adoption of measures with respect to their nationals for the conservation of the living resources of the high seas (Articles 116–120)

**Creation of regional or sub-regional fisheries organizations**
States have an obligation to cooperate in the adoption of conservation and management measures through the creation of regional or sub-regional fisheries organizations (Article 117)

**Use of best scientific evidence available to maintain and restore stocks to maximum sustainability**
States shall take measures, using the best scientific evidence available, designed to "maintain or restore populations of harvested species at levels which can produce the maximum sustainable yield" as qualified by relevant factors (Article 119.1[a])

**Maintain and restore associated species**
States shall take into consideration the effect of fishing on "species associated with or dependent upon harvested species" to maintain or restore the latter above levels at which their reproduction may become seriously threatened (Article 119.1[b])

**Regular exchange of scientific and fisheries information**
States shall contribute such information to one another and through competent international organizations on a regular basis (Article 119.2)

**Marine mammals**
States must protect marine mammals on the high seas (Article 120)

law governance of the high seas beyond the limits of national jurisdiction (beyond claimed EEZs and extended continental shelves).

## Main Ocean Governance Principles under Public International Law

Although the term "freedom" can be considered a misnomer in light of the many reservations applying to the traditional freedoms of the high seas, the sole jurisdiction of the flag-State over vessels flying its flag (in which the vessel actually carries the sovereignty of its nation to sea) has ensured that enforcement actions on the high seas have remained at a minimum. This is particularly relevant to enforcement of fisheries regulations of a regional, sub-regional and global nature. Reactive measures have prevailed over proactive ones to protect the high seas marine environment and conserve its living resources.

Over the last century, however, a number of fundamental conservation principles for governance of the high seas have been developed under major public international law instruments. These include the 1982 Convention on the Law of the Sea, the 1992 Convention on Biological Diversity (CBD 2001–2004), the 1993 Food and Agricultural Organization Compliance Agreement (FAO 1995a), the 1993 FAO Code of Conduct for Responsible Fishing (FAO 1995b) and the 1995 Agreement for the Implementation of the Provisions of the United Nations Convention on the Law of the Sea Relating to the Conservation and Management of Straddling Fish Stocks and Highly Migratory Species. While the major criticism of conservation groups and the international experts at the *DOE* Conference is that these principles are not implemented effectively, it is still critical to understand their scope, as summarized below:

### The United Nations Convention on Law of the Sea (UNCLOS) 1982

The Convention codifies and recognizes overarching duties for States to cooperate and consult with one another, in a non-discriminatory manner (Article 119.3) and without "unjustifiable interference" with other States' activities in conformity with the Convention (Article 194.4), with regards to the living resources (Table 1) of the high seas (Articles 116–120) and the protection and preservation (Table 2) of the marine environment (Articles 192–194).

### Convention on Biological Diversity (CBD)

The CBD Preamble declares that the conservation of biodiversity "is a common concern of humanity." It notes that "where there is a threat of significant reduction or loss of biological diversity, lack of full scientific certainty should

Table 2

**Governance principles in UNCLOS for protection and preservation of the marine environment of the high seas, including its biodiversity**

**General obligation to protect and preserve the marine environment** within and beyond the limits of national jurisdiction (Article 192)

**Take all necessary measures to prevent, reduce and control pollution from any source** (Article 194.1), beyond the limits of "where they [States] exercise sovereign rights" (Article 194.2)

**Prevent the release of toxic, harmful or noxious substances** from land, atmosphere or by dumping (Article 194.3[a])

**Prevent pollution from vessels, including accidents and/or discharges** (Article 194.3[b])

**Prevent pollution from installations and devices** used in the exploration or exploitation of natural resources of the seabed and subsoil (Article 194.3[c]) or any other installations or devices (Article 194.3[d])

**Legal basis for the establishment of marine protected areas (MPAs)**
Take measures necessary to "protect and preserve rare or fragile ecosystems" including "the habitat of depleted or endangered species and other forms of marine life" (Article 194.5), and a requirement for the Deep Seabed Authority Council to "disapprove areas for exploitation" when "substantial evidence indicates the risk of serious harm to the marine environment" (Article 162.2[x])

not be used as a reason for postponing measures to avoid or minimize such a threat." It also notes that the conservation and sustainable use of biological diversity "will strengthen friendly relations among States and contribute to peace for humankind," including present and future generations. It balances a recognition of the special needs of developing countries (access to additional financial resources and appropriate technologies to enable economic and social development and poverty eradication) while stressing that "substantial investments are required to conserve biological diversity" and reap the "broad range of environmental, economic and social benefits from those investments."

The first objective of the CBD, in Article 1, is

*the conservation of biological diversity, the sustainable use of its components and the fair and equitable sharing of the benefits arising out of the utilization of the genetic resources, including by appropriate access to genetic resources and by appropriate transfer of relevant technol-*

Table 3

**Governance principles in Convention on Biological Diversity: Legal basis for the establishment of high seas marine protected areas**

**General duty to ensure *in-situ* conservation**
Parties have a duty to ensure the conservation of ecosystems and natural habitats and the maintenance and recovery of viable populations of species in their natural surroundings (Article 2), and where the effects of coastal State processes or activities affect the high seas, to establish "a system of protected areas, or areas where special measures need to be taken to conserve biological diversity" (Article 8[a] through [m])

**Duty to prevent damage**
Parties are responsible for preventing activities within their jurisdiction or control from causing damage to the environment of other States or beyond the limits of national jurisdiction (Article 3)

**Duty to minimize effects on high seas**
Duty extends to areas within and areas beyond the limits of national jurisdiction, thus making States responsible for activities or processes carried out by them that may have effects "beyond the limits of national jurisdiction" (Article 4.2)

**Duty to cooperate in sustainable use of high seas biodiversity**
Parties have a duty, as far as possible and appropriate, to cooperate directly or through competent international organizations in the conservation and sustainable use of biological diversity beyond the limits of national jurisdiction (Article 5)

Table 4

**Governance principles in the United Nations Fisheries and Agriculture Organization's Compliance Agreement (FAOCA)**

**Application to all fishing vessels**
The FAOCA applies to "all fishing vessels that are used or intended for fishing on the high seas" (Article II.1)

**Flag-State responsibility**
Each State shall take measures with regard to fishing vessels flying its flag to ensure their compliance with international conservation and management measures on the high seas (Article III.1[a]); and shall not authorize its own or another party's fishing vessels that, *prima facie*, are unable to comply with such measures or over which the flag-State cannot exercise its responsibility effectively (Articles III. 3 and 5[a] through [d], and 6–8)

**Duty to cooperate and exchange information**
These duties relate to a party's own fishing vessels, vessels of another party and vessels of third parties voluntarily in the port of a State party. These duties apply among parties, non-parties and with FAO (Articles V and VI)

**Duty to encourage non-parties to join FAOCA**
States shall cooperate to prevent non-party vessels from engaging in activities that undermine the effectiveness of conservation and management measures adopted under FAOCA (Article VIII.1)

*ogies, taking into account all rights over those resources and to technologies, and by appropriate funding.*

Under CBD Article 2, sustainable use refers to:

*The use of components of biological diversity in a way and at a rate that does not lead to the long-term decline of biological diversity, thereby maintaining its potential to meet the needs and aspirations of present and future generations.*

CBD Articles 3, 4 and 5 also address nations' conservation obligations on the high seas (Table 3).

**The 1993 FAO Compliance Agreement (FAOCA)**

The FAOCA (Table 4) is the successful result of the Cancun Declaration and the United Nations Conference on Environment and Development (UNCED) Agenda 21 which calls on States to deter the "reflagging" of vessels for the purpose of avoiding compliance with UNCLOS conservation and management rules for high seas fishing and to pressure fishing nations to join Regional Fisheries Management Organizations (RFMOs).

The Preamble of the FAOCA balances the right of States to engage in high seas fishing with the duty to take measures with respect to their own nationals for the conservation and sustainable use of the living resources of the high seas. It calls on States to join RFMOs and to exercise effective jurisdiction and control over vessels flying their flag—including "fishing vessels and vessels engaged in the transshipment of fish"—and to cooperate for those effects.

**The 1995 U.N. Fish Stocks Agreement (UNFSA)**

The Preamble of the UNFSA responds to the call of Chapter 17, Agenda 21, of UNCED, for States to ensure the adequate management of high seas fisheries and their proper utilization. The UNFSA (Table 5) seeks to address unregulated fishing, over-capitalization, excessive fleet size, reflagging of vessels to escape controls, insufficiency of selective gear, unreliable databases and lack of sufficient cooperation among States. It aims to avoid adverse impacts on the marine environment, preserve biodiversity, maintain the integrity of marine ecosystems and minimize the risk

Table 5

**Governance principles in the United Nations Fish Stocks Agreement (UNSFA) 1995**

With regard to the duty to conserve and manage straddling fish stocks and highly migratory fish stocks, the UNFSA recognizes the following principles and duties:

**Optimum utilization**

Requires States to ensure long-term sustainability; prevention or elimination of overfishing and excess fishing capacity and fishing effort; use of the best scientific evidence available, assessing the impact of fishing and other human activities on target stocks and species belonging to the same ecosystems or associated species; and to minimize impacts on non-target species (especially endangered species) by minimizing pollution, waste, discards and catch by lost or abandoned gear, using selective, environmentally safe and cost-effective gear and techniques (Article 5 [a through h]); and to collect relevant information, perform scientific research and adopt effective measures for the monitoring, control and surveillance of fishing on the high seas (Article 5 [j] through [l])

**Precautionary approach**

States "shall be more cautious when information is uncertain, unreliable or inadequate" and these reasons shall not be used to postpone or fail to take conservation and management measures (Article 6.2 and as specified in Article 6.3[a] through [d] and 4 through 7)

**Duty to cooperate**

States must adopt very detailed and specific measures in order to fulfill their duty to cooperate (Articles 7 and 8) in a manner compatible with both conservation and management measures. This duty applies to non-party States (Article 7.8) and places restrictions on the traditional freedoms of the high seas. The duty to cooperate requires that "States shall enter into consultations in good faith and without delay" (Article 8.2), and "shall give effect to [it] by becoming members" of RFMOs or other arrangements, or by "agreeing to apply the conservation and management measures established" thereof (Article 8.3)

**Duties of flag-State**

The flag-State has extensive duties to comply (Article 18) and to enforce the UNFSA (Article 19–22)

of long-term or irreversible effects of fishing operations. UNFSA, Article 2, outlines its objective: "To ensure long-term conservation and sustainable use of straddling fish stocks and highly migratory fish stocks through effective implementation of the relevant provisions of the UNCLOS."

### The 1995 FAO Code of Conduct for Responsible Fisheries

This is a voluntary and non-binding agreement. Its principles are important (Table 6), however, and the recommendations from the *DOE* Conference incorporate their spirit.

## Fishing at Overcapacity within and beyond the 200-Nautical-Mile EEZ: Urgent Action Needed

In spite of the plethora of major governance principles, rules and duties identified above, fishing fleets around the world since World War II have grown to overcapacity through multinational financing from banks and government subsidies. Few governance principles have achieved the status of customary international law, and even those duties regarding marine resource conservation that have become binding have lagged behind the race to fish. The international law obligations to consult and cooperate have not been implemented effectively. Disputes relating to conservation of living marine resources under the sovereign rights of coastal States in their EEZs are not subject to a truly comprehensive regime of compulsory dispute settlement under the 1982 UNCLOS (Orellana 2004). And provisions of the 1995 UNFSA defer to UNCLOS in cases involving highly migratory species such as many commercially attractive—but overfished—species of tuna, Patagonian toothfish and swordfish.

Over the last century, international lawyers and ocean policymakers sought to pressure nations to abstain from conduct that leads to the irreparable damage of shared living marine resources. The failure over the last century of national authorities to comply with international law in compelling those with an economic interest in ocean fisheries to exercise restraint in their exploitive practices is well documented (Rayfuse 2003). These exploitive practices involve many industrial sectors within and across nations, including those engaged in fishing, the transshipment of endangered stocks, the processing of fish products and related industries and the opening of ports to the landing of highly depleted but especially valued fish destined to please the palate of an elite domestic consumer market.

States have adopted mechanisms for sharing stocks whose habitat spans more than one national EEZ through measures to conserve, manage and exploit those stocks jointly as transboundary resources. These mechanisms also cover the management of fish species that straddle several EEZs and the high seas beyond the limits of national jurisdiction, and highly migratory species that traverse vast expanses of ocean area during their life cycle. There are currently five Regional Fisheries Management Organizations (RFMOs) established to deal with highly migratory species whose life cycle covers several nations' EEZ and high seas areas; four

Table 6
**Governance objectives in the FAO 1995 Code of Conduct**

Under this Code of Conduct, States shall:

- Ensure sustainability of humanity's nutritional, economic and social well-being
- Ensure long-term sustainable use of fisheries resources
- Ensure safe and fair labor conditions for those employed in fisheries and related trades
- Implement education and training programs
- Conduct trade according to World Trade Organization rules
- Promote concerted conservation at national, bilateral, regional, sub-regional and global levels, ensuring transparency and consultation among interested sectors
- Implement effective mechanisms for monitoring and control of fleets
- Protect habitats and ecosystems
- Develop adequate gear and practices
- Prevent and or eliminate overfishing and excess capacity
- Introduce ecosystem-based and precautionary management
- Use best scientific evidence available to guide management principles, taking into account relevant environmental, economic and social factors, and also using traditional and artisanal community knowledge in the formulation of such management decisions

RFMOs deal with straddling stocks, and four new RFMOs are being established in areas of the high seas where there has been no legal regime in place for the conservation and sustainable development of marine living resources (Table 7).

RFMOs have pressured fishing nations into restricting fishing in high seas areas close to or beyond the outer limits of the EEZ in some regional areas and curtailing the freedom of fishing in certain high seas areas where States have adopted a range of voluntary and mandatory measures to protect fish stocks. RFMOs have used the following measures thus far to restrict fishing effort:

- Area and time closures
- Compulsory registration of fishing vessels engaged in high seas fisheries
- Good standing status for vessels abiding by fishing restrictions
- Annual reporting requirements by member States

Table 7
**Selected regional fisheries management organizations**

**A. Highly migratory species covering several EEZs and the high seas:**

- Inter-American Tropical Tuna Commission (IATTC)
- International Convention for the Conservation of Atlantic Tunas (ICCAT)
- Indian Ocean Tuna Commission (IOTC)
- General Fisheries Commission for the Mediterranean (GFCM)
- Commission for the Conservation of Southern Bluefin Tuna (CCSBT)

**B. Straddling stocks across several EEZs:**

- North East Atlantic Fisheries Commission (NEAFC)
- Northwest Atlantic Fisheries Commission (NAFO)
- Commission for the Conservation of Antarctic Marine Living Resources (CCAMLR)
- Convention for the Conservation and Management of Pollock Resources in the Central Bering Sea

**C. High seas (where no previous regime existed)**

- Western Central Pacific Fisheries Commission (WCPFC)
- South East Atlantic Fisheries Organization (SEAFO)
- Galapagos Agreement—not yet in force
- South West Indian Ocean Fisheries Commission (SWIOFC)—under development

- Catch limits and quotas (set and/or reduced harvest limits and levels)
- Quota-sharing arrangements
- Reporting requirements for trade by exporters of frozen stocks
- Protection of juveniles
- Minimizing dead discards
- Prohibition of fishing in spawning areas
- Research on gear, equipment and fishing techniques
- Scientific research and tagging programs
- Limits on and evaluations of fishing capacity
- Prohibition against retention of landing and sale in domestic and foreign markets
- Monitoring and sightings
- Trade sanctions

However, these measures fall short of what is required to establish a proper management system for the protection of high seas marine life, biogenetic and other living resources. Gjerde and Breide (2003) consider RFMO efforts necessary but insufficient for several reasons. First, sector-based governance (having the user set the limits of his use) is incompatible with ecosystem-based management and does not address the complex nature of the issues that need to be integrated. Second, there remain a number of unregulated or "orphan" activities on the high seas (deepsea fishing, bioprospecting, commercial energy/sequestration projects and scientific experiments, including large-scale ecosystem manipulation) for which there is no environmental impact assessment carried out. Responsibility and liability for causing harm are implemented poorly and there are few mechanisms in place to raise claims. Third, it is not unusual for RFMOs and other sector-based governance bodies to apply inconsistent principles and mandates or fail to implement internationally agreed principles. And last, the rules on transparent decision-making and governance accountability are seen to be weak and poorly enforced, especially by RFMOs, and it is doubted that RFMOs can deal adequately with global fishery resources.

The *DOE* Conference recommendations—following the lead of Professor Michael Orbach, IUCN positions and those of many concerned scientists and conservation groups—call for a much more effective form of "enclosure" for the remainder of high seas fisheries than the ones achieved by RFMOs to date (Orbach 2002, Gorina-Ysern 2004). *DOE* recommendations included: designating fish populations not raised through aquaculture as "wildlife" rather than "marine living resources," and establishing—through the U.N General Assembly—a moratorium on bottom-trawl fishing on the high seas. This activity is regarded as unsustainable, destructive of ancient and non-recoverable habitats and inconsistent with UNFSA provisions and binding UNCLOS obligations for all States to protect rare and fragile ecosystems (Gjerde and Breide 2003, Gianni 2004).

In addition, *DOE* recommended longer-term action on development of ecosystem-based protection and management in high seas areas, including the establishment of strictly protected areas and other place-based management tools. The Malaga Workshop of 2003 and the World Parks Congress of 2003 recommended the development of these criteria for including tropical marine ecosystems for potential nomination as World Heritage Sites through a representative network of high seas marine protected areas (HSMPAs).

With regard to habitats, networks and ecosystems, the successful implementation by 2012 of a representative set of marine and coastal protected areas (MCPAs), as called for by the WSSD and reinforced by the CBD advisory body, would signify a considerable improvement over current MCPAs. Existing MCPAs are described as "*extremely* deficient in purpose, numbers and coverage" (Kimball 2003). The following actions are urgently needed:

- Restricting high seas activities at the national level through flag-State approaches
- Harnessing the cooperation of like-minded States that are threatened by a similar activity
- Establishing competent RFMOs where they do not exist, or where those in place are insufficient or inefficient
- Requesting the International Sea Bed Authority to exercise its full mandate to prevent mining where deep seabed ecosystems may be endangered
- Demanding that RFMOs implement the widely established precautionary principle and the ecosystem-based management approach under the UNFSA and/or implement stronger measures under existing agreements, including in continental shelf areas beyond 200nm
- Pressuring more non-member States to enlarge the membership of RFMOs and to become parties to existing multilateral agreements and regional arrangements
- Using the UNESCO World Heritage Convention criteria as a basis for negotiating consensus criteria among States on most urgent as well as longer-term protection needs
- Conducting extensive mapping of specific areas and scientific study of their relationship to surrounding marine ecosystems

Implementation of these measures could result in greater RFMO accountability in compliance, enforcement and emergency coordination powers between States and relevant local, regional, sub-regional and international institutions and stakeholders. It could foster a more effective integration of existing legal regimes and better protection of marine life, biogenetic and other living resources through higher quality information and assessment available for decision-making. Kimball suggests that these measures can be complemented with a number of effective tools: greater port-State enforcement (if a flag-State fails to perform its obligations); more trade control over illegally harvested marine species through the application of the 1973 Convention on International Trade in Endangered Species of Wild Flora and Fauna (CITES); use of catch documentation, certification schemes, import bans and Vessel Monitoring Systems (VMSs).

Views within the conservation movement vary (Redgwell 1998). The above measures are not meant to result in "a prohibition on all high seas fishing any more than [they mean] a free-for-all for fishers: it means finding the right balance between fishing rights and conservation obligations, reinforced by effective enforcement" (Kimball 2003).

However, some at the *DOE* Conference sought to "push the envelope" by considering whether a prohibition on all high seas fishing could be possible, so that—in the vision of Sylvia Earle—consumers would no longer eat tuna and swordfish, just as no consumer seeks to feed on land animals protected as wildlife, such as tigers, elephants and lions. This notion of considering uncultivated high seas fish as wildlife rather than living marine resources is a very controversial proposition that would require, as a first step, an in-depth discussion of the nature of the high seas and the nature of the rights of the international community over marine life for the purpose of protecting it for its own sake, as well as for the sake of current and future generations.

## World Ocean Public Trust (WOPT): A New Ocean Ethos

The WOPT embraced by the *DOE* Conference is a necessary concept formulated to put pressure on the United Nations and agencies, national governments, non-governmental organizations (NGOs) committed to marine conservation and industry to adopt a precautionary approach to all uses of high seas marine life, biogenetic and other living resources, habitats and ecosystems, and to protect them through the integration of existing regulations, regimes, programs and objectives. The WOPT is promoted through a "soft law" approach that, through repetition methods, techniques and strategies seeks to create a dialogue and achieve a shift from a reactive (tort/fault-based) approach to a proactive (precautionary-based) approach to ocean uses, ocean habitat and marine life conservation.

The WOPT is based on public trust doctrine that is useful only to a certain extent, because—unlike national parks and wilderness areas—the high seas do not fall under the domestic sovereignty of any single nation. The development of a WOPT is controversial because, in the process of promoting ocean enclosures and access restrictions, it impinges on the traditional rights of nations. The nature of the title that States may have over high seas living resources has been examined extensively by ocean policymakers and social scientists (Table 8).

**Open access *(res nullius)*** refers to a resource that belongs in the public domain—not owned by anyone—and is therefore regulated or controlled only in general terms, without being subject to any specific system or mechanism. The "rules of capture" or "first-come, first-served" apply to high seas fisheries, subject to the governance principles identified earlier as general duties.

**Common property *(res communis)*** refers to the joint control exercised by an identified group of people with a legitimate right to exclude others through restraints on access and use of the resource. The practices of RFMOs and MPAs can be integrated in this system.

**Private property *(res privatae)*** refers to a resource that is controlled by private individuals under generally agreed and socially acceptable rules relating to use. This regime applies to private property ownership in areas under sovereignty/sovereign rights of nations. Although this type of property rests in private hands, the State may still exercise control and place restraints on use on the basis of moral, ethical, cultural and religious values.

**State or government property *(res publicae)*** refers to a situation where the State or its citizens hold title over the asset and control and regulate its use according to social objectives and widely agreed limits imposed on a non-discriminatory basis, even if the government may retain exceptional rights over the resource. This regime could correspond to the WOPT concept, if an analogy can be drawn between parks and wilderness areas within the national territory and high seas protected areas beyond the limits of national jurisdiction.

The establishment of a WOPT, according to the recommendations of the *DOE* Conference, would result in all of the high seas being placed under an internationally managed regime with some areas designated as limited-access or

Table 8

**Four types of property rights regimes, with their associated rights and duties**

The following table provides high-level insight into the types of legal arguments, title claims and consequences that arise from a discussion of the World Ocean Public Trust (WOPT) concept in the context of property rights as they relate to high seas living resources.

| Regime type | Owner | Owner right | Owner duties |
|---|---|---|---|
| Open access *(res nullius)* | No owner free-for-all subject to no rules; or subject to International law rules and Community of States | Capture versus general duties of protection, cooperation and preservation | No rules; or general rules of UNCLOS, CBD, FAO, UNFSA, IMO, UNEP, RFMOs, CITES, CMS, ICRW and other regimes |
| Common property *(res communis)* | Condominium; Common ownership; (RFMOs) | Exclusion of non-owners, unless they accept standard rules | Maintenance; conventional rules of a regional, sub-regional and international law restricting uses |
| Private property *(res privatae)* | Individual nations (EEZ, EFZ, etc.) | Socially acceptable uses, control of access | Avoidance of socially unacceptable uses; domestic/international law |
| State or public property *(res publicae)* | All Citizens–All States; Custody, stewardship | Determined rules; Custody, Stewardship; Common Concern | UNCLOS, CBD, FAO, UNFSA, IMO, UNEP, RFMOs, CITES, CMS, ICRW and other regimes |

**Source:** Montserrat Gorina-Ysern (modified from Hanna, Folke & Maller 1995; Eggertsson 1993; Jodha 1993; Mohd *et al* 2004; Bromley & Cochrane 1994).

marine protected areas. The specific areas recommended by *DOE* are discussed below.

## Integrating Governance Mechanisms

The *DOE* Conference identified eleven large marine ecosystems (LMEs) for which protection should be of the highest priority for governments and marine conservation institutions (see Chapter 6). These areas include the Benguela Current, Bering Sea, Baja California System (Baja and Sea of Cortés), Caribbean, central West Pacific (Palau to Tuvalu), Antarctic waters, Coral Triangle (tropical Indo-Pacific), eastern tropical Pacific (Costa Rica basin), Humboldt System (Chilean/Peruvian), Patagonian and the Western Indian Ocean/East Africa. Parts of some of these areas fall outside national EEZs and include high seas areas.

There is also some overlap between the LME areas proposed by *DOE* for high-priority attention (Chapter 6) and some existing RFMOs, so the RFMOs are an existing management mechanism that would need to be engaged in order to implement the *DOE* recommendations. For example, protection of an LME in the Bering Sea would require involvement with the mandates of the International Pacific Halibut Commission (IPHC) and the North Pacific Marine Sciences Organization (PICES). Fisheries issues in the Caribbean LME area are overseen by the Western Central Atlantic Fishery Commission (WECAFC). Work in the Baja California System LME would involve interaction with the Inter-American Tropical Tuna Commission (IATTC) and the Latin American Organization for the Development of Fisheries (OLDEPESCA). Similarly, the Eastern Tropical Pacific LME would involve the Council of the Eastern Pacific Tuna Fishing Agreement (CEPTFA) and, possibly, International Convention for the Conservation of Atlantic Tunas (IATTC). The Humboldt System would engage the South Pacific Permanent Commission (CPPS). Work in the Benguela LME would involve the Commission for South East Atlantic Fisheries Organization (SEAFO), the International Commission for South East Atlantic Fisheries (ICSEAF), the Commission for the Conservation of Southern Bluefin Tuna (CCSBT) and, possibly, the International Commission for the Conservation of Atlantic Tunas (ICCAT). Protection of areas in the Western Indian Ocean LME would involve the Western Indian Ocean Tuna Organization (WIOTO) and, possibly, the Indian Ocean Tuna Commission (IOTC) and the Southwest Indian Ocean Fisheries Commission (SWIOFC). Work in the Coral Triangle LMEs might involve the Asia-Pacific Fishery Commission (APFIC) and IOTC; activities in the central West Pacific LME might involve the South Pacific Commission (SPC); in the Patagonian LME could involve the Joint Technical Commission for the Argentine/Uruguay Maritime Front (COFREMAR). And the Convention for the Conservation of Antarctic Marine Living Resources (CCAMLR) designates an area for some types of conservation management (see Chapter 3).

There is also potential overlap with Regional Seas Programs implemented by the United Nations Environmental Program (UNEP) for the same areas. UNEP has established regional seas programs in the Northeast and Southwest Pacific; the Wider Caribbean region, the upper Southwest Atlantic, off West and Central Africa, in the Mediterranean and Black Seas, Eastern Africa, the Red Sea and the Gulf of Aden, in the Regional Organization for the Protection of the Marine Environment (ROMPE) sea areas, in South and East Asian Seas and the Northwest and South Pacific. Other UNEP regional seas programs include the Arctic and Antarctic regions, Northeast Atlantic and the Baltic Sea (United Nations Environment Programme 2000).

In addition, the *DOE* Conference identified seamounts as supportive of highly unique (endemic) faunas (Chapter 2) and expressed concern that new technologies used by commercial fisheries to harvest deep-water species around seamounts are causing the collapse of certain species like orange roughy and causing massive collateral damage to "complex and poorly studied benthic (seafloor) communities." The *DOE* Conference recommended "establishing a comprehensive global reserve system to protect representative seamount habitats throughout the world."

## Conclusions

The Preamble of the CBD affirms the "intrinsic value of biological diversity" as well as its ecological, genetic, social, economic, scientific, educational, cultural, recreational and aesthetic values. Rooted in the Stockholm Declaration, the CBD and the WSSD principles, the *DOE* Conference recommended a range of approaches to high seas marine life, habitats and ecosystems. In the choice of strategy and tactics, the *DOE* recommendations satisfied a range of deepsea ecology and animal rights approaches to the environment, from the most traditional—such as those that regard the environment from an anthropocentric point of view of how "useful" the environment and its resources are to humans—to those whose purpose is to protect marine life on the high seas for its own sake and for the sake of current and future generations, irrespective of the near-term "useful" value these elements of nature may provide to fishermen, tourism and recreation industries, industrialists and other economic resource profit seekers.

The Ocean Governance theme group at the *DOE* Conference adopted a "zero tolerance" philosophy to environmentally unacceptable uses of the high seas. The success of *DOE*'s governance proposals would have considerable impacts on industrial and commercial activities encouraged by other sectors of international economic law. The dialogue between those holding conflicting interests is essential but likely to be confrontational, at least initially.

At the time of this writing, some practical and achievable actions resulting from the *DOE* Conference have been identified and are in-work, including the following:

- Working toward a U.N. General Assembly moratorium on high seas bottom-trawl fishing
- Developing criteria and mechanisms for high seas area-based conservation management
- Developing the scientific and legal bases for improved management, including a representative set of strictly protected and/or managed areas on the high seas
- Continuing to reform, evaluate and critique the performance of Regional Fisheries Management Organizations
- Establishing public trust finance and cost recovery mechanisms
- Continuing to pursue education and communications campaigns directed toward the general public and governmental decision-makers
- Resorting to original thinking and other stepping stones to highlight the need for new mechanisms/frameworks to achieve sustainable and equitable high seas governance

## Literature Cited & Consulted


Balton, D. 2003. State department official describes fisheries legal framework. Testimony before the Subcommittee on Fisheries Conservation, Wildlife and Oceans, House Committee on Resources, Sept. 11, 2003. *http://www.state.gov/g/oes/rls/rm/2003/24725.htm* (accessed September 25, 2004).

Boyle, A. 2003. The world trade organization and the marine environment. In *The Stockholm Declaration and the Law of the Marine Environment,* ed. M. H. Nordquist, J. N. Moore and S. Mahmoudi, 109–118. Boston: Martinus Nijihoff Publishers.

Bromley, D. W. and J. A. Cochrane. 1994. Understanding the global commons. Working Paper 13. EPAT/MUCIA Research & Training, University of Wisconsin-Madison. *http://www.wisc.edu/epat/.res-price/.global/.format/* (accessed August 24, 2004).

CBD. 2001–2004. *Convention on Biological Diversity,* 5 June 1992. United Nations Environment Programme. *http://www.biodiv.org/doc/legal/cbd-en.pdf* (accessed September 24, 2004).

Defying Ocean's End. 2003. *Executive Summary of Defying Ocean's End: An Agenda for Action,* ed. L.K. Glover, 12. Washington, D.C.: Conservation International.

Department of the Navy, Office of the Chief of Naval Operations. 1995–Current. *The Commander's Handbook on the Law of Naval Operations.* Naval Warfare Publication NWP 1-14M (Formerly NWP 9) COMDTPUB P5800.7. *http://lawofwar.org/naval_warfare_publication_N-114M.htm* (accessed September 26, 2004).

Edeson, W. 2003. Soft and hard law aspects of fisheries issues: Some recent global and regional approaches. In *The Stockholm Declaration and the Law of the Marine Environment,* ed. M. H. Nordquist, J. N. Moore and S. Mahmoudi, 165–182. Boston: Martinus Nijihoff Publishers.

Eggertsson, T. 1993. Economic perspectives on property rights and the economics of institutions. Beijer Discussion Paper no. 40. Stockholm: Beijer International Institute of Ecological Economics.

FAO. 1995a. *Agreement to Promote Compliance with International Conservation and Management Measures by Fishing Vessels on the High Seas,* approved on 24 November 1993 by Resolution 15/93 of the Twenty-Seventh Session of the FAO Conference. Rome: Food and Agriculture Organization of the United Nations. *http://www.fao.org/DOCREP/MEETING/003/X3130m/X3130E00.HTM#Top%20Of%20Page* (accessed September 27, 2004).

———. 1995b. *Code of Conduct for Responsible Fisheries.* Twenty-eighth Session of the FAO Conference on 31 October 1995. *http://www.fao.org/fi/agreem/codecond/codecon.asp* (accessed September 27, 2004).

Garibaldi, L. and L. Limongelli. 2003. *Trends in oceanic captures and clustering of large marine ecosystems: Two studies based on the FAO capture databases.* FAO Fisheries Technical Paper, No.435. Rome: Food and Agriculture Organization of the United Nations.

Gianni, M. 2004. *High Seas Bottom Fisheries and Their Impact on the Diversity of Vulnerable Deep-Sea Ecosystems.* Report prepared for IUCN–The World Conservation Union, Natural Resources Defense Council, WWF International and Conservation International.

Gjerde, K. M. and C. Breide. 2003. Towards a strategy for high seas marine protected areas. Proceedings of the IUCN, WCPA and WWF experts' workshop on high seas marine protected areas (IUCN, World Commission on Protected Areas and WWF International) at *www.iucn.org/themes/marine/pubs/pubs.htm* (accessed September 27, 2004).

Gorina-Ysern, M. 2004. World ocean public trust: High seas fisheries after Grotius—towards a new ocean ethos? *Golden Gate University Law Review* 34 (3): 645–714.

Hanna S., C. Folke and K. G. Maller. 1995. Property rights and environmental resources. In *Property Rights and the Environment,* ed. S. Hanna and M. Munangsinghe, 15–29. Washington D.C.: The Beijer International Institute of Ecological Economics and the World Bank.

Jodha, N. S. 1993. Property rights and development. *Beijer Discussion Paper no. 41.* Stockholm: Beijer International Institute of Ecological Economics.

Juda, L. 2002. Rio plus ten: The evolution of international marine fisheries governance. *Ocean Development and International Law* 33:109–144.

Kimball, L. A. 2001. International ocean governance: Using international law and organizations to manage marine resources sustainably. Gland, Switzerland: IUCN–The World Conservation Union. *http://www.iucn.org/themes/marine/pdf/IUCN%20book.pdf* (accessed September 27, 2004).

——. 2003. Governance of high seas biodiversity conservation: A framework for identifying and responding to governance gaps. Presentation at the Workshop on the Governance of High Seas Biodiversity Conservation, 16–19 June 2003, Cairns, Australia. CD available from National Oceans Office, Commonwealth of Australia.

Lizárraga, M. Medal award, code of conduct for responsible fisheries. *http://www.fao.org/fi/agreem/codecond/codecon.asp* (accessed August 26, 2004).

Marr, S. 2003. The precautionary principle in the law of the sea. Modern decision making in international law. Boston: Martinus Nijihoff Publishers.

Mohd, R., K. M. Noor, A. N. A. Ghani and S. Mohamed. 2004. Forest concession policy for sustainable forest management. Department of Forest Management, Faculty of Forestry, Unversiti Putra Malaysia. *http://www.forr.upm.edu.my/~awang/CONCES1.htm* (accessed August 24, 2004).

NPACI and SDSC Online. Seamounts: Window on ocean diversity. Volume 5, Issue 15, July 25, 2001. *http://www.npaci.edu/online/v5.15/seamounts.html* (accessed September 25, 2004).

Orbach, M. 2002. Beyond the freedom of the seas: Ocean policy for the third millennium. Fourth Annual Roger Revelle commemorative lecture, National Academy of Sciences, November 13, 2002.

Orellana, M. A. 2004. The law on highly migratory fish stocks: ITLOS jurisprudence in context. *Golden Gate University Law Review* 34 (3): 459–95.

Rayfuse, R. G. 2003. Enforcement by non-flag states of high seas fisheries conservation and management measures adopted by regional fisheries organizations. PhD dissertation, University of Utrecht.

Redgwell, C. 1998. Life, the universe and everything: A critique of anthropocentric rights. In *Human Rights Approaches to Environmental Protection,* ed. A. Boyle and M. Anderson, 71–87. Oxford: Clarendon Press.

Regional fisheries bodies—world ocean coverage. Map available at *http://www.fao.org/fi/body/rfb/Big_RFB_map.htm* (accessed August 26, 2004).

Rieser, A. 1999. Prescriptions for the commons: Environmental scholarship and the fishing quotas debate. *Harvard Environmental Law Review,* 393.

Roach, J. A. 2003. Particularly sensitive sea areas: Current developments. In *The Stockholm Declaration and the Law of the Marine Environment,* ed. M. H. Nordquist, J. N. Moore and S. Mahmoudi, 311–50. Boston: Martinus Nijihoff Publishers.

United Nations Environmental Programme, Regional Seas. 2000. A Survival Strategy for Our Oceans and Coasts. Nairobi, Kenya: United Nations Environment Programme. *http://merrac.nowpap.org/download/RSbooklet-E.pdf* (accessed September 27, 2004).

UNGA. 1995. *Agreement for the Implementation of the Provisions of the United Nations Convention on the Law of the Sea of 10 December 1982 relating to the Conservation and Management of Straddling Fish Stocks and Highly Migratory Fish Stocks.* United Nations General Assembly. UN Conference on Straddling Fish Stocks and Highly Migratory Fish Stocks. Sixth session, New York, 24 July–4 August 1995. *http://ods-dds-ny.un.org/doc/UNDOC/GEN/N95/274/67/PDF/N9527467.pdf?OpenElement* (accessed September 27, 2004).

WWF/IUCN. 2001. The status of natural resources on the high-seas. Gland, Switzerland: WWF/IUCN. *http://www.biodiv.org/doc/meetings/mar/temcpa-02/other/temcpa-02-high-seas-en.pdf* (accessed September 27, 2004).

### UN documents and other treaties cited

Agreement to Promote Compliance with International Conservation and Management Measures by Fishing Vessels on the High Seas. Adopted in 1993, reproduced at 33 I.L.M. (1994) 968, available at *http://www.fao.org/legal/treaties/012t-e.htm.* Entered into force April 2003 after receiving 25 ratifications.

Agreement for the Implementation of the Provisions of the United Nations Convention on the Law of the Sea of 10 December 1982 Relating to the Conservation and Management of Straddling Fish Stocks and Highly Migratory Fish Stocks. Reproduced at 34 I.L.M. (1995) 1542. In force 11 December 2001, 36 parties at August 2003.

The Declaration of the United Nations Conference on the Human Environment. U.N. Doc.A/CONF.48/14, 11 I.L.M. 1416. June 16, 1972.

*Oceans and the Law of the Sea.* Report of the Secretary-General, UNGA Doc. A/5757, March 7, 2002, at paragraphs 675-686.

United Nations Law of the Sea Convention. Done at Montego Bay, Jamaica, 10 December 1982, U.N. Doc. A/Conf.62/122, reproduced in 21 I.L.M. 1261 (1982). In force 16 November 1994, with 143 parties at August 2003.

CHAPTER 12

# *The Unknown Ocean*

Laurence P. Madin, *Woods Hole Oceanographic Institution*
Fred Grassle, *Rutgers University*
Farooq Azam, *Scripps Institute of Oceanography*
David Obura, *Coral Reef Degradation in the Indian Ocean (CORDIO)*
Marjorie Reaka-Kudla, *University of Maryland*
Myriam Sibuet, *Institut Francais de Rechercher pour l'Exploitation de la Mer (IFREMER)*
Gregory S. Stone, *New England Aquarium*
Karen Stocks, *University of California, San Diego*
Anne Walls, *British Petroleum*
Gerald R. Allen, *Conservation International consultant*

© SYLVIA A. EARLE

## Introduction

One hundred and twenty years ago the eminent English biologist Thomas Huxley declared that ". . . the cod fishery, the herring fishery, the pilchard fishery, the mackerel fishery, and probably all the great sea-fisheries, are inexhaustible; that is to say that nothing we do seriously affects the number of fish . . . given our present mode of fishing. And any attempt to regulate these fisheries consequently . . . seems to be useless" (Smith 1994). Indeed, throughout most of human history, the seas have been considered so vast and bountiful that they would forever be immune to the impacts of civilization. We have learned very different lessons in the last 50 years, particularly in the coastal ocean where fishes, turtles, marine mammals and even beds of turtlegrass and kelp have been drastically diminished as a result of the removal of resources, the addition of pollutants or other environmental modifications made by human society. The seriousness of the state of the "known" oceans nearshore is now evident to almost anyone who goes to a beach or a fish market, but it has been comforting to assume that those parts of the open sea and the deep ocean far from human endeavors would still remain in their natural state.

We know now that the effects of human civilization can be seen in almost every part of the Earth, from the ozone layer of the upper atmosphere to the glaciers of Greenland and the depths of the ocean. The composition of our atmosphere has been altered by industrial production in the last century to a state never seen over the last 400,000 years. Given the air/sea interaction, it seems inescapable that even remote and seemingly pristine parts of the ocean will experience previously unanticipated human impacts in the next few decades. The "unknown ocean" includes those places and ecosystems that we know exist and have characterized to some extent (Figures 1 and 2) but which are still so remote and unexplored that we lack sufficient information to fully assess the effects of our activities on them. We are at a point now when additional timely research can help us understand how these ecosystems function, forecast how they may be affected by human actions and plan conservation priorities before it is too late. Precautionary protective measures taken in advance of predictable future impacts in these remote areas will be easier to put in place before economic or political interests are organized to oppose them.

In this chapter, we consider several poorly known ocean environments with respect to the need for additional research and possible conservation action. There are many possible candidates—even the waters of our bays and harbors remain largely unknown with respect to many microbial organisms and ecological processes, and little is known even about the life histories and behaviors of common marine species. At the other end of the continuum there are undoubtedly places in the ocean that are still truly unknown—as undiscovered now as the chemosynthetic communities that "feed" on chemicals from hydrothermal vents were before 1977. Our unknown environments here fall between these extremes. They are places whose existence is known and about which we have enough preliminary information to guess at their ecological characteristics and vulnerabilities.

In comparing and ranking these unknown habitats we first considered their characteristic ecological structure and processes, insofar as these are known, to assess several criteria:

- Likelihood that the biological community is specialized (specially adapted) to the habitat
- Known or probable level of biodiversity (number of species) and endemism (species found in no other environment)
- Probable value of basic scientific knowledge to be gained from an undisturbed environment and community
- Known or anticipated human impacts that would have adverse effects
- Potential for irretrievable loss of ecological functions or species diversity from human disturbance

LEFT TO RIGHT: ART HOWARD, NAPRO; LIZ BAIRD; NOAA/U.S. DEPARTMENT OF COMMERCE; NOAA/U.S. DEPARTMENT OF COMMERCE

**Figure 1** *Left to right:* Sea urchin on a deep reef • Porcelain crab on the outer continental shelf • Rattail fish in the deep Atlantic • Deep-water anemone.

- Sufficient preliminary knowledge to form a basis for timely research planning and/or conservation action
- Feasibility of conservation actions with respect to technical capability, economic cost versus benefit and political limitations

For most unknown habitats we lack enough species-level data to evaluate whether some may be sufficiently threatened or endangered to qualify for inclusion on the Red List of IUCN–The World Conservation Union, or to qualify for protection under any nation's endangered species legislation. In the absence of this information, it seems most effective to design management or conservation efforts that protect entire habitats, defined by their special geographic and ecological characteristics. Even without detailed knowledge, we know enough in many cases to plan actions that are precautionary. Although further research is always a desirable component of effective conservation, current knowledge and understanding of ocean ecology and biodiversity is sufficient in some areas to justify taking precautionary steps now that may be refined in the future if additional information so warrants. Where there is the likelihood of severe ecological damage or irretrievable loss of biodiversity, conservation actions should not be held up by a desire for further basic research.

Our Unknown Ocean theme group at the *Defying Ocean's End* Conference distinguished two categories of unknown habitats. The first group of four includes those with evidence of immediate conservation needs because the ecosystem is unique and valuable and is either already exposed to damaging human impacts or at high risk of such impacts in the near future. These are also environments for which there is sufficient data to plan further steps and for which conservation actions are likely to be technically, economically and politically possible. These four "habitats with conservation priority" are: seamounts, deep and remote coral reefs, canyon and seep environments of the continental margin and marine anchialine (cave) habitats. For each of these, members of the working group prepared summaries that describe the habitats and current or anticipated threats to them and then recommended priorities for research and conservation, with timelines and estimated costs. In the discussion of recommended actions that follows, those concerning seamounts are most developed because this remote and little-known environment emerged as a high priority at the *DOE* Conference.

A second group of environments was designated "habitats with research priority." These are places in the ocean with significant ecological importance but for which there is some combination of insufficient data to assess conservation needs, no apparent human impact in the foreseeable future or no apparent mechanism to put conservation measures in place. At present the highest priority for these areas is continued or expanded scientific research, both to understand their basic structure and functioning and to ascertain whether they will need protective measures in the future. The six habitats in this category are: the open-ocean water column; the deepsea benthos (seafloor); ridge and vent ecosystems; polar seas and ice-edge zones; fronts and upwelling regions; and prokaryote (microbial) communities throughout the world ocean. There is high scientific value in all of these environments and continuing research is considered to be the highest priority for them.

## Unknown Habitats with Conservation Priority

### Seamounts

*The environment*

Seamounts are isolated submarine mountains that rise to an elevation of 1,000 meters or more above the ocean floor. They are distributed throughout the world's ocean. Although only about 1,000 seamounts have been named, it is estimated there may be tens of thousands of seamounts in the

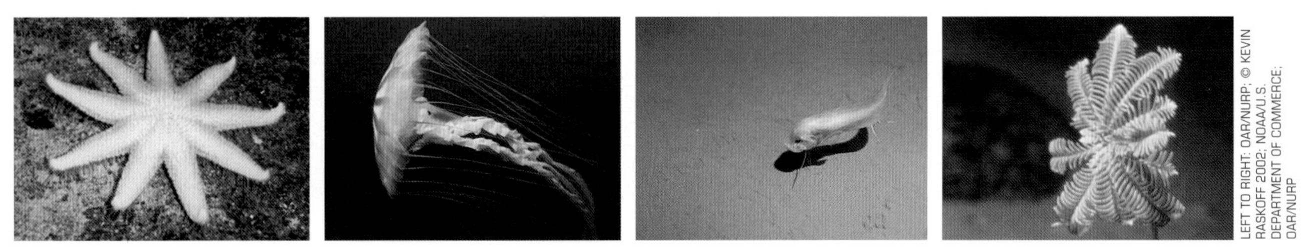

**Figure 2** *Left to right:* Sea star on a deep volcanic rock • Jelly under the Arctic ice • Unidentified fish in the Arctic • Crinoid in the deep Pacific.

LEFT TO RIGHT: OAR/NURP; © KEVIN RASKOFF 2002; NOAA/U.S. DEPARTMENT OF COMMERCE; OAR/NURP

Pacific Ocean alone. The biology and ecology of seamounts has been discovered and studied only in recent years. Only about 350 seamounts have been biologically sampled at all, and fewer than 90 have had quantitative, taxonomically broad surveys. They are of high interest for conservation, however, because this preliminary sampling has identified an unexpected level of highly unique and endemic faunas (Rogers 1994).

Seamounts represent potentially large pools of undiscovered biodiversity in the ocean. Average levels of endemism (the degree to which species are found on one seamount or seamount chain and nowhere else in the ocean) are probably at least 15% and have been documented to be as high as 30–50% on some seamounts. In addition, seamounts may be sites of speciation (evolution of new species), refuges for rare species and stepping-stones for the distribution of species across wide areas of the open ocean. While these functions are not fully understood, they indicate that seamounts may have important roles in the speciation and distribution of species in the ocean. For a more detailed description of seamount ecology and biodiversity, see Chapter 2.

*Threats*

Seamounts are increasingly under threat from deepsea fishing activities. Some characteristic species on seamounts are very long-lived and their biological communities, including deepsea corals, can take hundreds or even thousands of years to develop. Newly developed fisheries for deepsea species are now altering these habitats faster than they can be studied. Trawling on deep seamounts became possible with the constant development of more capable fishing gear technology. Large-scale deepsea commercial fishing began in the 1960s for orange roughy, pelagic armorhead, hoki, oreos and related species, leading to the collapse of many of these populations by the 1990s. It is likely that many of the world's shallow seamounts have experienced some degree of damage from fishing already. Current management efforts in Australia and New Zealand include closure of parts of the seamount fisheries, but developing fisheries in other parts of the world are largely unregulated. Collateral damage from deepsea trawling is a significant effect, destroying complex, long-lived and poorly known benthic seafloor communities in the process of catching fish. One study done on underwater Tasmanian hills indicates that unfished hills had twice the biomass of fished ones and only 10% bare rock compared to 95% bare rock on fished sites (Koslow *et al* 2001).

Research and conservation needs for seamounts are summarized below, and the proposed timetable and costs are presented in Table 1.

*Research needs*

Only a fraction of 1% of the 30,000 to 100,000 seamounts estimated to exist worldwide have been sampled biologically, and this sampling is often very incomplete, documenting only a few species. While it is not feasible to sample all the seamounts in the ocean, targeted sampling is urgently required to fill in current gaps in scientific knowledge so we have a sound basis for predicting conditions in the unsampled seamount areas of the ocean. Existing gaps that should be addressed by future research include:

UNDERSTUDIED REGIONS. Areas such as the southern Pacific/ Southern Ocean, the Indian Ocean and near-equatorial regions have high densities of seamounts, and virtually no biological research has been done on seamounts in these areas. Indonesian-area seamounts are also little explored, but they are likely to support highly endemic and diverse communities given the generally high marine biodiversity of this area. A comprehensive research effort is needed to get representative sampling from a range of latitudes, ocean regions and distances from land. In addition, certain types of seamounts, including very deep and soft-sediment seamounts, are not well known and should be studied.

POORLY KNOWN BIOLOGY. Fishes have been by far the largest focus for biologic sampling, yet seamounts support a diverse array of invertebrate species (animals without backbones) (Figure 3). Top-to-bottom sampling to study the entire seamount community—from bacteria and viruses through the larger species—is needed for several seamounts. These studies should include distribution information and population genetics research to determine the relative isolation and evolutionary history of seamount species.

TIME-SERIES MONITORING. To assess how seamount environments and communities change over time, time-series monitoring of several representative seamounts should be initialized *before* they are impacted by bottom-trawl fishing.

FISHING IMPACT STUDIES. Assessments of bycatch and overharvesting in seamount fisheries, as well as habitat damage and recovery times, should be initiated as soon as possible.

TECHNOLOGY DEVELOPMENT. New tools are needed for exploring and sampling remote, deep and difficult-to-access seamounts. Remotely operated and autonomous vehicles

Table 1

**Research and conservation plans for seamounts**

| Objectives | Actions within one year | Actions within 2–10 years |
|---|---|---|
| Map known ranges of seamount species and begin assessment for IUCN Red Listing of threatened or endangered species | • Convene a workshop for standardizing IUCN marine habitat files and mapping standards | • Continue mapping species distributions, including identifications<br>• Revise species systematics as needed<br>• Collect genetic data to improve understanding of genetic relationships among species<br>• Digitize and enter all data, including maps and images, into appropriate databases<br>• Continue evaluation of species for possible IUCN Red List inclusion<br>• Publish in appropriate research and conservation journals |
| | **Estimated costs: $175K** | **Estimated costs: $450K annually** |
| Create global maps of estimated species diversity, endemism and fishing pressure | • Create preliminary maps to guide conservation and research planning efforts<br>• Analyze current state of seamount fisheries and prospective growth | • Continue refining maps with new data collected |
| | **Estimated costs: $175K annually** | |
| Establish a Seamounts Campaign | • Designate an executive committee, secretariat and working groups | • Continue coordination of research, conservation, legal and public relations aspects of seamount protection |
| | **Estimated costs: $500K annually** | |
| Expand the SeamountsOnline system (Seamounts Online 2004) to hold all available seamount data—biological sampling, environmental characteristics and publications—in an online database for ease of access | • Expand SeamountsOnline to hold all range map data produced by research efforts above<br>• Include EEZ and protected area status for all seamounts | • Create a web-based collaborative work environment to facilitate data sharing, group document editing and strategic planning<br>• Incorporate all existing seamount biological sampling information<br>• Extend SeamountsOnline capabilities to contain images and include mapping features<br>• Translate seamount data from non-English sources for database entry<br>• Establish data format standards and automated upload features for new expedition data |
| | **Estimated costs: $215K annually** | |
| Complete a program of globally representative ecological sampling on seamounts by 2015 | • Conduct two initial planning workshops | • Convene planning workshops twice each year<br>• Support 45 days of ship time for research cruises per year to study 50 representative seamounts<br>• Provide salary support for scientists on seamount research cruises, including data analysis after the cruise<br>• Support new technology development as needed for research |
| | **Estimated costs: $150K** | **Estimated costs: $27.2M total** |
| Create policies to protect seamounts | • Support development of a high seas protection mechanism in consultation with national governments, United Nations, NGOs and non-governmental conservation organizations | • Achieve protection from fishing and other impacts for seamounts within EEZs of three countries<br>• Have 70% of seamounts in high seas and national EEZ waters fully protected from fishing and all other extractive activities |
| | **Estimated costs: Determined by the *DOE* Ocean Governance theme group** | |

Table 1 (continued)
**Research and conservation plans for seamounts**

| Objectives | Actions within one year | Actions within 2–10 years |
|---|---|---|
| Raise international public awareness of seamount values and threats | • Develop an action plan for increasing public awareness of seamounts and the threats to them | • Launch a marketing campaign to discourage catch and consumption of seafood from seamounts; introduce labeling for "seamount-safe seafood"<br>• Enlist cooperation of major seafood wholesalers and seafood restaurant chains<br>• Decrease consumer demand for deepsea fishes from seamounts and create a consequent drop in the related fishing activity |
| | **Estimated costs: Determined by *DOE* Communications and Ocean Governance theme groups** | |
| Design and install a system for monitoring fishing activity on seamounts | • Design a global surveillance system to detect bottom-trawl fishing boats over seamounts<br>• Prepare plans for an international mechanism to monitor wholesale fish markets for seamount species, using DNA-based probes | • Develop and deploy surveillance technology for fishing boats over seamounts<br>• Establish international agreements for monitoring and regulating fishing activity and commerce in seamount fish species |
| | **Estimated costs: Initial development $10M, annual operating costs $200K** | |

that can do sonar and video mapping and collect biological samples are promising approaches.

INFORMATION AGGREGATION. Large-scale patterns on seamounts are currently difficult to assess because data are spread among many sources and are hard to access. A comprehensive effort to aggregate global data and create predictive models will help guide future research and conservation efforts on seamounts. The Web-based program "SeamountsOnline" is a good initial effort to integrate global data from seamounts for easier analysis and synthesis, but it must be expanded.

*Conservation actions*

Commercial fishing is the single largest threat to seamounts and has already resulted in significant habitat destruction in these unique underwater habitats. Reducing or eliminating these impacts, especially from bottom-trawl fishing, is the highest priority for seamount conservation. Recommended conservation actions include:

- Evaluate the status of seamount species for addition to the IUCN Red List of endangered species.
- Create a new framework for high seas marine protected areas or other fisheries controls. There is currently no policy framework for protecting seamounts as habitats, but efforts to persuade the United Nations General Assembly to adopt a moratorium on bottom-trawl fishing on seamounts in international waters (Deep Sea Conservation Coalition 2004) should be supported.
- Encourage the creation of marine protected areas (MPAs) for seamounts within EEZs as the best mechanism for protecting some of these unique ecosystems.
- Develop new monitoring systems to enforce these marine protected areas or seamount fisheries restrictions, including:
  - Monitoring the presence of fishing vessels over seamounts with satellite sensors
  - Installing acoustically triggered underwater monitoring devices that report the presence of vessels within a given range of seamounts via satellite
  - Establishing international arrangements to monitor wholesale fish markets for seamount endemic species, perhaps using a DNA-based probe system
- Establish public education campaigns, similar to the "dolphin-safe tuna" campaign, to decrease demand for seamount species that are long-lived and slow to reach reproductive age and thus very vulnerable to overfishing.

- Establish a regulatory framework and appropriate restrictions on future mining, oil exploration or other commercial activities that will be potentially damaging to seamounts.

## Deep and remote coral reefs

*The environment*

Coral reefs have been studied and well characterized in many parts of the world, adding significantly to our understanding of marine biology and ecology (Wilkinson 2000, 2004). The beauty, complexity and vulnerability of coral reefs have made them a primary focus of marine conservation efforts throughout their range (Bryant *et al* 1998). Many reefs and some reef zones have not been studied to any extent, but they appear to be critical for the long-term conservation and sustainability of marine ecosystems. We focus here on the deeper parts of reefs beyond the reach of bright sunlight or SCUBA-diving researchers, and on remote reefs and shallow oceanic banks that are too distant or inaccessible for easy exploration. In this discussion, "deep reefs" refers to reef zones below about 50 meters but contiguous with shallow reefs associated with islands or continental margins. Much deeper coral beds and reefs occur on seamounts and continental slopes, but these habitats are considered in the preceding section on seamounts and also in Chapter 2. The focus in this section is on the "twilight zone" reefs between about 50 and 150 meters, and on remote and unexplored shallow reefs less than 50 meters deep.

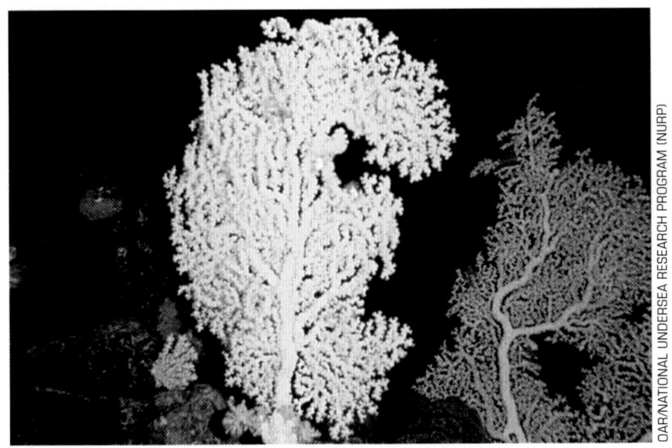


**Figure 3** Deepsea coral found on the side of a seamount more than 300 meters deep off Hawai'i.

Most coral reef studies have been constrained by the depth limits of SCUBA-diving technology—about 70 meters. The region below this level is often called the twilight zone because of the decreasing sunlight at this depth caused by attenuation of light through the water column (Pyle 1996). Because it is below ordinary SCUBA-diving depths, this zone around tropical coral reefs remains largely unexplored, but preliminary work has shown that it contains many new species (Randall & Pyle 2001, Colin *et al* 1986) and offers new insights into the biological and ecological properties of coral communities (Porter 1973, Fricke & Knauer 1986, Thresher & Colin 1986). Twilight zone reefs may also be subject to many of the same threats as the shallow-water ecosystems.

Although remote reef ecosystems have been visited periodically by scientific expeditions over the last 200 years, research on them is not systematically planned in any current program due to limitations of funding and accessibility. Long distances for research vessels and lack of nearby logistical support make these areas particularly expensive to study. Many extensive open-ocean shallow bank systems are poorly studied for the same reasons. But the very remoteness of these areas promises completely unexpected discoveries. Some recent expeditions have discovered extraordinarily high levels of plant and animal biodiversity including many rare and/or undescribed species (Reaka-Kudla 1997, South, Skelton & Yoshinaga 2001, Bouchet *et al* 2002). These remote ecosystems grow in importance as repositories of biodiversity and refugia (safe havens) as more accessible locations are increasingly affected by human commercial activities and climate change.

*Threats*

Remote and deep coral reefs can be subject to overfishing, global warming, disease and sedimentation from coastal development, with the extent of each threat varying for shallow and deep regions. For example, coral bleaching is mainly a shallow reef problem, but the disintegration of shallow reef structures can cause increased sedimentation in deeper areas as the calcium carbonate debris rains down through the water column. This kind of increased sedimentation can also lead to increased turbidity (suspended sediment) at shallower depths, dramatically reducing the intensity of sunlight that reaches the deeper reefs and leading to potentially catastrophic impacts on the deep reef communities. Overfishing occurs at all depths, and sedimentation caused from deforestation and development of coastal areas is directly related to the density of nearby human settlements.

*Research and conservation priorities*

Conservation action for deep and remote reefs will require both research and communications efforts. We recommend a campaign to study, conserve and popularize these deep and remote reef ecosystems. The program should include systematic surveys of these reef systems, coupled with the development of conservation and long-term monitoring strategies, as well as a communications plan to showcase the uniqueness and vulnerability of these areas. An effective campaign would have a number of major elements summarized below, with costs and timelines highlighted in Table 2.

PREPARATION AND PRIORITIZATION. Planning for a global survey and conservation initiative will require a consultative and review process coupled with the identification and development of methods, opportunities and technologies to be used. The objectives will be to:

- Identify patterns of island, seafloor and ocean current interactions for areas to be surveyed
- Develop a strategy for using, adding to, coordinating and assessing global databases
- Develop technological capacity in remote areas, including training, vessels, vehicles and research equipment
- Select priority areas for the communications strategy and establish a timetable for implementation

EXPLORATION OF TWILIGHT ZONE REEFS. To date, most deep reef surveys have been conducted by divers using SCUBA rebreathers, mixed gases and other technical diving approaches (Colin 1999, Pyle 1998, 2000, Sharkey & Pyle 1992). This approach still has many advantages, but we recommend that future deep surveys be carried out using submersible vehicles whenever possible to maximize safety and time on the seafloor. Although large submersibles like *Alvin* are prohibitively expensive for research of this kind, many smaller vehicles that operate to depths of 1,000 meters or more have been used (Fricke & Schuhmacher 1983, McIntyre *et al* 1991) and can be available on a charter basis. A small submersible that can accommodate an observer and pilot, be supported from a fairly small

Table 2
**Research and conservation plans for deep and remote coral reefs**

| Objectives | Actions within one year | Actions within 2–10 years |
|---|---|---|
| Design exploration and research program | • Select 15 to 20 priority research sites<br>• Establish processes for integrating new data into appropriate databases<br>• Identify existing research and support capacity near priority sites and begin developing additional infrastructure and expertise | • Develop global set of monitored sites<br>• Assess results of initial surveys<br>• Design conservation plans for selected sites<br>• Completely assess survey results from all priority sites<br>• Establish a global set of sites, incorporating all monitoring and biodiversity data on OBIS |
| Survey priority areas | • Complete survey and assessment of two remote shallow reefs and two deep reef sites | • Install monitoring programs in first group of sites<br>• Complete survey and assessment of 15 remote shallow and 18 deep reef sites |
| | **Estimated costs: $200K for shallow, $700K for deep, total $900K** | **Estimated costs: $1.5M for shallow, $6.5M for deep, total $8M** |
| Develop and implement conservation strategies for surveyed sites | • Develop and implement conservation strategy at one site | • Develop and implement conservation plans at 49 additional priority sites |
| | **Estimated costs: $50K initially, then $30K annually** | **Estimated costs: per site, $50K initially, then $30K annually, total $4M** |
| Implement communications plan for remote shallow and deep reefs | • Magazine articles from one or two new sites<br>• Begin work on television and film projects | • Develop, maintain and update aquarium exhibit(s) on deep reefs<br>• Complete television/film program(s)<br>• Publish several pictorial and scientific books |
| | **Estimated costs: $100K** | **Estimated costs: $3M** |

vessel, and have an operating depth of at least 400 meters and dive duration of six to eight hours, would be a valuable exploratory tool. It should be equipped with still and video cameras, an exterior lighting system and biological collecting devices, including a suction sampler and manipulator for sampling fish and benthic (bottom-dwelling) invertebrates. The submersible program would be organized to develop an inventory of certain key indicator groups (fishes, crustaceans, corals and sponges) and an ecological assessment of the overall deep reef community, perhaps using protocols for rapid assessment similar to those used for shallow reefs by Conservation International (McKenna & Allen 2002, forthcoming, McKenna, Allen & Suryadi 2002) and other organizations.

Two other promising technologies for surveys of twilight zone reefs are remotely operated vehicles (ROVs) and autonomous underwater vehicles (AUVs). Small ROVs are easily deployed from boats and can be used to conduct video and photographic surveys at low cost. Although most small ROVs lack sampling capability, they can scout areas for later sample collecting by SCUBA divers or manned submersibles. Most currently available autonomous vehicles are specifically designed to collect oceanographic data in the water column, but one recent vehicle, SeaBED (Singh *et al* 2004), is designed specifically to "fly" over the seafloor and make video swaths or photo-mosaics at high resolution. This vehicle, which can also be deployed from a fairly small boat without a winch or handling system, can be programmed to cruise a few meters above the bottom—following contours and avoiding obstacles—to create a continuous video record of seafloor life and features.

Exploration of remote reefs and banks. Shallow reef survey teams for remote reefs (Figure 4) should include specialist taxonomists for key biological groups (such as soft and hard corals, fishes, crustaceans, sponges and echinoderms) and reef ecologists that can assess the overall "health" and ecological parameters of each survey site. Approximately eight to twelve scientists would participate

NEIL HAMMERSCHLAG

**Figure 4** Damselfish flutter around a coral colony on a remote atoll reef in the Indian Ocean.

in each survey, depending on the size of the reef and available vessels and facilities. Participation could be arranged through existing regional networks of coral reef monitoring teams (Linden *et al* 2002). As with the deep reefs, a standardized protocol would be used at each dive site, and related oceanographic data (temperature, salinity, oxygen profiles, primary production, currents) should be collected wherever possible. The main survey method would consist of direct underwater observations and video and photo records made by SCUBA divers. Relatively few specimens will be collected, except where necessary, to facilitate the identification of new species. Both deep and shallow reef surveys could be supported by chartered live-aboard dive vessels at far lower cost than using standard oceanographic research vessels.

ANALYSIS AND ARCHIVING OF DATA. For both deep and remote shallow reef survey programs, data in the form of survey records, video transects, digital photographs and mosaics, maps, preserved specimens and genetic material need to be catalogued, identified, annotated and entered into one or more centralized databases for later retrieval and analysis. Where possible, the Ocean Biogeographic Information System (OBIS) and its specialized component databases, like Fishbase and Cephbase (Zhang & Grassle 2003, Wood & Day 2004, Froese & Pauly 2004), should be used as the repository for new data from deep and remote reefs. Funding for exploratory expeditions needs to include adequate support for post-expedition curating of specimens and data.

DEVELOPMENT OF CONSERVATION STRATEGIES. Experience with well-studied shallow reef environments—coupled with new information from surveys of deep and remote reefs—will allow development of model conservation strategies and options tailored to specific remote and twilight reef areas. These plans should include the required research, financing mechanisms, institutional or governmental structures and the necessary national, regional or international agreements. Specific requirements include:

- Plan for long-term monitoring requirements, with sites, specific monitoring parameters or indices, appropriate technology, infrastructure and funding to provide the necessary continuity
- Assess and standardize survey methods for exploring and saving remote reefs and twilight zones
- Integrate newly explored deep and remote reefs into a coordinated global set of monitoring sites and protected areas
- Develop site-specific, long-term conservation strategies, including legal and regulatory agreements and multiple-zoned protected areas

ESTABLISHMENT OF A COMMUNICATIONS STRATEGY. A communications effort must be developed to highlight the uniqueness, biodiversity, vulnerability and global value of twilight zone and remote reef regions. The communications should be targeted at the public with a view to influencing policy and governments and might include photographic books and magazine articles in publications such as *National Geographic*, IMAX movies, public television programs and linked aquarium and museum exhibits.

## Continental margins, canyons and seeps

### *The environment*

Continental margins are the boundary zones between the shallow shelf regions that surround most continents and the deeper "abyssal plains" of the seafloor. Also called the continental slopes, these margins are regions of steep profile, often with deep canyons, other rugged topography and unusual geophysical features. Large underwater canyons—such as Monterey Canyon off California, the Gully off Nova Scotia and the Murray Canyons off Australia—are some of the largest and most dramatic topographic features on the planet, dwarfing the Grand Canyon in depth and extent. The flow of underwater ocean currents over slope and canyon regions can be strong compared to the abyssal deepsea, and seismic events that trigger turbidity currents (underwater avalanches of suspended sediments) have powerful impacts on the physical and biological features of slope habitats (Rhodas & Hecker 1994).

The margins are characterized by many species-rich deepsea communities, mostly dependent on food produced in the upper layers of the ocean (Hecker *et al* 1983, Hecker 1994, Vetter & Dayton 1998). Large coral colonies, sponges and their associates thrive on rocky outcrops and are especially abundant in submarine canyons and on carbonate mounds. These communities flourish where strong bottom currents carry food, supporting the growth of these large seafloor colonies and concentrating fish living close to the seafloor. The heads of submarine (underwater) canyons are often particularly productive nursery areas for fish (Yoklavich *et al* 2000).

Cold seep environments are a relatively new discovery on continental slopes (Paull *et al* 1984, Embley *et al*

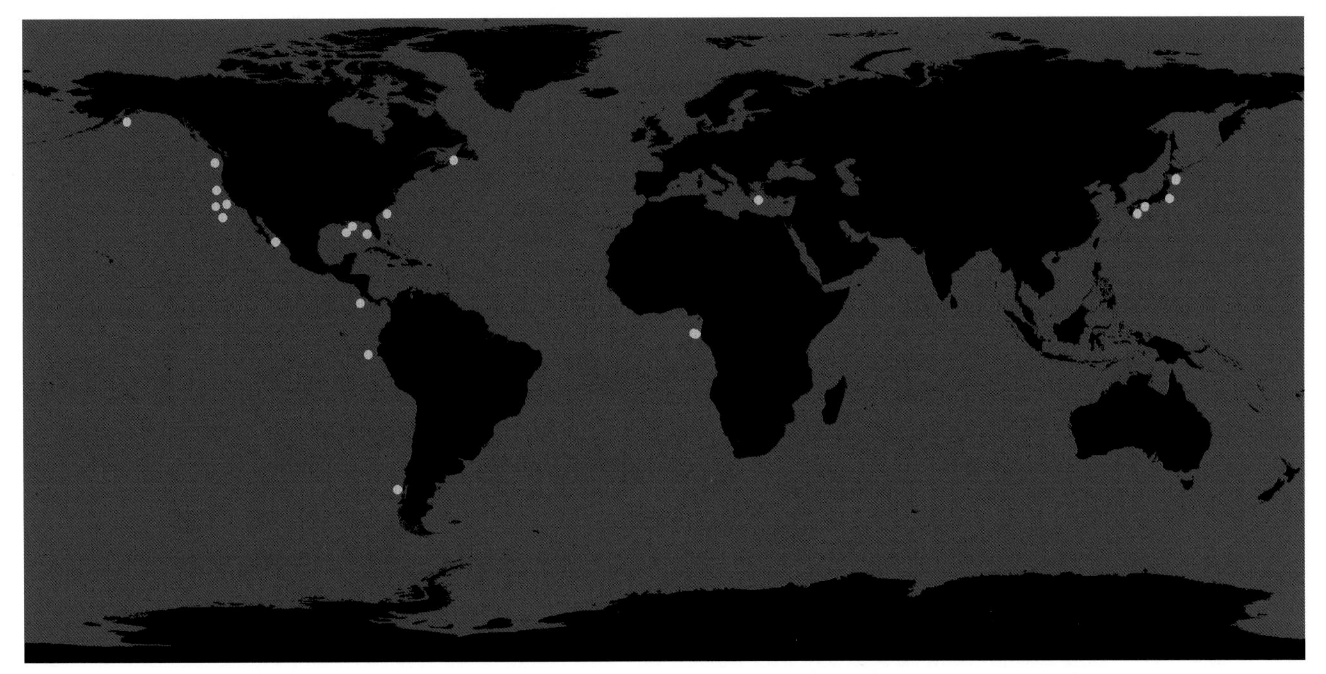

**Figure 5** Worldwide locations of known cold seep environments (adapted from Sibuet & Olu 1998 and Sibuet & Olu-LeRoy 2002).

1990) about which we are only beginning to learn. These are "chemosynthetic" ecosystems dependent on chemicals seeping from the seafloor for nutrients and life, rather than on photosynthesis that uses sunshine in shallower areas to produce food. The previously unknown species of large animals in these chemosynthetic sites have been discovered where liquid hydrocarbons, methane, salt brines and other compounds seep from subsurface geological formations providing energy—essentially food—for these strange organisms, analogous to deepsea hydrothermal vent communities. This form of life on Earth was completely unknown until the late 1970s. The relationships between the many newly discovered species, and bacteria and other microbe (microscopic animal) assemblages in sediments beneath the seafloor, are only beginning to be understood (Brooks *et al* 1987, Juniper & Sibuet 1987, Barry *et al* 1996, Sibuet & Olu 1998). The diverse environments of the continental margins, including canyons and cold seep habitats, are a priority for research and conservation because they support: a vast reservoir of undiscovered species on both soft sediments and hard rock bottoms; unique assemblages of microbes and invertebrates dependent on chemical energy from a variety of geological formations; and highly productive canyon systems that support commercial fisheries.

Biodiversity in the sediments of the upper continental slope is the highest of any known deepsea environment—each new region studied yields collections where 50% to 90% of the species found are new to science (Snelgrove 1998, Grassle & Maciolek 1992, Poore & Wilson 1993, Merrin 2004). In submarine canyons and seep environments (Figure 5), submersible research vehicles are able to explore dense clusters of species similar to, but different from, those found in hydrothermal vent systems. New higher taxa are continually being discovered in these unusual chemical environments (Hecker 1985), and the comparison of seep fauna with both deepsea hydrothermal vent forms and shallow water species promises to provide important insight into the evolutionary history of many types of animals (Black *et al* 1997, Goffredi *et al* 2003).

### *Threats*

Many of these continental margin environments are—or may soon be—under threat from pollution, oil exploitation, waste disposal and overfishing. Activities that have been prohibited on continental shelves are moving to deep-water environments where the ecosystems are poorly understood and monitoring is more difficult. Many areas of the continental margins are known or thought to contain valuable mineral deposits (Glover & Smith 2003, Rona 2003), and

Table 3

**Research and conservation plans for continental margins, canyons and seeps**

| Objectives | Actions within one year | Actions within 2–10 years |
|---|---|---|
| Create global maps of seafloor bathymetry and habitat features of the continental margins, species diversity and endemism, areas of mineral and oil leases, waste disposal and fishing activity | • Conduct a workshop to coordinate and standardize marine mapping standards and IUCN marine habitat authority files<br>• Begin assembly of existing data for mapping projects | • Assemble, edit and produce maps from scientists working in these areas<br>• Develop protocols for long-term monitoring of deepsea continental margin sites<br>• Develop a minimum set of standards for seep surveys<br>• Review and enter species data into OBIS database<br>• Continue the process of identifying species for Red List protection<br>• Begin the process for formal IUCN Red Listing of threatened species |
| | **Estimated costs: $200K** | **Estimated costs: $1.7M** |
| Implement ecological sampling and monitoring of priority continental margin sites | • Plan and fund initial surveys of priority seep and soft-sediment sites<br>• Begin survey programs including sampling, video survey and sample processing of species, including identifications, data analysis, public outreach, database entries and publications<br>• Initial sites possibly in Gulf of Guinea | • Make results from initial sites available as baselines against which to assess future changes<br>• Launch additional survey expeditions to other sites including the New Zealand continental margin, the Banda and Sulu Seas and others to be determined<br>• Establish a long time-series study of continental margin soft-sediment communities. Do initial sampling in year 2 at each of three priority regions, to be resampled in years 5 and 8 |
| | **Estimated costs: $3.5M per site** | **Estimated costs: 10 seep site expeditions at $3.5M each, total $35M; 9 soft-sediment site expeditions at $3.5M each, total $31.5M** |
| Build an expanded OBIS system to hold all available continental margin data (biological sampling, environmental characteristics and publications) as an online database available to scientists and the public | • Expand the Continental Margin Database to hold all maps of species ranges produced in the efforts described above and to include EEZ and protected area status for continental margin regions | • Create a web-based collaborative work environment with data sharing, group document editing and strategic planning functions<br>• Incorporate all existing sampling data, improve capability for holding images and provide mapping tools<br>• Translate data from non-English sources<br>• Develop a data format standard and automated upload feature for new expedition data |
| | **Estimated costs: $265K annually** | |
| Establish a continental margin communications campaign | • Establish an executive committee and appropriate working groups<br>• Hire a coordinator, secretary, program officer, legal expert and communications director | • Raise public awareness about continental margins via magazine articles, books, television programs, exhibits and a website |
| | **Estimated costs: $650K** | **Estimated costs: $750K annually for operating budget, salaries and expenses** |

some are leased for oil and mineral extraction. In many parts of the world, submarine canyons are heavily fished (Moore 1999, Moore & Mace 1999), and impacts on long-lived species are likely to be similar to the effects of overfishing on seamounts (see above and Chapter 2). Large deposits of methane hydrates on continental slopes have led to interest in possible future extraction for production of methane gas. Policy and regulatory issues associated with exploitation of methane hydrates and other hydrocarbons at cold seeps are only beginning to be addressed. This issue is of particular importance because, in addition to impacts of extraction techniques on the local biological communities, there is a risk of catastrophic instability of some hydrate structures if material is removed—such as a huge underwater avalanche, for example—that could impact not only large areas of the seafloor, but potentially the water column and atmosphere as well (Renssen *et al* 2004).

The rich biodiversity of canyon ecosystems has been recognized with varying degrees of protection. The Monterey Bay National Marine Sanctuary encompasses much of Monterey Canyon, and the Gully off Nova Scotia was declared a marine protected area in May 2004, with limitations on oil exploration and fishing. Efforts are in place to secure protection for other canyon regions, including the New England canyons (Natural Resources Defense Council 2001) and Murray Canyon in Australia. Research and conservation recommendations for these areas are presented below, and the timetable and costs are summarized in Table 3.

*Research needs*

The 200-mile Exclusive Economic Zones (EEZs) of most countries have canyons and seeps that are unexplored. There is only one area where long-time-series sampling of megafauna (large animals) has been conducted and another where the macrofauna (medium-sized animals) in sediments is sufficiently well known to assess change. Baselines need to be established in EEZs throughout the world and several long-time-series stations should be established or designated as centers for study of ecological processes in the deepsea.

UNDERSTUDIED AREAS. In addition to studies currently underway, the continental margins of Indonesia and New Zealand are in urgent need of study. New kinds of seeps are also being discovered in the Mediterranean, western North Atlantic, Southeast Atlantic and in the New Zealand EEZ—one in New Zealand is characterized by flows of mercury. Areas of active oil exploration or mining should become priority areas for research.

UNDERSTUDIED TAXONOMY. The training of the next generation of deepsea taxonomists who identify and classify newly discovered plant and animal species is of particular concern. For many plant and animal groups there is no longer an authority for confirmation of identifications. Use of new genomic methods to study seep organisms, especially seep microbial communities, should be a priority. Such studies have potential for major discoveries on the range of biodiversity on the planet and the origins of life on Earth.

PHYLOGEOGRAPHY OF COLD-SEEP SPECIES. Genetic studies of organisms from seeps and deepsea coral communities are needed to understand evolutionary processes in the deepsea. This information will be used in determining locations for marine protected areas.

LIFE HISTORY STUDIES OF DEEPSEA FISH. It is unlikely that commercial exploitation of any deepsea fish below 1,000 meters is sustainable because of their slow maturation to reproductive age (ten to fifteen years). Scientific studies are needed to unequivocally define limits on fishing of deepsea species.

*Conservation recommendations*

NEW LEGISLATION TO SUPPORT EXPLORATION AND PROTECTION OF THE LIVING RESOURCES OF THE CONTINENTAL MARGINS WITHIN NATIONAL EXCLUSIVE ECONOMIC ZONES (EEZS). Biodiversity of continental margin areas must be assessed and cataloged urgently—in an accessible format such as the Ocean Biogeographic Information System (OBIS) (Zhang & Grassle 2003). Waste disposal (including garbage, sewage and sludge, dredge spoils and industrial waste) should be prohibited or at least strictly regulated in these environments. Immediate consideration should be given to establishing some large marine protected areas (MPAs) in continental margin depths below 1,000 meters. Unlike the open-ocean seafloor, continental margins fall mainly within national EEZs. This should remove the complexity of negotiating international agreements on high seas protection but does require many nations to develop an understanding of the importance and vulnerability of their continental slopes and take appropriate action at a national level.

DEEPSEA TIME-SERIES MONITORING. The bottom of the ocean is warming and the changes in life on the seafloor must be watched as an important indicator of global change over large areas of the planet. Time-series monitoring from ocean floor observatories should include temperature profiles, current patterns, methane flux and other chemical signatures, and biological parameters. These data will provide essential information on the role of these systems in the planet's carbon cycle and their responses to global climate changes.

A CAMPAIGN FOR AWARENESS OF LIFE IN THE DEEPSEA. There are currently no accurate and publicly accessible depictions of the richness of life in the canyons, seeps, rocky outcrops or sediments of the continental margin seafloor. While the public is increasingly aware of hydrothermal vent communities through films, television, books and magazine articles, a plan to foster understanding and appreciation of the wonders of deep canyons and newly discovered seep sites must be developed to provide a broad basis of support for conservation measures.

## Anchialine (marine cave) habitats

### *The environment*

Anchialine habitats are underground and underwater cave environments containing saltwater and are formed either by the dissolving of limestone (karst) platforms by seawater during periods of low sealevel or by lava flows from underwater volcanic eruptions. Anchialine limestone caves may occur worldwide in locations where carbonate formations are dissolved through exposure to seawater, although conditions for their formation seem most common along coastlines in tropical and subtropical areas. Anchialine habitats include both caves having only underwater openings and blue holes (cenotes) that have openings on land. According to a resolution of the International Congress of Marine Cave Biology held in Bermuda in 1984, anchialine habitats are defined as

> *flooded inland marine caves and groundwaters that lack any direct surface connection with the open sea, having a wide range of different salinities and showing the following subterranean features:*
> - *horizontal zonation of salinity and oxygen*
> - *relative lack of food*
> - *limited accessibility for marine fauna*
> - *darkness*
> - *presence of stygomorphic organisms (those that live in complete darkness)*

Many of the best known anchialine limestone caves are on the Yucatan Peninsula, which has the world's largest anchialine cave—the 130-kilometer-long Ox Bel Ha cave system—along its Caribbean coast, Bermuda, the Bahamas,

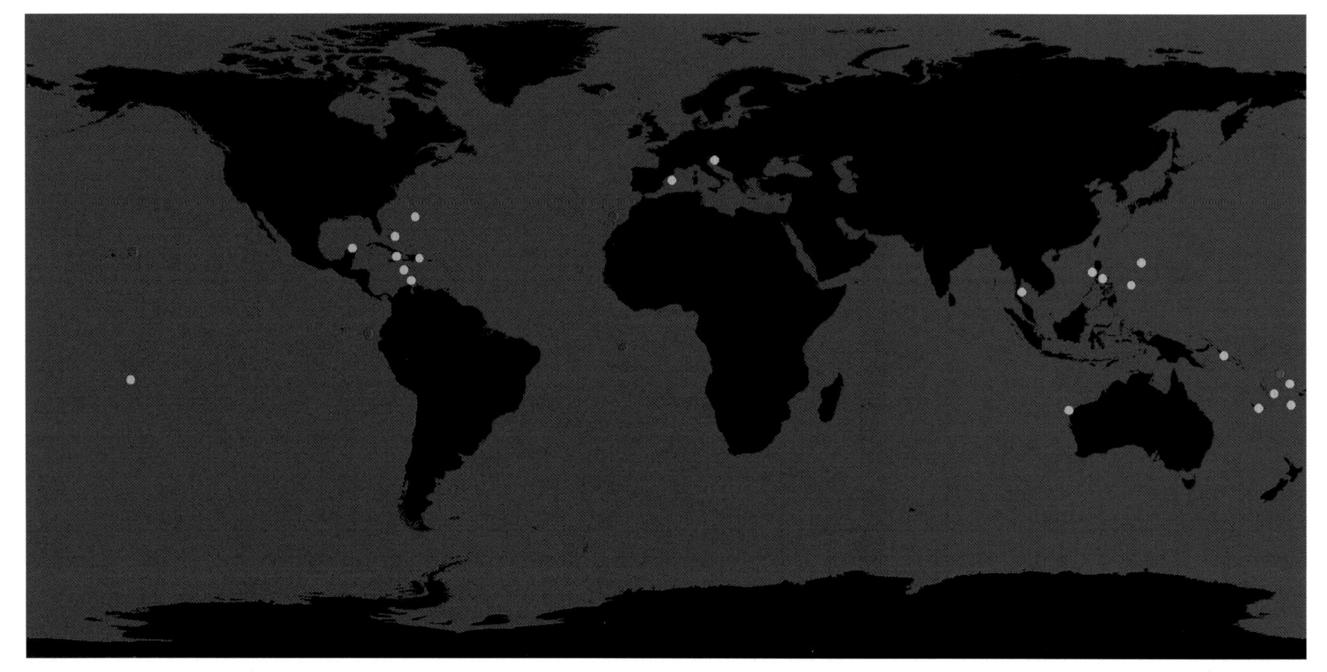

**Figure 6** Location of known marine caves. Limestone (karst) caves indicated in gold; volcanic caves in red (Iliffe 2000).

the Balearic Islands and Italy. Notable anchialine lava tube caves are found in the Canary Islands, Hawai'i and the Galápagos (Figure 6). These caves were first described in the 1960s (Reidl 1966), and subsequent researchers have documented the unique fauna that live in these isolated environments (Iliffe 1992, 2000, Yager 1994, Sket 1996, 1999).

The organisms isolated in marine caves are adapted to extremely low nutrients, low oxygen, no light and no disturbance. The unique assemblages of marine animals in anchialine environments include an extremely high percentage of previously unknown species and even higher levels of biological taxonomy (genus, family, order, class, etc.). Recent investigations in Bermuda and the Bahamas have cataloged more than 300 new species, 50 new genera, ten new families, two new orders, and discovered a whole new class of living invertebrates, the Remipedia (Yager 1981). These "living fossils," which represent one of the most basic early types of Crustacea, had been previously known only as fossils from the Paleozoic era, 230 to 600 million years ago (Figure 7).

Crustaceans make up the bulk of cave-dwelling species and include representatives of both common (copepods, mysids, decapods) and rare (thermosbenaceans, remipedes) groups (Kornicker & Iliffe 2000). The known occurrence of these species in caves around the world is summarized in Table 4. These organisms are highly endemic (unique) to particular cave systems, and species or higher taxonomic levels are often restricted to a single island or a single cave. Although these unique animal assemblages have been documented primarily in caves in Bermuda, the Bahamas, Cuba, Hispaniola, the Yucatan Peninsula, the Balearic Islands, the Canary Islands, southern Italy and western Australia, they may well be present in isolated and still unexamined caves wherever limestone or volcanic cave systems are present, like other areas of the Mediterranean, southeastern China, Indonesia and islands in the Indo-South Pacific.

Animals previously thought to be limited to inland anchialine caves have recently been discovered in completely submerged caves on the seafloor. The unusual genetic and morphological (shape) characteristics of the anchialine animals raise questions about their origins. One theory holds they are derived from deepsea species, with similar adaptations to low food levels and darkness (Iliffe, Jickells & Brewer 1984, Hart, Manning & Iliffe 1985). The discovery of copepods in Bermudian caves that are closely related

**Figure 7** Possible new species of Remipedia from Big Fountain Blue Hole in the Bahamas, collected by T. Iliffe.

to deepsea species supports this idea (Boxshall 1984). An alternative hypothesis (Stock 1986, Danielopol 1990) is that they evolved from shallow-water species that entered caves from surrounding coastal waters, evolving over time into new species adapted to their highly specialized anchialine environments. The very unusual distribution of some of these species with a broad range in tropical cave locations worldwide ranging from western Australia to both sides of the Atlantic suggests they may be relicts from the ancient globe-girding Tethys Sea known to exist between about 50 and 135 million years ago (Humphreys 2000).

*Threats*

Both anchialine caves and the extremely specialized, often ancient and highly endemic fauna that inhabit them can be at great risk from a number of human activities. Because of the special nature of their environments and limited food supply, cave-dwelling species are inherently rare and vulnerable. Their high degree of specialization to cave habitats and the vulnerability of those habitats to human influences makes all cave species—some surviving for millions of years—highly susceptible to stress or rapid extinction. In Bermuda alone, 25 cave-dwelling species are on the 2003 IUCN Red List of Threatened Species (IUCN 2003). From a research perspective, the rare and often primitive types of crustaceans and other organisms found in caves provide important insights into the evolution of these groups and the mechanisms of their dispersal into anchialine environments around the world (Humphreys 2000).

Table 4

**Number of species inhabiting anchialine caves from the Bahamas, Bermuda, Yucatan Peninsula (Iliffe 2000), northwestern Australia (Humphreys 2000), Lanzarote in the Canary Islands (Iliffe 2000) and worldwide (Pesce 2001)**

| | Bahamas | Bermuda | Yucatan | NW Australia | Lanzarote | Worldwide |
|---|---|---|---|---|---|---|
| Ciliates | | 2 | | | | 2 |
| Sponges | 3 | | | | | 4 |
| Turbellarians | | | | 1 | | 1 |
| Annelids | 1 | 2 | | | 5 | 10 |
| Molluscs | | 2 | | | | 5 |
| Mites | | 5 | | | | 5 |
| Tantulocarids | | | | | | 1 |
| Copepods | 18 | 20 | 8 | 6 | 10 | 55–60 |
| Ostracods | 14 | 19 | 2 | 1 | 4 | 40–45 |
| Phyllocarids | 1 | 1 | | | | 2 |
| Decapods | 6 | 6 | 10 | 2 | 2 | 95–100 |
| Isopods | 7 | 5 | 6 | 1 | 2 | 35–40 |
| Amphipods | 6 | 8 | 5 | 12 | 3 | 95–100 |
| Tanaidaceans | | 2 | | | | 2 |
| Mictaceans | 1 | 1 | | | | 2 |
| Cumaceans | 9 | 6 | | | | 15 |
| Mysids | 3 | 2 | 2 | | 1 | 35–40 |
| Thermosbaenaceans | 2 | | 1 | 1 | 1 | 30–35 |
| Remipedes | 8 | | 1 | 1 | 1 | 11 |
| Chaetognaths | 1 | | | | | 4 |
| Fishes | 1 | | 2 | 2 | | 10 |
| **Total** | **81** | **81** | **37** | **27** | **29** | **459–494** |
| Percent crustaceans | 92.6 | 86.4 | 94.6 | 88.9 | 82.8 | 91.5 |

**Source:** Table courtesy of T. Iliffe.

In highly developed and populated places like Bermuda, the threats include vandalism and souvenir collecting of cave minerals like stalactites, dumping of trash into cave openings, quarrying, cesspit leakage, deep-well injection of wastes, construction work and drilling water wells above or near caves. Church Cave in Bermuda contains the largest underwater lake (1,500 square meters) and, on an island nearby, Bitumen Cave is the deepest underwater cave (25.5 meters). Both contain endemic species, of which nine are on the IUCN Red List of Threatened Species. These caves are located on the grounds of a resort hotel, where construction of luxury condominiums was proposed above the caves. Concerns were raised about possible collapse of part of the cave system and nutrient pollution from treated wastewater used to irrigate the adjacent golf courses. The project was allowed after developers threatened to close down their hotel and put 400 local people out of work, but after the project was approved, the hotel was shut down anyway and the employees dismissed. More recently in Bermuda, a large cave with exquisite natural beauty was discovered by blasting in a privately owned limestone quarry. Quarry workers who entered the cave smashed many delicate cave crystals, and blasting operations shattered meter-thick flowstone formations within the cave. Despite the discovery of a fifteen-meter-deep lake in the cave found to contain at least four endangered species on the IUCN Red List, the quarry operator is now petitioning the government for permission to quarry this cave out of existence.

In almost any developed region, pollution of the groundwater from human sewage, agriculture or injection of sewage waste into deep wells can lead to organic pollution of cave waters (Iliffe, Jickells & Brewer 1984). Chemical, microbial or nutrient pollution can create toxicity or lead to "blooms" of microbes that consume available oxygen and endanger anchialine species that are adapted to low nutrient and low food conditions. Lacking sunlight, the cave ecosystem has no photosynthesis to replenish oxygen, and any reduction in its levels due to excess microbial growth can seriously affect

COLIN KLIEWER

**Figure 8** Sampling zooplankton in a Yucatan marine cave.

Table 5
**Research and conservation plans for anchialine (marine cave) habitats**

| Objectives | Actions within one year | Actions within 2–10 years |
|---|---|---|
| Compile existing data and research plans | • Develop comprehensive catalogs and maps, from literature and museum collections, of all known anchialine species and all known underground and underwater marine karst and volcanic cave systems where anchialine biota may exist<br>• Develop preliminary estimates of relative threats to all global marine cave areas<br>• Establish coordination with existing karst environment researchers and conservation groups, including cave-diving organizations<br>• Develop a comprehensive and prioritized plan for documentation of the fauna<br>• Develop a comprehensive and prioritized global conservation action plan<br>**Estimated costs: $500K** | |
| Field surveys of cave environments | | • Complete field investigation, including geological, chemical and biological characterization, of at least 90% of all cave environments in every major coastal or marine karst region of the world<br>• Add new data on cave biodiversity to the Ocean Biographic Information System (OBIS) database<br>**Estimated costs: $750K annually** |
| Develop and implement conservation plans where needed | | • Identify threats to specific anchialine environments<br>• Develop conservation plans to address each situation<br>• Work toward establishing and enforcing protective regulations at local and national levels<br>• Promote awareness of the unique qualities and vulnerability of anchialine habitats, and work with diving and conservation organizations to protect caves from over-exposure to recreational diving<br>**Estimated costs: $250K annually** |

cave species. Although it is an essential tool for exploring and understanding marine cave geology and ecology (Figure 8), cave diving as a recreation also has the potential to disturb, contaminate and degrade these habitats. Finally, evidence is now emerging that cave communities can be affected by changes in temperature related to global climate shifts (Chevaldonne & Lejeusne 2003). Research and conservation priorities, and costs, for these threatened environments are presented in Table 5.

## Unknown Habitats for Continued Research

The working group identified and discussed several other ocean regions for which the biodiversity and ecological structures are also relatively unknown. They are all areas that have great significance to the overall stability of the global ocean ecosystem and to the planet as a whole. Because of their great extent and/or their remoteness from human activity, these regions are judged to be less threatened with disturbance or disruption, however, and therefore of a lower priority for conservation action than the four regions described above. While there are potential threats that can be identified for many of these six areas, in general we lack sufficient information to predict the severity or timing of the threats or to specify protective actions.

However, all of these regions command a high priority for continued and accelerated research into their basic physical, chemical and biological properties and functions. There is much fundamental knowledge about the origin, adaptation and diversity of life to be gained from modern exploration and investigation of these parts of the ocean. Gaining a fuller understanding of the functioning of these ecosystems is also a basis for assessing their vulnerability to damage from human or natural causes and planning conservation actions in the future.

© RICHARD HERRMAN/SEAPICS.COM

**Figure 9** Very little is known about the life history and ecology of some fishes, such as the very large ocean sunfish, in the open-ocean upper water column.

### Open-ocean water column

With a mean depth of 3,700 meters and a total volume of approximately 1.3 billion cubic kilometers, the open-ocean water column of the world is the largest habitat for life on the planet, actually the largest in the solar system. It is the source of most of the primary food production in the ocean and nearly half of the oxygen in the atmosphere and is an absolutely vital link in the global cycles of carbon, heat, water and nutrients. The diversity of invertebrates and fishes in the upper waters is reasonably well known and surprisingly high for such a relatively homogeneous environment (Figure 9). In the mesopelagic waters (200 to 1,000 meter depths) and bathypelagic regions (depths greater than 1,000 meters) organisms are sparser and poorly known, with new species being discovered regularly through exploration with submersible vehicles. Recent investigations have revealed complex microbial communities throughout the water column, including small microscopic animals (bacteria, archaea and protists) with largely unknown characteristics (Karl 2002, Karner, Delong & Karl 2001). Although the environment is vast and seemingly well buffered for stability, shifts in the composition or functioning of its biological communities could affect nutrient and carbon cycles worldwide. There is recent evidence of "regime shifts" in the community structure in some parts of the open ocean, which may be due to global climate change (Karl, Bidigare & Letelier 2001). The ocean plays a significant role in removing carbon dioxide from the atmosphere, but this is now beginning to change the pH balance (acidity) of ocean waters, which could have widespread effects on phyto- and zooplankton with calcareous (calcium carbonate) shells that will dissolve in acidic waters (Feely *et al* 2004). Removal of top predator fish by high seas fishing may disrupt the lower trophic structures (predation and feeding relationships).

### Deepsea benthos

The soft-sediment environment of the abyssal plains (the flat seafloor in the deepest parts of the ocean) is another vast and seemingly stable habitat that is remote from land and human impacts. Despite its relatively homogenous nature, it has exceptionally high species diversity, particularly of "infaunal" organisms that live within the seafloor sediments (Grassle & Maciolek 1992). Although the deepsea benthos (seafloor habitats) hold few economically viable biological resources, there is the potential for future impacts from exploitation of minerals like manganese nodules (Rona 2003). Ocean dumping of radioactive and other wastes is currently prohibited, but there is a possibility of future disposal of carbon dioxide ($CO_2$) from power plants into the deepsea in either liquid or hydrate form, either of which would have severe effects wherever they were dumped. At present neither seafloor mining nor $CO_2$ disposal are economically feasible (Hoagland 2003) but could become so at some future time.

### Ridge and vent systems

The remarkable chemosynthetic communities of hydrothermal vents (Figure 10) were one of the great scientific discoveries of the 20th century. These ecosystems offer clues to the origin of life on Earth as well as unique adaptations of both microbial and larger life forms to extreme environments. Their remote location makes them difficult and expensive to find and visit, so the only significant human impact currently is research activity (although limited tourism is starting to develop). Hydrothermal vents are ephemeral by nature, appearing and disappearing as a result

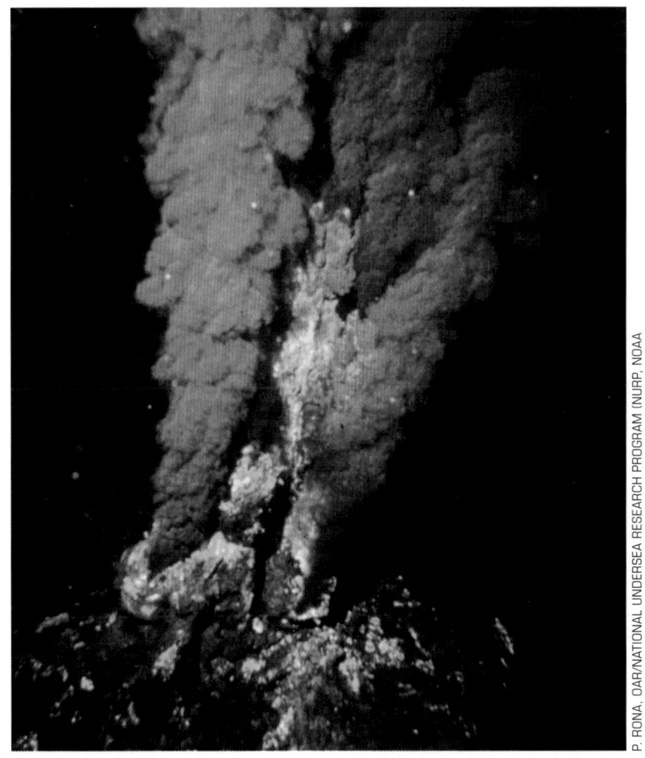

P. RONA, OAR/NATIONAL UNDERSEA RESEARCH PROGRAM (NURP, NOAA

**Figure 10** Black smokers, such as this one at a mid-ocean ridge hydrothermal vent, supply clouds of hot chemical and mineral "food" for animals that live by chemosynthesis.

of changes in the supply of magma from below the ocean floor. Research activities probably have minimal impact on the habitats or organisms, especially compared to the research value of the discoveries. There is a possibility of future mining operations for the rich metallic content of vent chimneys and surrounding deposits, but like manganese nodule mining, this is not likely to be economically viable in the foreseeable future (Glover & Smith 2003).

### Frozen seas and ice edges

These specialized and seasonally variable habitats play important roles in polar ecology. The seasonal and interannual variations in ice cover affect primary production (the transferring of sunlight energy into food) and the population dynamics of krill and other zooplankton (small drifting animals) that are the main food resource of penguins and marine mammals. New under-ice and ice-slush communities of microbes and invertebrates have been discovered in the past few decades that are likely critical to the functioning of the ecosystems. There are already changes in ice cover and thickness at both poles that may be related to climate warming (Comiso *et al* 2003), and continued reduction of ice cover may have drastic and long-term consequences for these highly productive marine environments (Burghart *et al* 1999).

© SYLVIA A. EARLE

**Figure 11** Large mats of *Sargassum* seaweed congregate along middle Atlantic and Gulf of Mexico oceanic fronts where different watermasses come together.

### Fronts and upwelling zones

These zones of enhanced primary and secondary production are often located in the coastal ocean but can also occur at the boundaries of major currents in the open sea. The concentrations of life relative to the surrounding waters (Figure 11) make these ocean fronts important feeding areas for large pelagic fishes, reptiles and marine mammals that often congregate along these strong oceanic features (Sherman 1987, Seki *et al* 2002). This aggregation of fishes in narrowly defined locations increases their vulnerability to commercial fishing pressure, especially on the high seas outside of national jurisdiction and regulation.

### Microbial ecology

It is now abundantly clear that microbial diversity and physiological processes are fundamental and indispensable to the functioning of all ecosystems, whether marine or terrestrial. The apparent diversity of microscopic organisms collected by new explorations and revealed by new molecular techniques is staggering and scarcely understood (Azan & Worden 2004). While microbes may have unparalleled capacity for adaptation and survival, shifts in their relative abundance and activity could trigger significant consequences for other parts of the ecosystem. We know far too little about these interactions and balances. Molecular, biochemical and genomic tools promise fundamental discoveries of microbial diversity and ecosystem function, thus enabling the incorporation of microbial processes into future conservation models.

## Summary and Recommended Actions

The Unknown Ocean theme group at the *Defying Ocean's End* Conference considered several poorly known ocean environments to determine needs for conservation action and/or research. We lack species-level data for many unknown habitats and further basic research in these areas is important to conservation, but in many areas we already know enough to plan precautionary actions. In some areas, swift action is critical to avoid losing forever many unknown ocean species before they can even be discovered. We recommend actions in three general areas.

**Unknown habitats with conservation priority**

First, we must initiate concerted research and conservation action in four high priority unknown habitats, chosen based on: human impacts; potential irreversible loss of habitat type or biodiversity; adequate initial data; and economic feasibility of conservation actions.

*Seamounts*

Invest in a minimum of 45 ship days per year *for ten years* for seamount sampling. Study 50 new seamounts—chosen to be representative of global seamounts in depth, latitude, bottom-type and geologic age; describe and document the thousands of new species that will be discovered. Create an international framework for protecting seamounts from destructive fishing practices, and develop a global satellite-based surveillance system to enforce fishing restrictions on seamounts.

*Deep "twilight zone" coral reefs and remote shallow reefs*

Identify priority areas, conduct initial reef area surveys and develop a pilot project for conserving one site. *Within ten years,* survey an additional 28 or more sites and develop conservation strategies where needed.

*Continental slope, canyons and cold seeps*

Complete ecological surveys of three seep sites *within three years,* and determine changes there *within ten years.*

*Anchialine habitats—underwater caves and blue holes*

Complete field studies of at least ten underwater cave environments in each major region of the world where they occur *within three years* and survey at least 90% of all cave environments *within ten years.*

**Unknown habitats for continued research**

Initiate research efforts in a second group of habitats with lower priority for immediate conservation attention but warranting further research based on: high scientific or ecological value; inadequate initial data; no obvious immediate threats; and opportunities for long-term conservation planning.

*Open-ocean water column*

This largest marine habitat is the source of most primary production and a vital link in the global cycles of carbon, heat, water and nutrients. Changes in these communities may have significant effects on global carbon budgets and climate.

*Deepsea benthos*

A vast habitat with exceptionally high biodiversity but poorly known ecology. Future impacts from mining or deep-sea fishing are possible.

*Ridge and vent systems*

These chemosynthetic communities have unique adaptations to extreme temperature and pressure and offer clues to the origin of life on earth and possible life forms on other planets.

*Frozen sea and ice edges*

These areas have important roles in seasonally productive polar ecosystems and may show early effects of climate warming. The role of polar regions in sustaining ocean circulation patterns is an important regulator of global climate.

*Fronts and upwelling zones*

These are highly productive regions generated by ocean current and nutrient patterns that are vital to the survival of large pelagic fishes, reptiles and marine mammals.

*Microbial ecology*

Microbial communities are the biogeochemical foundation of all ecosystems on earth, without which no other life forms would survive. Although microbes are ancient and ubiquitous, their diversity and ecology are poorly understood. New genetic and molecular tools can provide unprecedented insights into microbial characteristics and functions and may help alert us to possible harmful human impacts.

**Data infrastructure for research and conservation**

We must also invest in data infrastructure to provide improved support to both research and conservation:

- Expand the capabilities and scope of the Ocean Biogeographic Information System (OBIS) of the Census of Marine Life Program (O'Dor 2004)
- Build the OBIS database to 25 million records *within 10 years*
- Establish new taxonomic authority groups for marine species as needed
- Evaluate additional marine species from unknown regions for inclusion on the IUCN Red List of threatened or endangered species

The environments described in this chapter are remote and poorly understood. Their biologic inhabitants appear to be ancient, unique and vulnerable to change. There is a good chance that a significant percentage of the Earth's biodiversity—and an understanding of the Earth's ecosystem—still lie undiscovered in these ocean environments.

## Literature Cited & Consulted


Azam, F. and A. Z. Worden. 2004. Microbes, molecules, and marine ecosystems. *Science* 303:1622–1624.

Barry, J. P., H. G. Greene, D. L. Orange, C. H. Baxter, B. H. Robison, R. E. Kochevar, J. W. Nybakken, D. L. Reed and C. M. McHugh. 1996. Biologic and geologic characteristics of cold seeps in Monterey Bay, California. *Deepsea Research, Part I.* 43 (11-12): 1739–1762.

Black, M. B., K. M. Halanych, P. A.Y. Maas, W. R. Hoeh, J. Hashimoto, D. Desbruyères, R. A. Lutz and R.C. Vrijenhoek. 1997. Molecular systematics of vestimentiferan tubeworms from hydrothermal vents and cold-water seeps. *Marine Biology* 130:141–149.

Bouchet, P., P. Lozouet, P. Maestrati and V. Heros. 2002. Assessing the magnitude of species richness in tropical marine environments: Exceptionally high numbers of molluscs at a New Caledonia site. *Biological Journal of the Linnean Society* 75:421–436.

Boxshall, G. A. 1984. The functional morphology of Benthomisophria palliata Sars, with a consideration of the evolution of the Misophrioida. *Crustaceana Supplement* 7:32–46.

Brooks, J. M., M. C. Kennicutt II, C. R. Fisher, S. A. Macko, K. Cole, J. J. Childress, R. R. Bidigare and R. D. Vetter. 1987. Deepsea hydrocarbon seep communities: Evidence for energy and nutritional carbon sources. *Science* 238:1138–1142.

Bryant, D., L. Burke, J. McManus and M. Spalding. 1998. *Reefs at Risk: A Map-Based Indicator of Potential Threats to the World's Coral Reefs.* Washington, D.C.: World Resource Institute.

Burghart, S. E., T. L. Hopkins, A. G. Vargo and J. J. Torres. 1999. Effects of a rapidly receding ice edge on the abundance, age structure and feeding of three dominant calanoid copepods in the Weddell Sea, Antarctica. *Polar Biology* 22:279–288.

Chevaldonne, P. and C. Lejeusne. 2003. Regional warming-induced species shift in north-west Mediterranean marine caves. *Ecology Letters* 6:371–379.

Colin, P. L. 1999. Palau at depth. *Ocean Realm* Summer 1999:77–87.

Colin, P. L., D. M. Devaney, L. Hills-Colinvaux, T. H. Suchanek and J. T. Harrison, III. 1986. Geology and biological zonation of the reef slope, 50-360m depth at Enewetak Atoll, Marshall Islands. *Bulletin of Marine Science* 38 (1): 111–128.

Comiso, J. C., J. Yang, S. Honjo and R. A. Krishfield. 2003. Detection of change in the Arctic using satellite and in situ data. *Journal of Geophysical Research* 108 (C12): 3384.

Danielopol, D. L. 1990. Ground water ecology—more basic research needed. *Ground Water Monitoring Review* 10 (4): 5–10.

Deep Sea Conservation Coalition. 2004. *Urgent Action Needed to Protect Seamounts, Cold Water Corals and Other Vulnerable Deep Sea Ecosystems.* Deep Sea Conservation Coalition (DSCC) position paper. *http://www.highseasconservation.org/publicdocs/DSCC_Position_US.pdf* (accessed October 3, 2004).

Embley, R. W., S. L. Eittreim, C. H. McHugh, W. R. Normark, G. H. Rau, B. Hecker, A. E. DeBevoise, H. G. Greene, W. B. Ryan, C. Harrold and C. Baxter. 1990. Geological setting of chemosynthetic communities in the Monterey Fan Valley system. *Deepsea Research* 37:1651–1667.

Feely, R. A., C. L. Sabine. K. Lee, W. Berelson, J. Kleypas, V. F. Fabry and F. J. Millero. 2004. Impact of anthropogenic CO sub(2) on the CaCO sub(3) system in the oceans. *Science* 305:362–366.

Fricke, H. W. and B. Knauer. 1986. Diversity and spatial pattern of coral communities in the Red Sea upper twilight zone. *Oecologia* 71:29–37.

Fricke, H. W. and H. Schuhmacher. 1983. The depth limits of Red Sea stony corals: An ecophysiological problem (A deep diving survey by submersible). *PSZNI Marine Ecology* 4:163–194.

Froese, R. and D. Pauly, eds. 2004. *FishBase.* World Wide Web electronic publication. *www.fishbase.org,* version 06/2004.

Glover, A. G. and C. R. Smith. 2003. The deepsea floor ecosystem: Current status and prospects of anthropogenic change by the year 2025. *Environmental Conservation* 30:219–241.

Goffredi, S. K., L. A. Hurtado, S. Hallam and R. C. Vrijenhoek. 2003. Evolutionary relationships of deepsea vent and seep clams (Mollusca: Vesicomyidae) of the *pacifica/lepta* species complex. *Marine Biology* 142:311–320.

Grassle, J. F. and N. J. Maciolek. 1992. Deepsea species richness: Regional and local diversity estimates from quantitative bottom samples. *American Naturalist* 139:313–341.

Hart, C. W. Jr., R. B. Manning and T. M. Iliffe. 1985. The fauna of Atlantic marine caves: Evidence of dispersal by sea floor spreading while maintaining ties to deep waters. *Proceedings of the Biological Society of Washington* 98:288–292.

Hecker, B. 1985. Fauna from a cold sulphur-seep in the Gulf of Mexico: Comparison with hydrothermal vent communities and evolutionary implications. *Bulletin of the Biological Society of Washington* 6:465–473.

———. 1994. Unusual megafaunal assemblages on the continental slope off Cape Hatteras. *Deepsea Research, Part II* 41:809–34.

Hecker, B., D. T. Logan, F. E. Gandarillas and P. R. Gibson. 1983. Megafaunal assemblages in canyon and slope habitats. In *Canyon and Slope Processes Study,* Vol. III, Chapter I. Final Report prepared for U.S. Department of the Interior, Minerals Management Service. Washington, D.C.: U.S. Department of the Interior.

Hoagland, P. 2003. Exploring sea-floor resources. *Science* 300:1093.

Humphreys, W. F. 2000. The hypogean fauna of the Cape Range Peninsula and Barrow Island, Northwestern Australia. In *Ecosystems of the World. 30. Subterranean Ecosystems,* ed. H. Wilkens, D.C. Culver and W. F. Humphreys, 581–601. Amsterdam: Elsevier Science.

Iliffe, T.M. 1992. Anchialine cave biology. In *The Natural History of Biospeleology,* ed. A.I. Camacho, 613–636. Madrid: Museo Nacional de Ciencias Naturales.

———. 2000. Anchialine cave ecology. In *Ecosystems of the World. 30. Subterranean Ecosystems,* ed. H. Wilkens, D. C. Culver and W. F. Humphreys, 59–76. Amsterdam: Elsevier Science.

Iliffe, T. M., T. D. Jickells and M. S. Brewer. 1984. Organic pollution of an inland marine cave from Bermuda. *Marine Environmental Research* 12:173–189.

IUCN–The World Conservation Union. 2003. *2003 IUCN Red List of Threatened Species. http://www.redlist.org* (downloaded December 2003).

Juniper, S. K. and M. Sibuet. 1987. Cold seep benthic communities in Japan subduction zones: Spatial organization, trophic strategies and evidence for temporal evolution. *Marine Ecology Progress Series* 40:115–126.

Karl, D. M. 2002. Microbiological oceanography: Hidden in a sea of microbes. *Nature* 415:590–591.

Karl, D. M., R. R. Bidigare and L. M. Letelier. 2001. Long-term changes in plankton community structure and productivity in the North Pacific subtropical gyre: The domain shift hypothesis. *Deepsea Research, Part II* 48:1449–1470.

Karner, M. B., E. F. DeLong and D. M. Karl. 2001. Archaeal dominance in the mesopelagic zone of the Pacific Ocean. *Nature* 409:507–510.

Kornicker L. S. and T. M. Iliffe. 2000. Myodocopid Ostracoda from Exuma Sound, Bahamas, and from Marine Caves and Blue Holes in the Bahamas, Bermuda, and Mexico. *Smithsonian Contributions to Zoology* 606:1–98.

Koslow, J. A., K. Gowlett-Holmes, J. K. Lowrly, G. C. B. Poore and A. Williams. 2001. Seamount benthic macrofauna off southern Tasmania: Community structure and impacts of trawling. *Marine Ecology Progress Series* 213:111–125.

Linden O., D. Souter, D. Wilhelmsson and D. Obura. 2002. *Coral Reef Degradation in the Indian Ocean.* Status report 2002. CORDIO. *http://www.cordio.org/* (accessed September 5, 2004).

Macintyre, I. G., K. Ruetzler, J. N. Norris, K. P. Smith, S. D. Cairns, K. E. Bucher and R. S. Steneck. 1991. An early holocene reef in the western Atlantic: Submersible investigations of a deep relict reef off the west coast of Barbados, W.I. *Coral Reefs* 10 (3): 167–174.

McKenna, S. A. and G. R. Allen. 2002. Coral reef biodiversity assessment and conservation. In *Implications for Coral Reef Management and Policy: Relevant Findings from the 9ICRS,* ed. B. A. Best, P. S. Pomeroy and C. M. Balboa, 92–94. Washington, D.C.: USAID.

———. Forthcoming. A Rapid Marine Biodiversity Assessment of the Coral Reefs of Northwest Madagascar. *Rapid Assessment Program RAP Bulletin of Biological Assessment* 31. Washington, D.C.: Center for Applied Biodiversity Science, Conservation International.

McKenna, S. A., G. R. Allen and S. Suryadi, eds. 2002. A marine rapid assessment of the Raja Ampat Islands, Papua Province, Indonesia. *Rapid Assessment Program RAP Bulletin of Biological Assessment* 22. Washington, DC: Center for Applied Biodiversity Science, Conservation International.

Merrin, K. 2004. The unknown marine asselote ispod crustaceans of New Zealand. *Water & Atmosphere* 12 (1): 24–25.

Moore, J. A. 1999. Deepsea finfish fisheries: Lessons from history. *Fisheries* 24:16–21.

Moore, J. A. and P. M. Mace. 1999. Challenges and prospects for deepsea finfish fisheries. *Fisheries* 24:22–23.

Natural Resources Defense Council. 2001. *Priority Ocean Areas for Protection in the Mid-Atlantic.* Washington D.C.: Natural Resources Defense Council.

O'Dor, R. 2004. A Census of Marine Life. *BioScience* 54:92–93.

Paull, C. K., B. Hecker, C. Commeau, R. P. Feeman-Lynde, C. Neumann, W. P. Corso, G. Golubic, J. Hook, E. Sikes and J. Curray. 1984. Biological communities at Florida escarpment resemble hydrothermal vent communities. *Science* 226:965–967.

Pesce, G.L. 2001. The Zinzulusa Cave: An endangered biodiversity "hot spot" of south Italy. *Natura Croatica* 10 (3): 207–212.

Poore, G. C. B. and G. D. F. Wilson. 1993. Marine species richness. *Nature* 314:731–732.

Porter, J. W. 1973. Ecology and composition of deep reef communities off the Tongue of the Ocean, Bahama Islands. *Discovery* 9:3–12.

Pyle, R. L. 1996. The twilight zone. *Natural History* 105 (11): 59–62.

———. 1998. Use of advanced mixed-gas diving technology to explore the coral reef "twilight zone." In *Ocean Pulse: A Critical Diagnosis,* ed. J. T. Tanacredi and J. Loret, 71–88. New York: Plenum Press.

———. 2000. Assessing undiscovered fish biodiversity on deep coral reefs using advanced self-contained diving technology. *Marine Technology Society Journal* 34 (4): 82–91.

Randall, J. E. and R. L. Pyle. 2001. Three new species of labrid fishes of the genus Cirrhilabrus from islands of tropical Pacific. *Aqua* 4 (3): 89–98.

Reaka-Kudla, M. L. 1997. The global biodiversity of coral reefs. In *Biodiversity II: Understanding and Protecting Our Natural Resources,* eds. M. L. Reaka-Kudla, D. E. Wilson and E. O. Wilson, 83–108. Washington, D.C.: Joseph Henry/National Academy Press.

Renssen, H., C. J. Beets, T. Fichefet, H. Goosse and D. A. F. Kroon. 2004. Modeling the climate response to a massive methane release from gas hydrates. *Paleoceanography* 19:PA2010.

Rhodas, D. C. and B. Hecker. 1994. Processes on the continental slope off North Carolina with special reference to the Cape Hatteras region. *Deepsea Research, Part II* 41:965–80.
Riedl, R. 1966. *Biologie der Meereshöhlen. Topographie, faunistik und okologie eines unterseeischen lebensraumes. Eine monographie.* Berlin: Verlag Paul Parey.
Rogers, A. D. 1994. The biology of seamounts. *Advances in Marine Biology* 30:305–50.
Rona, P. 2003. Resources of the sea floor. *Science* 299:673–674.
SeamountsOnline. 2004. SeamountsOnline: An Online Information System for Seamount Biology. *http://seamounts.sdsc.edu/* (accessed September 5, 2004).
Seki, M. P., J. J. Polovina, D. R. Kobayashi, R. R. Bidigare and G. T. Mitchum. 2002. An oceanographic characterization of swordfish *(Xiphias gladius)* longline fishing grounds in the springtime subtropical North Pacific. *Fisheries Oceanography* 11:251–266.
Sharkey, P. and R. L. Pyle. 1992. The twilight zone: The potential, problems, and theory behind using mixed gas, surface-based SCUBA for research diving between 200 and 500 feet. In *Diving for Science, Proceedings of the American Academy of Underwater Sciences Twelfth Annual Scientific Diving Symposium.* Costa Mesa, CA: American Academy of Underwater Sciences.
Sherman, K. 1987. *Large marine ecosystems as global units for recruitment experiments.* Collected Papers, ICES Council Meeting 1987.
Sibuet, M. and K. Olu. 1998. Biogeography, biodiversity and fluid dependence of deepsea cold-seep communities at active and passive margins. *Deepsea Research, Part II* 45:517–567.
Sibuet, M. and K. Olu-LeRoy. 2002. Cold seep communities on continental margins: Structure and quantitative distribution relative to geological and fluid venting patterns. In *Ocean Margin System,* eds. G. Wefer, D. Billet, D. Hebbeln, B. B. Jorgensen and T. J. van Weering, 235–251. Heidelberg: Springer-Verlag.
Singh, H., R. Armstrong, F. Gilbes, R. Eustice, C. Roman, O. Pizzaro and J. Torres. 2004. Imaging coral I: Imaging coral habitats with the SeaBED AUV. *Subsurface Sensing Technologies and Applications* 5:25–42.
Sket, B. 1996. The ecology of anchihaline caves. *TREE* 11 (5): 221–225.
———. 1999. The nature of biodiversity in hypogean waters and how it is endangered. *Biodiversity and Conservation* 8 (10): 1319–1338.
Smith, T. 1994. *Scaling Fisheries.* Cambridge, England: Cambridge University Press.
Snelgrove, P. V. R. 1998. The biodiversity of macrofaunal organisms in marine sediments. *Biodiversity and Conservation* 7:1123–1132.
South, G. R., P. A. Skelton and A. Yoshinaga. 2001. Subtidal Benthic Marine Algae of the Phoenix Islands, Republic of Kiribati, Central Pacific. *Botanica Marina* 44:559–570.
Stock, J. H., T. M. Iliffe and D. Williams. 1986. The concept "anchialine" reconsidered. *Stygologia* 2:90–92.
Thresher, R. E. and P. L. Colin. 1986. Trophic structure, diversity, and abundance of fishes of the deep reef (30–300m) at Enewetak, Marshall Islands. *Bulletin of Marine Science* 38 (1): 253–272.
Vetter, E. W. and P. K. Dayton. 1998. Macrofaunal Communities within and adjacent to a detritus-rich submarine canyon system. *Deepsea Research, Part II* 45:25–54.
Wilkinson, C. 2000. *Status of Coral Reefs of the World.* Global Coral Reef Monitoring Network (GCRMN). Townsville: Australian Institute of Marine Science.
Wood, J. B. and C. Day. 2004. Cephbase. *http://www.cephbase.utmb.edu/* (accessed September 5, 2004).
Yager, J. 1981. Remipedia, a new class of Crustacea from a marine cave in the Bahamas. *Journal of Crustacean Biology* 1:328–333.
———. 1994. *Speleonectes gironensis,* new species (Remipedia: Speleonectidae), from anchialine caves in Cuba, with remarks on biogeography and ecology. *Journal of Crustacean Biology* 14 (4): 752–762.
Yoklavich, M. M., H. G. Greene, G. M. Cailliet, D. E. Sullivan, R. N. Nea and M. S. Love. 2000. Habitat associations of deep-water rockfishes in a submarine canyon: An example of a natural refuge. *Fishery Bulletin* 98:625–641.
Zhang, Y. and J. F. Grassle. 2003. A portal for the ocean biogeographic information system. *Oceanologica Acta* 25:193–197.

CHAPTER 13

# *Business Plan*

Benjamin A. Vitale, *Conservation International*
Larry Linden, *Goldman Sachs & Company*
Ivan Barkhorn, *Redstone Strategy Group, LLC*
Dietmar Grimm, *Redstone Strategy Group, LLC*

NEIL HAMMERSCHLAG

## Introduction

"What will it cost and how much time will it take to protect the biological treasures and economic value of the world's ocean in perpetuity?" This was one of the questions posed to a group of world experts in ocean science, policy, law, economics, conservation and business at the *Defying Ocean's End* Conference. This effort was reminiscent of another occasion when teams of experts were summoned to plan, prioritize and cost one of the most momentous challenges humankind has faced. To achieve President Kennedy's vision of landing a man on the moon, early cost estimates ranged from US $5 to $7 billion. These estimates rose to more than $20 billion over a ten-year period (Compton 1989), but the "impossible" was achieved on July 20, 1969, when Neil Armstrong and Buzz Aldrin set foot on the moon and brought back samples for scientific study. The effort eventually required over US $25.4 billion, roughly equivalent to $140 billion adjusted to 2003 dollars, and required more than 300,000 experts in fields ranging from logistics to rocket science (Marshall, Kreps & Chriss 1989). But despite all of the literally ground-breaking innovations required, this historic feat was accomplished in less than ten years.

Estimates of the cost to conserve global biodiversity vary greatly, but one credible study suggests that an effective network of mostly terrestrial protected areas would require US $45 billion annually in operational costs (Balmford *et al* 2002). At the 2003 World Parks Congress in Durban, South Africa, many speakers noted that marine conservation must be strengthened and that it will require significant additional funding. Marine conservation problems require a broader business planning approach because of the scale and immediacy of the ocean's problems:

- Many marine species roam vast areas of ocean between their larval, nursing, feeding and breeding stages
- The ocean lacks comprehensive ownership rights like those applicable on land
- The high seas are marked by freedoms that have hampered marine conservation efforts
- Only about 0.5% of global ocean areas are protected compared with nearly 10% of the world's land areas (Kelleher, Bleakley & Wells 1995)
- Fishing and other commercial activities have dramatically reduced the numbers of large fish worldwide—by 90% (Meyers & Worm 2003)
- Trawl fishing is irreversibly damaging many ancient deepsea corals and habitats (Deep Sea Conservation Coalition 2004)

In this chapter we describe the method used at the *DOE* Conference to determine the costs and time required to protect roughly five percent (5%) of the world's ocean over ten years—the same timeframe set for landing a man on the moon. The overall costs for the *DOE* action agenda were estimated to be between US $18.6 and US $30 billion over ten years. The details, and items included or omitted, are explained further in the results section below.

The vision for *Defying Ocean's End* was to combine the collective expertise of world-recognized scientists, policy experts, economists and conservationists to develop a multifaceted plan for conservation of the world's ocean. *DOE* also assembled a team of business professionals (Figure 1) to work with its seven theme groups over the course of five days to develop a business plan, including a timeline and implementation costs, for the final *DOE* recommendations. While the results are quite preliminary, the resulting integrated action plan, cost schedule and three-year and ten-year timelines constitute a credible first order marine conservation business plan. In this chapter we describe the approach, results and some potential next steps for marine conservation business planning efforts.

## What is a Marine Conservation Business Plan?

A business plan is used primarily to attract capital to launch or expand a new or existing commercial enterprise, but it is also an important tool to create "mind share" to attract high-caliber employees, strategic partners and customers. Commercial business plans describe products and services targeted at particular market segments to achieve a profit, and they develop financial statements that quantify the amount of financing expected from business revenues, private equity, debt and public equity markets for funding. A typical business plan first describes its vision in an executive summary—particularly important for a "new to the world" business—then includes descriptions of:

- The management team and their experience
- The market and planned products or services
- The competition and unique position of this company
- Associated risks
- The financial statements that project revenues and costs for a three- to five-year period

The global agenda for action from *Defying Ocean's End* is

indeed a "new to the world" effort that has not been undertaken before at such a scale. In contrast to a commercial enterprise, however, this global effort involves a complex combination of non-profit, for-profit and governmental organizations each with a particular perspective on how to conserve the ocean. Perspectives are varied and range from biodiversity conservation approaches, public policy views, activist outreach and achieving optimal commercial use of marine resources. Marine science, policy and conservation efforts are usually funded by governments and international development agencies, multilateral and bilateral funding institutions such as the World Bank and Asian Development Bank, and private organizations such as foundations, individuals and corporate philanthropic efforts.

BENJAMIN A. VITALE, CONSERVATION INTERNATIONAL

BENJAMIN A. VITALE, CONSERVATION INTERNATIONAL

**Figure 1** The *Defying Ocean's End* business team at work. Larry Linden (top) presents interim results of his business team. A member of the business team, Jim Hauslein (bottom right), consults with the team's staff at the *DOE* Conference with the Gulf of California as a backdrop.

Our *DOE* marine conservation business plan briefly addresses the five categories common in business plans, but each section has been adapted to accommodate the purposes, complexities and global breadth of *DOE*'s global, cross-discipline conservation recommendations. The primary purpose of our business plan is to enable a broad assemblage of organizations, governments, communities, financing institutions and other stakeholders to align behind a framework and practical set of actions to defy ocean's end. Our plan is sufficiently broad to indicate how a particular scientific program, marine policy or conservation action relates to the much broader scope, timeline and costs of a global plan.

We have developed top-down cost estimates and a timeline for *DOE* marine conservation recommendations, but given the timeframe of the Conference, there are many constraints on our effort. The main shortcomings are included in a "Limitations" section at the end of this chapter.

All of the recommendations and ideas incorporated into this *DOE* business plan rely heavily on the significant body of scientific, policy and conservation expertise provided before, during and after the conference. Of particular importance are two practical approaches widely used in the conservation community: business plans for establishing sustainable marine protected areas, and a set of tools incorporated into a discipline called conservation finance. It is important to describe how the plan developed at the Conference differs from both of these approaches.

The breadth of the *DOE* recommendations is significantly greater and more cost-focused than the typical site-based marine protected area business planning approaches, such as those detailed in *Financing Marine Conservation: A Menu of Options* (Spergel 2004). The approaches outlined in this paper and noted in other literature are typically applied to establish and sustain networks of marine protected areas (MPAs). Although developing networks of marine protected areas is one of the most effective tools for conserving marine biodiversity and functioning marine ecosystems (Palumbi 2003), it is only a subset of the recommendations offered at the *Defying Ocean's End* Conference. The scope of the plan developed at *DOE* is global in nature and includes not only the support required to establish networks of protected areas, but also the scientific, policy and communications efforts required for effective marine conservation worldwide.

"Conservation finance," on the other hand, is a discipline that has developed over the past twenty years into a creative and practical set of tools and methods to finance both terrestrial and marine conservation efforts. Two examples of these methods are the creation of debt-swaps with indebted countries to pay for conservation and the development of trust funds to finance the long-term management costs for protected areas. Another significant area in conservation finance is the "sale" of ecosystem "services" in protected areas, such as providing credits for the reduction of greenhouse gases or carbon dioxide, or for preserving the ecological integrity of watershed areas by conserving threatened forests within them. In the marine realm, ecosystem service values—such as the maintenance of productive fisheries enabled by no-take zones and the protection of coastal shorelines from storm surge and erosion through maintaining intact mangrove areas (Figure 2)—will be increasingly important during the timeframe of this business plan.

It has been estimated that marine ecosystems account for about two thirds of the Earth's ecosystem services value. For the Earth's entire biosphere, the value is estimated to be in the range of US $16 to $54 trillion per year, with an average of US $33 trillion per year. This is considered a minimum estimate given the nature of inherent uncertainties. The monetary value of most ecosystem services have not been identified, so they exist and operate outside the established trading and commercial markets. As Costanza *et al* (1997) remark, "We can think of ecological values under risk-avoidance as translating into insurance premiums that would willingly be paid to protect against the risk of destabilization." As the ecological services are more rigorously quantified and valued by governments, conservation organizations and local communities—and integrated into global and local policy processes—the justification and sources of a significant percentage of the funding required to sustain marine ecosystems (Wildlife Conservation Society 2004).

GETTY IMAGES

**Figure 2** The maintenance of intact mangrove forests, such as these red mangroves, provides economic benefits since the forests serve as a natural buffer against storm surges, help to reduce coastal erosion and provide an essential habitat for aquatic birds and coastal fishes.

## Methodology

Before presenting results of the business plan, it is important to explain the methods employed at the Conference to gather, organize and estimate the cost inputs. The goal of the business team at *DOE* was to assist in clarifying the specific recommendations for action being developed by the theme groups, and to gather and assess their estimated costs over a ten-year period. The business planning function at the Conference was unique and important in four key ways:

First, the business team underscored the need for urgent action and encouraged the theme groups to prioritize immediate actions. Each of the theme groups had performed significant background research before the Conference and gathered much of the quantifiable data required to support their recommendations at the event. To more easily document and integrate these recommendations across the seven thematic working groups, the business team distributed a data-gathering format to collect standardized inputs in the following areas:

- Definition of ten-year outcomes
- Description of supporting activities for years one, three and ten
- New global capacity requirements needed to achieve the outcomes
- Estimated costs per unit activity
- Key dependencies among recommendations of different theme groups
- Listings of current funding amounts and sources, if known, for comparable marine conservation initiatives

The theme groups provided increasingly sophisticated business planning inputs several times during the Conference.

Second, the individual theme groups often recommended similar programs of action, and the business planning process enabled the groups to identify areas of overlap and efforts that required integration across groups. In particular, marine protected areas were discussed from various perspectives in recommendations from three theme groups: Ocean-Use Planning and Protected Areas (reported in Chapter 6), Land/Ocean Interface (Chapter 8) and Maintaining and Restoring Marine Ecosystems (Chapter 9).

We reconciled the overlap and cost aspects among the recommendations of these groups as much as possible to establish a global approach to networks of marine protected areas and seascape-level conservation. Other key dependencies across theme groups at the Conference were shared through electronic exchange of information, by business team attendance at theme group meetings, and at special meetings of the theme group chairs. This integrating function of the business team's identified recommendations that could benefit from collective action by many parties. A striking example of this was the on-site beginnings of a seamount conservation campaign; this idea coalesced into a top priority for broad-based immediate action that required the integration of policy, law, governance, communications and science recommendations across several of the *DOE* theme groups.

Third, the *DOE* Conference design included the development of a cross-cutting global communications plan to support the major recommendations emerging from the Conference. The Communications theme group was asked to review all other theme group outcomes and milestones as input to their global communications plan to support the *DOE* agenda for action. They were able to use the theme groups' timelines to better prioritize and cost their global ocean communications strategy.

Lastly, it was generally acknowledged at the Conference that the terrestrial conservation movement has been more effective than the marine conservation movement at articulating the scale of ecosystem declines, global effects on species populations and the required scale of investment needed to implement improvements worldwide. The business planning team aided the theme groups in organizing and documenting their high-level cost estimates to better articulate these needs to various stakeholder groups and potential funding sources.

After assessing the range of recommendations emerging from the theme groups, the business team developed a skeleton cost model that included seven broad areas:

- Global science
- Large marine ecosystem (LME) programs
- Marine protected areas (MPAs)
- Fisheries reform
- Global governance
- Communications
- Program management

Each theme group reviewed their recommendations and provided estimated cost drivers, as requested by the business team, to improve the cost model. At the Conference the team worked to document missing detail and prepare questions for the theme groups to consider as input to the cost model. The business team also attended the theme group meetings to obtain detailed information and further refine the cost model estimates.

The cost estimation exercise identified a number of critical aspects of the global recommendations where major costs could not be sufficiently addressed. First, the investments required for sustainable economic use of marine areas may be far greater than the costs of protecting them. One study estimates that US $300 billion is required annually to truly conserve a representative group of protected areas worldwide, which includes roughly US $50 billion for compensating communities and providing incentives for protection (James, Gaston & Balmford 1999). Second, a reallocation of the global annual fisheries subsidies estimated at US $14–20 billion worldwide could provide incentives to restore degraded areas, improve profits due to better management and reduce overall costs (Milazzo 1998). Third, the costs of addressing a full range of land-based threats to marine areas will require land-use changes to reduce soil erosion, nutrient and pollution runoff from urbanization, deforestation and agricultural activities that are the source of 44% of the ocean's toxic contaminants (Campbell & Lawrence 2004). Fourth, our plan considered only the highest priority near-term actions needed, so our cost estimates span only a ten-year timeframe and do not include the costs in perpetuity of managing marine protected areas. Our methods of estimating costs could be strengthened by aggregating the costs of ongoing marine conservation programs and by addressing the constraints of our approach. But despite these limitations, our results provide an indication of the overall costs required to implement the *DOE* recommendations.

## Business Plan

This section reviews the high-level components found in typical business plans—the management team, market for planned products or services, competition, risks and the financial cost projections—and their application to the *DOE* recommendations.

**The management team**

The *DOE* action agenda has a very informal follow-up structure as a "management team" for executing the global plan. It is the combined senior management and their assigned staff of multiple organizations participating in Conference follow-on activities: World Wildlife Fund, The Ocean Conservancy, The Nature Conservancy, Natural Resources Defense Council (NRDC), The World Conservation Union—IUCN, SeaWeb, Conservation International and many others. Some *DOE* follow-on efforts have been quite successful, creating significant programs such as an expansion of early deepsea coral protection efforts into the Deep Sea Conservation Coalition (Figure 3). This group was formed principally to pursue the *DOE* recommendations of seamount ecosystem preservation and the proposed United Nations moratorium on destructive bottom-trawl fishing on the largely unregulated high seas. In an unusual approach to conservation funding, this group of institutions began implementing *DOE* recommendations without any outside funding, and two foundations came to the table to support the effort. The management team and governing structure for individual programs should be considered when determining the level and timing of financial support for a major coordinated program.

**Planned products and services**

The "products and services" of this business plan are the program recommendations developed by the seven theme groups at *DOE* and presented in Chapters 6 through 12. The costing of services from establishment of MPAs, such as coastal protection and fisheries benefits provided by mangroves (Sathiratha 2000), have been considered as products of this program.

There are also less tangible products and results of the proposed action agenda, such as improved governing policies and management structures to oversee key marine resources within large marine ecosystems. Each of the various products, services or results has a different mix of "market segments" or interested stakeholders with different willingness and ability to financially support these programs. An environmentally positive result of improved fisheries management must be balanced against the short-term costs to fishermen, increased long-term profits to the fishery and the effects of redirecting government subsidies. Ecotourism fee structures are an element often used in MPA business models to support the maintenance of

© 2004, IISD/FRANZ DEJON

**Figure 3** A coalition emerged from *DOE* that is seeking a United Nations moratorium on high seas bottom-trawl fishing and the protection of seamount habitats. Lisa Speer of the Natural Resources Defense Council represents the new Deep Sea Conservation Coalition at the United Nations in June 2004.

protected areas (Spergel 2004, Morris 2002, Phillips 2000). For long-term full financing to be secured, however, many new revenue streams must be established that go beyond single-issue projects and focus on reducing or mitigating multiple threats to marine areas (Gravestock 2002). Each of the stakeholders affected by a specific program—particularly those whose economic livelihoods are decreased—should be considered "market segments" for any resulting products and services from marine conservation programs.

**Competition**

Competition for a *DOE* plan comes from two areas: competition from other proposed conservation programs in a resource-limited "market," and broader competition between conservation programs and profit-based marine resource uses. While some competition among conservation programs is healthy, the immediacy and global nature of ocean conservation concerns require a more coordinated approach and combined application of scarce resources by all interested organizations.

The competition from fishing, oil and gas development and other industries is direct and as strong as any competing market force in commercial industry. The fisheries, fisheries management and conservation communities have traditionally held competing views and policies for managing access to and protection of the limited living marine resources. This competition is expected to continue and possibly intensify over time, given the increasingly drastic reductions in available resources. There is hope, however, that the increasing collapse of commercially viable fish stocks will bring these historic competitors to a more cooperative approach in the near future.

The competitive advantage of any commercial enterprise or conservation plan must be compared with all directly and indirectly competing offerings. In addition, the advantages of commercial efforts, such as oil and gas development, in competition with conservation, are considerable because of increased jobs provided to local communities and increased management revenues to governments. The weakness in this comparison, however, is that most commercial activities will have a "conservation cost," and more companies are being required to quantify these costs, develop environmental action plans to minimize their impacts and/or provide mitigation funds (Figure 4). The inherent competitive advantage of conservation is that it may increase biodiversity, ecosystem services and overall economic benefits. To strengthen the position of conservation versus commercial activity in a combined ecological/economic system, the conservation community must better quantify, compare and communicate long-term threats and benefits of competing alternatives.

### Risks

There are two general types of risk associated with the *DOE* marine conservation recommendations: execution risks—whether or not a given project will be implemented well; and "business-as-usual" risk—whether or not humankind can collectively improve current policies and practices affecting the world's ocean.

Providers of financial support can mitigate the project-specific risks by developing a portfolio approach to conservation investments and developing specific criteria for marine projects that receive investments. The more difficult risks to address are those regarding "business-as-usual" concerns. The economic risks of inaction in fisheries can be quite severe, as seen in the 1992 collapse of the Canadian cod fishery resulting in economic losses in the billions of dollars. The potential annual losses from the destruction of coastal mangrove and coral reef systems that protect islands and coastlines from severe storm events are estimated to be nearly US $50 billion (Munich Re Group 2003). There are also many unexplored and difficult to quantify economic benefits that may come to light as a result of increased marine research expenditures. The scope of risks associated with the business-as-usual scenario regarding global marine conservation is likely to be financially material to the business case and should be integrated into financial decisions to support the *DOE* action agenda recommendations.

### Financial costs

The theme groups, with input from the business team, estimated high-level costs for each activity that drives our cost model. At the Conference, each theme group developed a detailed matrix of recommended activities and their associated cost drivers, and shared ideas and final inputs with the other theme groups. Major costs were attributed to:

- Large marine ecosystem management, and the reduction of threats from sea-based industrial activities
- Establishment of networks of marine protected areas and the required infrastructure
- Local capacity-building and community outreach
- Global science, including dramatically increased resources for marine species and ecosystem assessments
- Global fisheries reform policy efforts and the redirection to positive economic incentives
- Efforts toward global and local policy enclosure for the ocean
- Communications in key marine resource demand and supply countries
- Management of these programs

© WOLCOTT HENRY 2001

**Figure 4** The offshore oil and gas industry, with oil rigs like this one in the Gulf of Mexico, provides jobs and commercial profits but has conservation costs that must also be considered.

Table 1

**Estimated costs for a global, ten-year, multi-faceted campaign for the protection of high seas seamounts**

| Activity | Costs |
|---|---|
| Map known ranges of seamount species, and identify new and endangered species | $3.4 million |
| Create global maps of estimated species diversity, endemism and current fishing trends | $1.6 million |
| Establish a Seamount Campaign, including international public awareness of seamount value and threats | $6.3 million |
| Develop an expanded SeamountsOnline database to hold all available seamount biological data | $1.8 million |
| Complete a globally-representative biological sampling on seamounts | $21.5 million |
| Design a system for monitoring fishing activity on seamounts | $3.8 million |
| Develop and create support for international policies to protect seamounts globally | N/A |
| **Total for 10 years** | **$38.4 million** |

Given the time constraints of the *DOE* Conference, there was a wide range of uncertainty in the cost estimates. Cost models for some recommended programs—such as the creation and operation of marine protected area networks and development of a seamounts protection campaign—are well founded. Protection of ancient habitats and newly discovered biodiversity on open-ocean seamounts emerged as a high-priority action from many of the *DOE* theme groups, and costs were estimated for a multi-faceted campaign which includes science, communications, advocacy, policy and law (Table 1).

Cost models for other programs—such as creating a high seas ocean public trust or changing the global mindset on ocean uses—have never been approached, so the costs and timeframes have a high degree of uncertainty. Costs of programs similar in scope, such as the United Nations Fish Stocks Agreement (for global policy efforts), were used where possible to estimate costs for comparable *DOE* recommended actions.

In the preliminary business plan, the overall cost estimate for the *DOE* marine conservation action agenda over ten years ranges from US $18.6 billion to more than US $30 billion (Table 2). This investment is quite small when compared to the much greater economic costs required for potential ecosystem restoration, decreases in economic livelihoods due to fisheries closures, and the $30 billion annually paid out by governments for "perverse subsidies" of various kinds that are actually supporting the catastrophic decline of ocean life. The immediate priorities are estimated to require an investment of approximately US $4.8 billion over the next three years. Some of these activities can be executed as individual programs, but the power of the *DOE* agenda is that key programs will support each other if implemented in a coordinated fashion.

**Limitations of the *DOE* business plan**

The *DOE* integrated cost model is not comprehensive and the financial estimates should be considered reasonable only as an order of magnitude. The plan focuses only on the highest priority actions required within three and ten years.

This initial model does not fully account for several significant cost factors. First, the model includes programs to protect approximately 20 large marine ecosystems (LMEs) and 400 marine protected areas (MPAs) that would together cover about 5% of the world's ocean, but it includes only limited support for existing MPAs which likely require additional funds (Gravestock 2002). Our business plan assumes that all sustainable economic and community development costs of establishing MPAs and LMEs and improving coastal land-use practices will be financed by current international global aid programs. The recommended fisheries reforms will be piloted in only a few fisheries, and our business plan assumes global fisheries programs will be largely self-funded by redirection of current perverse subsidies and expected economic benefits resulting from the reforms.

Our cost model has not included resources to cover perpetuity costs for MPA operations from trust funds or other potential revenue sources such as ecotourism. The *DOE* recommendations and resulting cost model do not allocate significant resources to address regional or global land-based threats, or the costs and revenue potential from aquaculture activities. The plan also assumes that enforcement costs, required to make MPA or LME management effective, will be covered by local government agencies. There are good

Table 2
**Minimal costs for the *DOE* action agenda**

| Conference action agenda | First 3 years | 10-year total |
|---|---|---|
| Global science | $0.32 billion | $0.65 billion |
| Large marine ecosystem (LME) programs | $2.38 billion | $9.32 billion |
| Marine protected areas (MPAs) | $1.18 billion | $5.42 billion |
| Fisheries reform | $0.17 billion | $0.49 billion |
| Global governance | $0.14 billion | $0.38 billion |
| Communications plan | $0.08 billion | $0.44 billion |
| Program communication | $0.51 billion | $1.93 billion |
| **Total** | **$4.78 billion** | **$18.63 billion** |

examples of MPA financing (Phillips 2000, Morris 2002), but the next step in financing marine protected areas must build on these models to include revenue streams from modifying or managing the offending land-use activities.

Lastly, a key limitation of our *DOE* cost model is that no substantial analysis of new funding sources for marine conservation programs was performed. The economic benefits of ecological services must be better documented to support increased funding for the *DOE* global action plan.

## Conclusions and Lessons Learned

Marine conservation is perhaps fifteen years behind terrestrial conservation efforts and has three additional challenges. The world's ocean lacks uniform and complete policy enclosure especially on the high seas; the threats posed by large-scale fishing for human livelihoods and consumption do not readily equate to the localized species loss from rainforest "bushmeat" trade; and conservation of marine ecosystems depends on improved practices occurring on land as well as in the ocean.

This business plan for global marine conservation was not meant to estimate the total costs for all actions ultimately needed to protect the 20% to 30% of the global ocean recommended by many of the *Defying Ocean's End* experts. Our plan focuses only on the most urgent actions that can be implemented in the next ten years. The *DOE* theme groups were asked to develop their conservation recommendations and to address which actions are feasible, during what timeframe, and what capacity and financial support is needed for implementation.

Urgent policy changes recommended by the Pew Oceans Commission and the U.S. Ocean Commission require significant systematic changes to governing laws and management authorities. The world must also act now to increase dramatically the scale and scope of marine conservation. Our business planning approach for marine conservation should be extended to include more stakeholders from governments, fishing communities and potential financing organizations. A more integrated business plan would simultaneously address cross-cutting issues, such as improving global economic incentives and disincentives, altering policy and management regimes, and improving marine communications across many cultures and countries.

The overall cost estimate for the *DOE* global action agenda ranges from a minimum of US $18.6 billion to more than US $30 billion over ten years. These estimated costs are required to protect only five percent (5%) of the world's ocean over the next ten years, and do not include the parallel financial support required for economic development in the affected communities. This *DOE* business plan is only a focused, high-priority subset of the estimated US $300 billion required annually to manage a system of protected areas globally that includes a representative subset of all types of marine habitats and ecosystems (James, Gaston & Balmford 1999). Roughly US $50 billion of this is required for compensating communities and providing positive incentives for protection. Our plan should be enhanced to include a more rigorous estimate of new approaches for addressing sustainable development in communities that rely heavily on marine resources for their livelihoods.

The marine community must take the next steps to more closely coordinate their collective plans into a global agenda for action, working from the *DOE* recommendations as a start. We should also work together on annual reviews of progress toward the global goals. The large-scale *DOE* recommendations for policy, scientific and conservation initiatives must be pursued collectively by governments, communities, financial providers, universities and conservation organizations. And we must implement a global communications plan—on the scale of the one crafted at *DOE*—to reach all audiences and explain why it is important for humankind to protect the health of our ocean.

There are many positive examples of conservation business planning and conservation finance that could be analyzed and used to improve this high-level approach developed at the *Defying Ocean's End* Conference. Many of the current approaches to conservation finance and business planning for marine protected areas rely exclusively on revenue opportunities within the delineated protected area such as ecotourism, with some supplemental funding from traditional philanthropic and international economic development sources. For MPA network financing to be viable for coastal areas in the long term, an enhanced business plan must include a detailed analysis linking terrestrial and marine ecological services to innovative scientific and economic approaches.

Unlike lunar exploration in the 1960s, marine conservation does not have the support of a Presidential challenge to save the world's ocean in ten years. However, it is critically important to life on Earth that the marine conservation community pursue this daunting challenge with the same immediacy, public attention, practical plan, financial support and sheer determination that developed in the early days of the space program.

## Acknowledgements

*We gratefully acknowledge the thoughtful and dedicated efforts of all members of the* DOE *theme groups we worked with during the conference. Their enthusiasm and excellent work in identifying and prioritizing the costs of their wide-ranging recommendations bodes well for expanding and refining this type of effort in the future.*

## Literature Cited & Consulted

Balmford, A., A. Bruner, P. Cooper, R. Costanza, S. Farber, R. E. Green, M. Jenkins, P. Jefferiss, V. Jessamy, J. Madden, K. Munro, N. Myers, S. Naeem, J. Paavola, M. Rayment, S. Rosendo, J. Roughgarden, K. Trumper and R. K. Turner. 2002. Economic reasons for valuing wild nature. *Science* 297 (5583): 950–953.

Campbell, D. and J. Lawrence. 2004. Ocean Facts. *http://www.marinebio.org/MarineBio/Facts/* (accessed September 5, 2004).

Compton, W. D. 1989. *Where No Man Has Gone Before: A History of Apollo Lunar Exploration Missions.* NASA Special Publication 4214.

Costanza R., R. d'Arge, R. de Groot, S. Farber, M. Grasso, B. Hannon, K. Limburg, S. Naeem, R. V. O'Neill, J. Paruelo, R. G. Raskin, P. Sutton and M. van den Belt. 1997. The value of the world's ecosystem services and natural capital. *Nature* 387:253–260.

Gravestock, P. 2002. Towards a better understanding of the income requirements of marine protected areas. *http://www.oceansatlas.org/cds_upload/1052753907869_Income_requirements_of_MPAs.pdf* (accessed September 5, 2004).

James, A. N., K. J. Gaston and A. Balmford. 1999. Balancing earth's accounts. *Nature* 401 (6751): 323–324.

Kelleher, G., C. Bleakley and S. Wells. 1995. *A Global Representative System of Marine Protected Areas.* Washington, D.C.: The World Bank.

Marshall, T., M. A. Kreps and N. C. Chriss. 1989. The eagle has landed: 20 years after Apollo 11. *Houston Chronicle.* 17 July.

Meyers, R. A. and B. Worm. 2003. Rapid worldwide depeletion of predatory fish communities. *Nature* 423 (6937): 280–382.

Milazzo, M. 1998. Subsidies in world fisheries. *World Bank Technical Paper WTP406.*

Morris, B. 2002. *Transforming Coral Reef Conservation in the 21st Century: Achieving Financially Sustainable Networks of Marine Protected Areas.* Washington, D.C.: The Nature Conservancy.

Munich Re Group. 2003. The economy of climate. *http://www.munichre.com* (accessed September 5, 2004).

Palumbi, S. 2003. Marine reserves, a tool for ecosystem management and conservation. In *America's Living Oceans: Charting a Course for Sea Change.* Arlington, Virginia: Pew Oceans Commission.

Phillips, A. 2000. *Financing Protected Areas.* IUCN–The World Conservation Union.

Sathiratha, S. 2000. Economic valuation of mangroves and the roles of local communities in the conservation of natural resources: Case study of Surat Thani, south of Thailand. *http://web.idrc.ca/uploads/user-S/10536137110ACF9E.pdf* (accessed September 17, 2004).

Spergel, B. and M. Moye. 2004. *Financing Marine Conservation: A Menu of Options.* WWF.

Wildlife Conservation Society. 2004. Conservation Finance Alliance. *http://www.conservationfinance.org* (accessed September 5, 2004).

CHAPTER 14

# Technology Support to Conservation

Robert J. Fine, *Conservation International*
Daniel A. Zimble, *Environmental Systems Research Institute (ESRI)*

© KIP F. EVANS, NATIONAL GEOGRAPHIC SOCIETY, 1998

## Introduction

Given the current threats to global ocean health, it is critical that we leverage all technology tools available to preserve the ecological processes and biodiversity of our oceans. Without a healthy ocean, there can be no healthy planet. The subject of "technology support to conservation" could be its own book, but this chapter focuses on technologies particularly relevant to recommendations of the *Defying Ocean's End* global plan of action for marine conservation.

Throughout this volume are syntheses of global themes, case studies and recommended actions in global governance, fisheries reform, communications, marine protected areas, large marine ecosystems and global ocean science. There are a number of technologies that can significantly enhance the likelihood of attaining the goals laid out in the *DOE* global agenda. Information technologies support this effort in many ways:

- Helping scientists better understand the conservation implications of their data
- Helping managers develop more sustainable management strategies
- Assisting policymakers in the design and implementation of more appropriate marine conservation policies

In the marine environment, understanding complex human/ecosystem dynamics within a given space across time has become critical to developing action plans. Within the last 30 years, the coinciding development of information technologies and a variety of sensors and measurement platforms are furthering our ability to provide much more useful information about our ocean and the planet as a whole. Within the highly dynamic marine environment, new technologies can assist scientists and policymakers in understanding the distribution and abundance of various species and the conditions of their environment. Only through understanding the ecology of community assemblages within their environment, plus human-induced or natural stresses on the system, will we be able to ensure the sustainability of the resources. The role these technologies play is becoming increasingly more important as human populations and subsequent demands on our ocean's resources continue to increase.

Animal tagging and tracking, one new technology for determining the location and behavior of a number of species, is contributing to the better understanding of these living resources. Some of the newer tracking systems are also sensing such variables as internal body temperature of the animal, plus the ambient (surrounding) temperature and pressure, and light attenuation (the degree of light penetration into the depths traveled by the animals) (Block *et al* 2003). Another means of collecting this type of information, although much less sophisticated, is through human observers stationed on fishing boats equipped with global positioning systems (GPS) that locate and log where and when particular species are caught (M. Hall personal communication).

In more remote settings, a number of technologies derive information about conditions of the marine environment. These may include remote sensors on a variety of platforms including:

- Earth orbiting satellites
- Aircraft
- Unmanned aerial vehicles (UAV)
- Sea-surface research vessels
- Moored data buoys
- Autonomous underwater vehicles (AUV)
- Underwater remotely operated vehicles (ROV)
- Manned submersibles

All these platforms can be equipped with specially tuned and configured sensors that can detect, store and transmit a great detail of environmental information such as bathymetry (water depth), coral reef distribution, wave height, air temperature, sea surface temperature (SST), chlorophyll-a (phytoplankton), dissolved organic nutrients and dissolved oxygen. A promising technology for sensing marine species is bioacoustics, which is increasingly able to distinguish one species from another based on their specific sound signatures, some of which indicate certain behaviors, such as spawning. While all of these sensing technologies are deriving a great deal of useful information, effective management and sustainment of healthy marine ecosystems requires processing, analysis and understanding of the interactions across and between different community assemblages in different environments, especially in determining how human pressures (both direct and indirect) are affecting any given system. This is the only way to design and implement truly sustainable management strategies that will ensure the long-term conservation of ocean resources.

A class of technology particularly suited to analyzing complex relationships across space and time is geographic information systems (GIS). GIS by nature involves multiple "layers" of information geographically located (called spatial databases) that assist in the understanding of complex relationships between and across datasets from a variety of information sources. GIS is an enabling technology completely dependent upon its data sources and must keep growing to develop ways to analyze, compare and display new types of data from sophisticated sensor systems. When data and other spatial information from multiple sources are put into a GIS system, it transforms into a decision support system (DSS) capable of providing responses to a whole host of complex questions. The output from a well-designed GIS can significantly help scientists, ecosystem managers, policymakers and the general public better understand how marine resources are impacted by a variety of stresses on any given ecosystem at any given time. In this way GIS systems effectively transition into DSS applications that are becoming increasingly important for ecosystem-based management. A good example of this type is NatureServe's Vista Decision Support System, currently being developed and implemented for land-based ecosystems (Riordan & Barker 2003). There is, however, very little effort currently being invested in developing these systems to support improved management of marine-based ecosystems.

Participants of the *Defying Ocean's End* Conference identified recommendations for global action in five areas:

- Global Governance
- Fisheries Reform
- Ocean Communications
- Marine Protected Areas & Large Marine Ecosystems
- Global Ocean Science

The technologies described and discussed in this chapter all have a significant role to play in supporting action recommendations in each of these areas.

## Remote Sensing Technologies

Remote sensing can be defined as follows:

> *The science and art of obtaining information about an object, area or phenomenon through the analysis of data acquired by a device that is not in contact with the object, area or phenomenon under investigation* (Lillesand & Kiefer 2000).

Remote sensing instruments that provide environmental information about marine resources range from very small-scale (low resolution) globally derived variables to extremely remote but large-scale (high resolution) samples of very specific locations. An example of small-scale remote sensing products include satellite-derived variables such as sea surface temperature, which can be globally averaged and displayed in monthly, yearly or even decadal GIS layers (Figure 1). Large-scale (high resolution) remote sensing products vary greatly depending on the sensor and its platform. In the spaceborne context, such systems are capable of measuring phenomena on the Earth in enough spatial detail to resolve objects the size of individual cars or boats. At all resolutions, information from remotely sensed data is fundamentally beneficial for users to better understand, manage and enforce good ocean governance. Remote sensing provides information critical to the following classes of activities:

- Providing a contextual backdrop or framework for other existing geographical information to be geolocated
- Providing an information resource in which features of interest can be extracted, measured and mapped
- Conducting spatial analysis and/or generating modeling scenarios regarding the locations and movements of marine resources and human impacts
- Providing a single large area snapshot of information over the geographic area of interest at a single moment in time to capture baseline conditions and/or to determine fundamental changes

Remote sensing includes many technologies in a data lifecycle from the collection stage, through processing to the development of final products.

### Platforms: airborne, space-based, waterborne

In this section we review the major approaches to remote sensing—including platforms, sensor characteristics and respective capabilities—to emphasize the importance of all of these technologies to marine conservation. Remote sensing is often considered to be primarily space-based, but many useful remote sensors are operating on a wide variety of platforms. As far back as the 1840s, balloonists used cameras to aid in land surveys, mapping, weather observations, urban planning and military reconnaissance of the battlefield. Remote sensors have been deployed on balloons, airplanes, ships, helicopters, unmanned aerial vehicles (UAV), autonomous underwater vehicles (AUV), remotely operated vehicles (ROV) and spacecraft. These platforms can generally be broken down into three classes: *airborne, space-based,* and *waterborne.*

Comparison of these major classes will help differentiate the value of the platforms for varying conditions. Waterborne and airborne platforms have limits for effective deployment time and coverage (e.g., limited range and

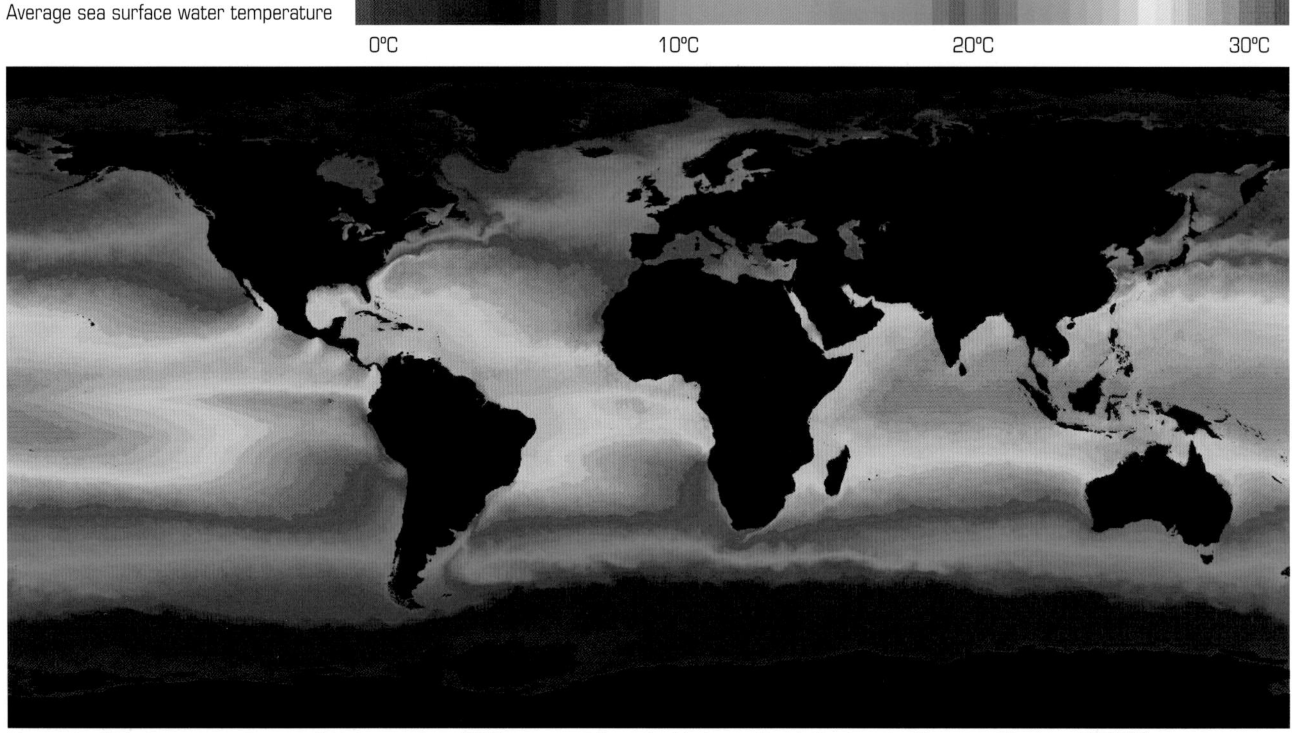


**Figure 1** Global averages of sea surface temperature (SST) measured from the Moderate Resolution Imaging Spectroradiometer (MODIS) sensor on NASA's Terra satellite.

limits on usable conditions due to adverse weather conditions, etc). Despite these limits, they do provide unique high resolution capabilities not yet achievable with most spaceborne sensors—such as extremely high resolution videography—which can facilitate the identification of known and unknown marine species and assist in detailed habitat analysis (Veisze & Karpov 2002). Another important aspect of remote sensing is the ability to image the same geography repeatedly over time, thus detecting change in an observable phenomenon. Satellite-based sensors in repeatable orbits with a mission lifetime of a few years can provide continuous coverage of the same geography over time. But the quality and utility of an image derived from a satellite-based sensor is often limited by atmospheric conditions (e.g., clouds, smoke, dust and haze) over the geography of interest. An airborne or waterborne sensor, on the other hand, can operate at altitudes/depths that may avoid many of the atmospheric/oceanic disturbances (smoke, haze or oil slicks). Use of a variety of remote sensing platforms is crucial for collecting the range of information needed for marine conservation and management in both the short-term and long-term.

*Airborne sensors* are useful as a lower cost option for high-resolution imagery. Another benefit of airborne sensors is their ability to focus on a small area of interest. In a time-sensitive situation, one can rely on airborne sensors to study a specific area quickly. A downside of this low-cost option is that it generally relies on film rather than digital sensors and the processing time from planning the flight to delivery of the data may take days, sometimes weeks. As more airborne sensors (cameras) take advantage of digital technology, the data delivery time will be significantly reduced.

*Space-based sensors* now collect digital, rather than film-based, images of the Earth environment and send them to a ground station in near real time. Many do not measure the full spectrum of light reflected from the marine environment, but pass the returned light through filters that break it into distinct wavelength bands. Each band is optimal for discrimination of different features, such as vegetation, water or biologic matter in the water. The multi-spectral scanner and thematic mapper on the Landsat satellite work in this way.

*Waterborne sensors* are very different from spaceborne or airborne sensors due to extreme differences in their operating environments. Many ocean-based sensors such as singlebeam or multibeam water-depth mappers use sonar, or sound devices, because the marine environment absorbs light so much that it cannot penetrate very far below the ocean surface in most areas. Without light penetration, information about the marine environment below the ocean surface is not accessible to most of the sensors used on airborne and space-based platforms. Many ROVs and manned submersibles carry powerful light sources into the ocean depths that can be used to support video, digital and film imaging of habitats, creatures and their behavior in the ocean's depths. This approach has been used to identify new species and to discover much information about their behavior that is just not available from traditional methods of studying dead creatures brought up from the depths in nets.

## Sensor characteristics

### *Active vs. passive*

Within these three classes of platforms for remote sensors, sensors are either *active* or *passive*. *Active sensors* use an active energy pulse—the most common of which are radio detection and ranging (radar), light detection and ranging (lidar) or sound navigation and ranging (sonar). The satellite sensor sends a pulse of energy toward its target area of interest and collects data based on the character of the energy spectrum reflected back to the sensor. *Passive sensors* receive energy from the sun or the earth, similar to the way a camera collects light energy. Multispectral sensors, such as those carried on Landsat satellites or the French Système Pour l'Observation de la Terre (SPOT) satellites, are some of the most commonly used passive remote sensing systems. The major distinction between active and passive sensor types is the manner in which they detect information about the Earth. Passive sensors are dependent upon collecting energy radiated from the land or ocean, mostly light or heat energy reflected from the sun. Active sensors provide their own light, heat, sound or other energy source to generate a reflection from the surface being sensed (Figure 2).

### *Resolution*

Active and passive sensors have a number of common characteristics that can describe their capabilities:

- *Temporal resolution, or how often an object, area or phenomenon is revisited and remeasured*. Depending on the application, the repeat cycle can be critical, especially as it pertains to monitoring the changes in or movement of various phenomena over time. Furthermore, this capability allows for *change detection*. GIS spatial data layers are particularly useful in identifying and displaying the

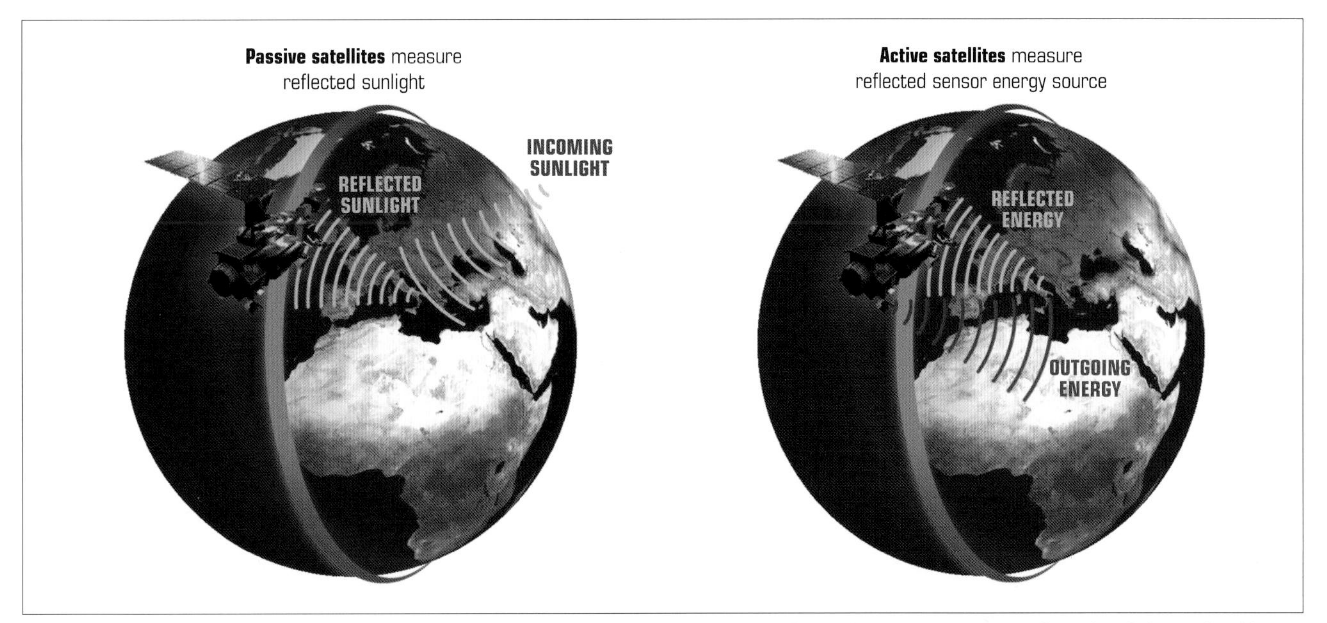


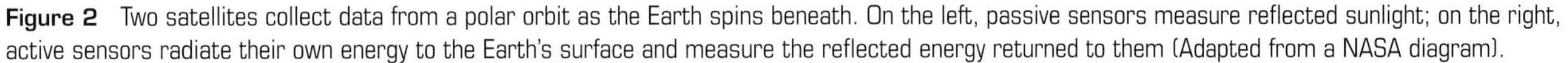
**Figure 2** Two satellites collect data from a polar orbit as the Earth spins beneath. On the left, passive sensors measure reflected sunlight; on the right, active sensors radiate their own energy to the Earth's surface and measure the reflected energy returned to them (Adapted from a NASA diagram).

**Figure 3** *Left to right:* MODIS 1-kilometer resolution image of the Nile River delta (courtesy NASA); Landsat 30-meter resolution image of the southwest Nile River delta with pyramids barely visible in the center (courtesy NASA); QuickBird 0.6-meter resolution image of the Great Pyramid of Giza, Egypt.

changes over time in objects, areas or phenomena of interest.

- *Spatial resolution, or how capable the sensor is in detecting small objects and details of information*. Spatial resolution is most often used to describe the fidelity of a remotely sensed image, since this characteristic determines what type of object, area or phenomenon is detectable within an image (Figure 3). With high-resolution commercial remote sensing photo-imaging now available—from Space Imaging's IKONOS satellite, Digital Globe's QuickBird satellite and Orbimage's OrbView-3 satellite —it is possible to monitor and assess an object, area or phenomenon of interest at scales appropriate to support marine ecosystem management activities.
- *Spectral resolution, or how much of the electromagnetic spectrum (wavelengths of energy) a sensor can detect*. Spectral resolution describes the range of the electromagnetic spectrum (EM) detectable by a sensor. Spectral ranges of photo-imaging sensors such as those on Landsat or SPOT typically cover the same visible portion of the spectrum that the human eye sees—wavelengths between 0.4 (blue) and 0.7 (red) nanometers. There are many space-based sensors, however, that sense information at lower and higher EM wavelengths, measuring Earth information from both the ultraviolet (UV) and infrared (IR) regions of the spectrum. Infrared sensors can measure non-visible phenomena, such as temperature, even in darkness or through cloud-cover. By combining both visible and non-visible portions of the EM spectrum, image-processing specialists can distinguish among different types of land cover, geology, algal blooms in coastal waters and other phenomena of interest. Active sensors (radar, lidar and sonar) also operate outside the visible range of the spectrum. Radar sensors operate in the microwave portion of the spectrum and can be used to measure the elevation of the land or wave height. For example, the TOPEX/POSIDEON (a joint mission between the U.S. and France) or the Jason-1 (U.S.) missions use radar altimeters to measure sea surface height. Such information is critical for understanding global oceanic climate patterns and phenomena such as El Niño Southern Oscillation (ENSO) events and ocean currents.
- *Radiometric resolution, or how sensitive a sensor is to the amount of transmitted information reflected from an object, area or phenomenon*. Radiometric resolution is the measure used for digital sensors to describe the number of difference values (or precision) that can be detected within any one spectral channel (band of wavelengths) per pixel (picture cell, the smallest resolution element). A channel (band) can be thought of as a slice of the EM spectrum. Sensors use separate channels to collect different types of information over a particular geographic area of interest. Each resulting data layer, or image, consists of rows and columns of digital information that together form a raster (gridded) data structure (Figure 4). At the cross-section of any row and column within a digital image or data layer is a cell (or pixel, picture element) of the image. It is in this cell that a quantitative value is recorded that represents a reflectance (brightness) value of the object, area or phenomenon of interest. Radiometric resolution is therefore a term used to describe how many values can be contained in a pixel and are usually described in units of bits (e.g., 8-bit, 16-bit or 24-bit). The greater the EM energy reflected in that wavelength, the greater the

value. Most images used today are typically 3-channel red, green, blue (RGB) composite images.

Sensor characteristics such as those mentioned above become more important when considering the process generally referred to as image classification or image processing. In this process, digital image information is processed and analyzed statistically to show comparison among images. An example of this is the Millennium Coral Reef Mapping project, managed through the Institute for Marine Remote Sensing (IMaRS) at the University of South Florida (USF) and funded by the Oceanography Program of the National Aeronautics and Space Administration (NASA). Its purpose is to provide an exhaustive worldwide inventory of coral reefs using high resolution satellite imagery. By using a consistent dataset of high resolution (30-meter) multispectral Landsat 7 images acquired between 1999 and 2002, USF is characterizing, mapping and estimating the extent of shallow coral reef ecosystems in the main coral reef provinces (Caribbean-Atlantic, Pacific, Indo-Pacific and Red Sea). The program highlights similarities and differences between reef structures at a scale never possible with traditional data

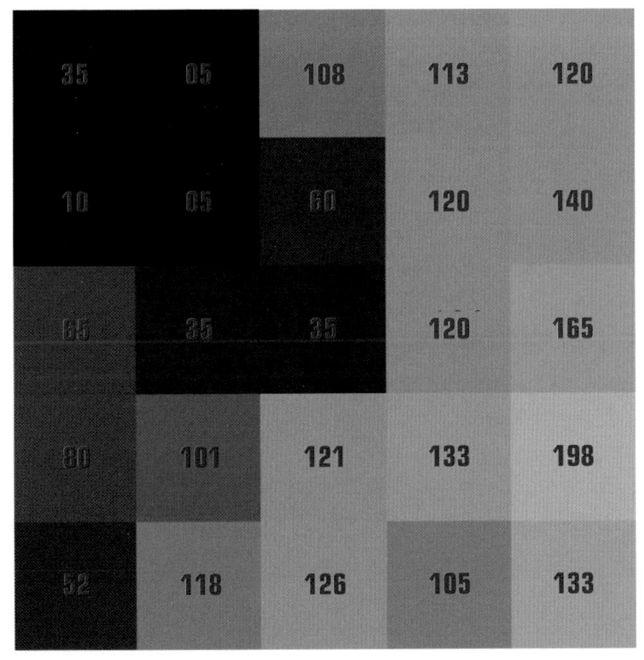


**Figure 4** An image consisting of rows and columns of information that together form a raster (gridded) data structure. At each intersection of a row and column is a cell (pixel) that stores the brightness value or intensity (a digital measure of the reflectance value of an object, area or phenomenon of interest) within a given sensor's frequency band.

sources based on field studies. It is providing critical information for reef managers in terms of reef location, extent and distribution.

Another example is the United States government's Coral Reef Task Force (2004) established in response to the growing global environmental crisis in the loss of coral reef ecosystems. Within the context of this initiative, the need for primary mapping data on all the coral reefs around the world is critical. It is in such contexts that geographic information systems and remote sensing have a fundamental role to play.

## Geographic Information Systems

Along with the development of remote sensing systems, geographic information systems have also matured over the last 30 years. GIS is an increasingly important tool for conservation, and we will review the basic concepts and discuss relevant past, present and future initiatives.

GIS has many definitions, which vary depending on how it is used and by whom (Longley *et al* 2001). While some people think of GIS only as software, it is really a system that can be described as

> *an arrangement of computer hardware, software and geographic data that people interact with in order to integrate, analyze and visualize that data and to further identify relationships, patterns and trends in order to find solutions to a given problem* (Kennedy 2001).

While the above definition is useful for understanding the technical aspects of GIS, there are critical institutional and political ramifications that affect how useful a GIS can be. The term "Enterprise GIS" is now taking hold, where organizations and even nations are implementing a rigorously structured and enforced geospatial infrastructure designed around their own data, information needs and sharing policies (Tomlinson 2003). The emergence of Enterprise GIS could be a major benefit for ocean conservation. Such systems would allow one to imagine and develop an entire framework around GIS-based decision-making processes specifically for oceans.

GIS is an approach designed to capture, store, update, manipulate, analyze and display geographically located information that can be utilized by numerous users. It is useful to think of a GIS as a container of digital maps (geographically located spatial databases) that can be overlain as data layers and then integrated and used to perform a variety of analyses (Figure 5). The most powerful ability of a GIS is often depicted as a series of multiple layers that overlie each other in the same geographic space, since it is this integration of spatial information from which all of the planning, change detection and other analyses can be made.

### Digitizing maps

The development of computer technologies—such as powerful processors, storage devices, digitizers, scanners and high-end graphics displays—has revolutionized the art of map-making and has intertwined the disciplines of remote sensing and GIS. The process of digitizing maps into a GIS involves using existing maps with a digitizing tablet to trace the feature of interest onto a map, so it can be sorted, overlain and compared with other data already in the GIS. This digitization process facilitated the growth of data repositories, which promoted the sharing of data among and across projects, yielding many new uses. GIS also has the ability to assign spatial references to each piece of information so it can be displayed at any scale and in any map projection (view of the Earth) such as spherical projections looking down on one of the poles, or a Mercator projection which shows the Earth spread into a flat rectangle. This spatial marking of the data allows for the efficient and accurate overlay of multiple data sets from different sources and allows for the accurate calculation of distances, direction, areas and comparison of many characteristics of a given geographic area. This ability to integrate multiple map layers from different data sources and sensors adds to the powerful ability of GIS to perform sophisticated analytical instructions quickly, providing analytical output that can significantly help with the marine conservation decision-making process.

**Figure 5** A geographic information system (GIS) presents overlays of different types of information on the same map base for analysis and comparison (courtesy of NOAA).

## GIS Output and Digital Maps

In addition to GIS capabilities available on desktop computer workstations, the development of the World Wide Web and Internet technology has provided the means for communicating interactive map layers and spatial information across the Internet to a wider audience. There are numerous online applications disseminating information about various aspects of marine environments. Online tools, coupled with high performance workstation GIS applications, are an incredibly powerful combination for supporting the demanding and dynamic information needs of marine ecosystem scientists, managers and policymakers.

### Institutions and applications

Over time, GIS has evolved from individual project-oriented efforts into larger department-wide or web-based global information sharing efforts. The culture of sharing is a major aspect of GIS practitioners that enables more than the simple sharing of data—it makes possible the sharing of experience and knowledge. ReefBase (ReefBase 2004) is one of the largest online information systems concerning coral reefs and provides information services to coral reef professionals involved in management, research, monitoring, conservation and education (Figure 6). The ReefBase program has provided the United Nations Environmental Programme/World Conservation Monitoring Centre (UNEP-WCMC) data on coral reefs and mangroves on the Web with online mapping tools that allow individuals to download information for their own projects and initiatives.

GIS is also increasingly acting as a clearinghouse, where users can find the data and information they need and then obtain the tools required for accessing that information. A great example of such a GIS portal is Geospatial One Stop,

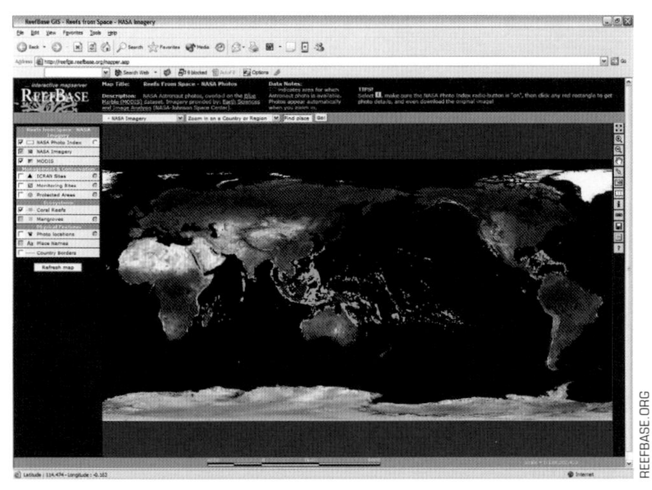

REEFBASE.ORG

**Figure 6** ReefBase is an example of a data-rich GIS system that is freely available on the Internet. Photographs and locations of coral reefs, mangrove areas, protected areas and monitoring sites can all be accessed and printed out in map form through this system.

currently provided by the U.S. federal government (Geodata.gov 2004). A new GIS portal specifically focused on marine species information is taking shape under the auspices of the Census of Marine Life (CoML). In particular, CoML's Ocean Biogeographic Information System (OBIS) is an information clearinghouse for marine species information that includes a database-driven online mapping component. This project continues to evolve as a leader in providing long-term, highly accessible information resources on marine biology species information (Census of Marine Life 2004).

Data on demographics, streets, property boundaries and weather stations are commonly shared and sold in value-added products that are used for a variety of GIS-enabled applications, including conservation planning, species distribution analysis, demographic analysis, routing services (like "Mapquest"), local municipality governance (e.g., tax enforcement) and local/regional weather forecasts. Typically the GIS decision-making process involves a series of steps that document the methodology for attacking a particular problem.

Though the history of GIS is a long and complex one, it is important for appreciating the level of maturity that currently exists in the geospatial community (Foresman 1998). We are now at an all-time high with regard to geospatial capacity. According to a U.S. Department of Labor study of market growth, the geospatial market currently is estimated to be worth $5 billion annually and is expected to expand to a $30 billion industry over the next several years (Sietzen 2004). The significance of this to marine conservation is considerable. Successfully sustaining ecological processes that foster healthy living marine ecosystems requires the command of a heightened level of awareness about human/ecosystem dynamics at multiple scales. How will developing a new residential area affect sedimentation rates, and where will any resulting increase in sedimentation end up? What are the multi-use conflicts among marine protected areas, migratory species and fishing fleets? GIS can be very effective for answering such questions, but data is the key to unlocking this potential. Another exciting aspect of this technology is that the network of users and related technologies continues to grow rapidly. This market growth will drive down the cost of GIS technologies, making them more accessible for marine conservation decision-makers. Growth in the geospatial community will provide expertise and a heightened level of awareness about our ocean world that can be leveraged to support marine conservation decision-making.

## Telemetry and Tagging

Widely used methods for collecting information on the behavior and habitat of marine species in the open ocean include two discrete technologies known as telemetry and tagging, which are used together to study the habits of species under water (Figure 7).

Telemetry can be defined as the science of conveying information from one location to another. There are two

TOPP

**Figure 7** Dr. Barbara Block, researcher in the Tagging of Pacific Pelagics (TOPP) program, places an archival tag on a tuna prior to its release.

main types of telemetry used for sending information in the marine realm: *radio telemetry* and *acoustic telemetry*. Radio telemetry is generally used in the air—including transmission from a data buoy or sensor at the ocean surface up to a data-relay satellite in space. A very low frequency (VLF) radio telemetry signal—from 10 to 30 kilohertz—can be used to create a "listening net" under water, but only over short distances. Acoustic telemetry is typically used underwater, and a set of "hydrophones" placed on the seafloor or suspended above the seafloor can create an acoustic telemetry listening net. Acoustic telemetry uses a wide range of frequencies, and extremely low frequency (ELF) signals can be heard under water for very long distances—even all the way around the world.

Radio telemetry uses transmitters to emit radio signals and receivers to detect and record them. When transmitting information, the data can be sent as *pulsed* signals or *digitally encoded* signals. Conventional radio telemetry involves the use of pulsed signals, providing a familiar "beep-beep-beep" signal. These pulsed systems can provide unique identification for individual subjects through the use of slightly different frequencies and/or a variety of pulse rates. A digitally encoded transmitter emits a unique numerical code that allows it to be differentiated from all others. Through this methodology, today's electronics allow for more than 200 transmitters to share one frequency. Because radio frequencies are a valued and scarce resource, this allows researchers to use the frequencies they are allocated optimally and increase the number of research subjects they can track. An added benefit to using a digitally encoded system with fewer frequencies is the decreased time required to scan the radio spectrum while monitoring a large number of subjects. A digitally encoded signal is also more distinguishable within a noisy radio environment and is easier to detect when the subject (or emitter) is mobile, providing more relevant data to the project. The transition from pulsed to digitally encoded signals within marine research is not unlike the transition in the cellular telephone industry from analog to digital. Many of the marine telemetry applications are based on a similar set of principles and technologies.

There are two basic ways the receivers function: by using either a manual tracking receiver or a fixed station receiver. Traditional manual tracking has been used to study large fish population movements, mammal migration patterns and the location of spawning grounds. When tracking is done manually, a search is performed on every frequency within the project for any emitted signals. A scanning receiver automatically "listens" on each frequency for a short period of time for an emitted signal. A fixed station receiver, on the other hand, can be deployed in the field and left to automatically scan all the frequencies in use. Once a valid signal has been detected, the receiver will automatically record the date, time, frequency, pulse rate and the signal strength of the emitter (Lotek Wireless Inc. 2003). With far-ranging and deep-diving marine species, the sensors and trackers are often used together. One example is the underwater acoustic sensors established in the water column in arrays lined up perpendicular to the U.S. West Coast. The acoustic sensors detect migratory species movement past them, and radio transmitters on the surface buoys transmit the data home for researchers to use.

There are also two basic types of tags attached to the animals. An *archival tag* collects data for a given time and is then mailed to the researcher when it is found by a fisherman catching the animal, or when it is automatically released from the animal and found floating at sea or washed up on shore. A *transponder tag* is a transmitter attached to a marine animal in order to study and track its behavior and movements; the radio transmitter sends location and other data—usually to a satellite—when the individual surfaces. A number of different *sensors* can be embedded within the tag to collect varying forms of information such as water depth, swimming direction, speed and heart rate. The sensors themselves are broken down into two categories: *physical sensors* and *physiological sensors*. Physical sensors measure temperature, pressure, light level, salinity, tilt angle, speed and acceleration, compass heading, magnetic field strength and sound. Physiological sensors measure characteristics that may provide more information about the health of the research subject including heart rate and tail-beat frequency. Some characteristics of individual sensor types include:

- *Temperature sensor*. This type of sensor uses a thermistor to measure ocean water temperature. One concern that needs to be taken into account when using this type of sensor is minimizing the amount of direct sunlight the sensor is exposed to, since this could cause a false reading.
- *Pressure sensor*. Pressure sensors are able to measure water depths accurately up to full ocean depth, but correct calibration is needed.
- *Light sensor*. A light sensor is comprised of a silicon diode and is normally about 5–10 square millimeters in size.

It is used to measure ambient light intensity in order to estimate the subject's geographical position. These sensors are highly sensitive and able to take light readings several hundred meters below the ocean surface.

- *Salinity sensor*. These sensors measure the concentration of saltiness in the water and can help identify when a subject has moved between an open sea area into a freshwater or brackish estuarine area.
- *Speed and acceleration sensors*. This sensor is comprised of a small turbine, hot film, an electromagnetic sensor or a small Doppler shift instrument. Alternatives to using these sensors would be to measure speed indirectly by interpreting the tail-beat frequency of the subject (Arnold *et al* 2002).

### TOPP (Tagging of Pacific Pelagics) Program

One program using telemetry and tagging technology to collect behavior information of marine species is a pilot program known as TOPP (Tagging of Pacific Pelagics). It is one of six programs under the Census of Marine Life (CoML) program umbrella and aims to provide a better understanding of habitat use by marine vertebrates and large squid within the North Pacific region. CoML is a ten-year program that will provide researchers and policymakers a unique historical perspective on the type, amount, health and behavior of life in the ocean. TOPP is using electronic tags to monitor various species to better understand their distribution and movements. The TOPP researchers are consciously using their tagged subjects as autonomous ocean profilers, covering sections and areas of the ocean where very little research has been done. The temporal and spatial data generated from the project will provide researchers an "organism-eye" view of the North Pacific. The program is addressing the following questions with its research:

- How do top predators respond to the varying physical environment?
- How does habitat selection take place?
- Where are the migration corridors, and why?
- What are the topographical and other physical features that organisms use to chart their courses or locate food concentrations?
- What are the areas of potential conflicts with human activities?

The information gathered from this project is vital to marine conservation efforts, and for popular commercial species such as tuna and shark, a better understanding of habitat structure will provide a better fisheries management plan for these species. Key areas to understand include: spawning sites, foraging locations, migratory routes and temporal and spatial movements within the ocean. In some cases, these "ocean profilers" will be able to provide data about the water column that cannot be currently ascertained from remote sensing. The TOPP program plans to tag and monitor over 5,000 animals of 22 different species (Table 1) (Block *et al* 2003, R. Kochevar personal communication).

Table 1

**List of species tagged and tracked in the Tagging of Pacific Pelagics (TOPP) program**

| | |
|---|---|
| **Sea turtles** | **Tunas** |
| Leatherback turtle | Bluefin tuna |
| Loggerhead turtle | Yellowfin tuna |
| | Albacore tuna |
| **Sea birds** | **Sharks** |
| Black-footed albatross | White shark |
| Laysan albatross | Mako shark |
| Pink-footed shearwater | Salmon shark |
| Sooty shearwater | Blue shark |
| **Pinnipeds** | Common thresher shark |
| Elephant seal | |
| California sealion | **Other species** |
| **Cetaceans** | Humboldt squid |
| Blue whale | Ocean sunfish |
| Fin whale | |
| Humpback whale | |
| Sperm whale | |

The TOPP program is using the most advanced technology available today in telemetry and tagging, including archival tags (where data are retrieved when the tag is returned to the research center for a reward), pop-up satellite tags (where the tag rises to the surface and radio-transmits data), satellite-linked data recorders (where data are radiotransmitted when the animal reaches the ocean surface) and Global Positioning System tags for determining more precise locations from the ocean surface. Some animals will be tracked for days to weeks, while others can be monitored for months or years. A linked research project from 1999 to 2000 involved the monitoring and tracking of six adult white sharks off the coast of California. The satellite tags used in that project recorded data every two minutes—including water depth, temperature and light

(Monterey Bay Aquarium 2002). The tags were programmed to automatically detach from the animals on a given date and rise to the ocean surface to transmit their data to an ARGOS satellite, which relayed them to the Hopkins Marine Earth Station. The TOPP program is a three-year effort at a cost of $15–$20 million (TOPP 2002).

## Submersibles

According to the *American Bureau of Shipping*, a submersible is defined as:

> *Any vessel or craft capable of operating under water, submerging, surfacing and remaining afloat under weather conditions not less severe than Sea State 3 (gentle breeze, 7–10 knots), without endangering the life and safety of the crew or passengers.*

Submersibles rely on direct support from a tending vessel and are hoisted on deck when transiting between research locations.

There is an array of distinct types of submersibles used for marine exploration and conservation work:

- DSVs, or Deep Sea Vehicles, are untethered, human-occupied, free swimming undersea vehicles.
- ROVs, or Remotely Operated Vehicles, are tethered, self-propelled vehicles with direct real-time control from a surface support ship.
- AUVs, or Autonomous Underwater Vehicles, are untethered, undersea vehicles that can be completely pre-programmed and equipped with decision aids to operate autonomously, or the operation may be monitored and revised by control instructions transmitted through a data link.

Breakthroughs are being made every day in the development of research submersibles. Very recently, Australian researchers announced the development of the world's smallest submarine at a mere 40 centimeters long, called *Serafina* (BBC News 2004). With a plastic hull, five propellers and rechargeable batteries, it can dive to a depth just over 5,000 meters and travel at about one meter per second. At a cost of US $700, it could be a very affordable research item for shipwreck discovery, search and rescue missions and marine conservation research.

With the advances in undersea robotic vehicles—equipped with fiber optic data transmission cables, high-definition television (HDTV) cameras and increasingly dexterous manipulator arms—some have asked why we should continue to build research submersibles for human use. Frequently, the expense and risk of sending humans hundreds or thousands of meters below the ocean surface are cited as reasons not to. Consider applying the same argument, however, to the study of terrestrial forests. What field biologist would be satisfied with lowering a robot camera or dragging a net from a helicopter through the forest canopy to study the rainforest? Imagine using the hodgepodge of samples and information this would yield to try to glean details on animal behavior and biology critical to laying out a solid conservation strategy. Yet, traditionally, this is how marine science has been done.

With a submersible craft, scientists can spend extended periods of time at depths unreachable using SCUBA and without the worry of decompression illness. The atmospheric pressure inside the submersible is equal to that at the surface, and carbon dioxide "scrubbers" clean the air to permit many hours of observation time. Much can be learned about a given ecosystem by simply setting the sub down on the seafloor and watching the action unfold over a span of several hours, or by methodically following transects. Simply having the luxury of time to explore and collect algae, rocks or new species can lead to new and unexpected finds.

Small, one- to three-person submersibles (Figure 8) are well suited to studies in the littoral zone, from near shore to depths of 1,000 meters. These compact and versatile craft are not dependent upon costly dedicated support ships and can be launched from vessels of opportunity. Many can be cost-effectively shipped anywhere in the world by air or in standard 20-foot sea containers.

Often deployed in pairs—like SCUBA-diving "buddies"—or in conjunction with a remotely operated vehicle providing real-time video to support personnel on the surface, these systems are ideal for both rapid assessment of areas and long-term observation and sampling missions. Both types of studies are important for understanding migration, feeding, predation and reproduction, and they permit the gathering of related physical oceanographic and geologic data. Beyond their raw data-gathering capability, submersibles provide for direct human observation and opportunities for discovery of things that might otherwise go unnoticed.

While diving in a one-person submersible at 400 meters off the coast of Hawai'i, a scientist was preparing to ascend following an exploratory dive. What appeared to be a bag or other human debris was glimpsed moving by in the current. When the scientist turned the sub for a closer look, the debris turned out to be a large deep-water octopus. In a "reverse aquarium" moment the octopus approached, running its suckered arms over the dome, around the lights and cameras and gazing at the human inside with its large eyes. While the cameras recorded the event, it was direct human observation that noted the large egg mass carried by the animal and the methodical, inquisitive fashion in which she investigated the submersible. Had the science party been using only an ROV, it is unlikely that the octopus would have been spotted at all. Later dives to similar depths led to the documentation of a new depth record for the ocean sunfish *Mola mola* at more than 300 meters and a breathtaking encounter with a fast moving billfish also at considerable depth.

Over the past five years, the number of scientists trained to operate one-person submersibles has grown into the hundreds. Those who have used both multi-person and one-person craft describe the experience as the difference between driving themselves and trying to tell a taxi driver where they want to go. With the excellent safety record and established protocols, even conservative government research vessels have achieved success in handling one-person, micro-submersible operations.

For all the recent success in the littoral zone, however, much remains to be done in the deep ocean. Only a handful of aging deep-diving research submersibles exist today and none of them can reach full ocean depth. Nearly all are large and cumbersome, requiring the expense of a dedicated "mothership," which drives up overall operational costs. Because so few exist and costs are high, few scientists have the opportunity to use this technology. Yet our experience on land tells us that the most successful conservation

© KIP F. EVANS, NATIONAL GEOGRAPHIC SOCIETY, 1998

**Figure 8** Many small one- to three-person submersibles can be easily transported worldwide and operated fairly inexpensively from the decks of ships of opportunity.

plans begin with understanding entire systems. While we have an ongoing strategy for removing great quantities of deep-water fish like orange roughy, we know almost nothing about their life cycle, natural behavior or other factors that contribute to a healthy population, and thus we have no real conservation strategy.

Deep cold-water corals are also poorly understood but are constantly under threat of destruction from deep-water trawling activity. Seamounts, which act as submerged islands with high levels of endemic species, are being trawled and fished bare before they can even be studied. Though small submersibles can reach the tops of some seamounts, the slopes and bases may extend below their rated depth.

The technology exists to build reasonably sized innovative craft capable of excursions to full ocean depth, yet the deepest parts of the ocean remain out of reach. The United States has a 4,000-meter submersible, Russia and France have 6,000-meter craft and Japan has one 6,500-meter depth submersible. The United States and China currently have plans to build new or replacement systems rated to 6,500 meters and 7,000 meters, respectively. All are widely touted to provide access to 95% of the world's ocean. But we won't really understand the whole ocean, and how it affects the Earth's environment, until we can see and study all of it.

## Summary

A general trend is now emerging in terms of the unification of knowledge about the Earth's environment. All of the above technologies have begun to be recognized through international agreements and informal international scientific cooperation, and a constellation of existing and new observation systems is becoming united into a common observation network for the planet Earth and its ocean. A highly recognized component of the Group on Earth Observation (GEO) and the Global Earth Observation System of Systems (GEOSS) is the Global Ocean Observing System (GOOS). In essence, we are slowly moving towards a more unified vision of our planet, logically structured in order to meet the needs of various sectors, including those of global marine conservation.

## Acknowledgements

*The authors wish to acknowledge the sharing of valuable information on marine science technologies by Randy Kochevar of the Monterey Bay Aquarium, Frank Herr of Lotek Wireless and Elizabeth Taylor of Deep Ocean Engineering and Research.*

## Literature Cited & Consulted

Annulis, H., J. Carr and C. Gaudet. 2003. Building the geospatial workforce. *URISA Journal Special Education Issue* 15 (1): 21–30.

Arnold, G., S. Gudbjornsson, M. Holm, T. Heggberget, N. Ó. Maoiléidigh and J. Sturlaugsson. 2002. Electronic tagging, telemetry and data-logging tags. In *Tagging Methods for Stock Assessment and Research in Fisheries*. Report of Concerted Action FAIR CT.96.1394 (CATAG), ed. V. Thorsteinsson, 179. Reykjavik, Marine Research Institute Technical Report (79). *http://www.hafro.is/Bokasafn/Timarit/catag.pdf* (accessed August 2, 2004).

BBC News. July 30, 2004. Tiny submarine makes big splash. *http://news.bbc.co.uk/2/hi/science/nature/3939355.stm* (accessed August 18, 2004).

Block, B. A., D. P. Costa, G. W. Boehlert and R. E. Kochevar. 2003. Revealing pelagic habitat use: The tagging of Pacific pelagics program. *Oceanologica Acta* 25:255–256.

Census of Marine Life. 2004. *http://www.iobis.org/* (accessed September 26, 2004).

Foresman, T., ed. 1998. *The History of Geographic Information Systems: Perspectives from the Pioneers*. Upper Saddle River, NJ: Prentice Hall.

Geodata.gov. 2004. *Geodata.gov: U.S. Maps and Data,* also known as Geospatial One Stop. U.S. Federal Office of Management and Budget. *http://www.geodata.gov/* (accessed September 26, 2004).

Kennedy, H. 2001. *ESRI Press Dictionary of GIS Terminology.* Redlands, CA: ESRI Press.

Lillesand, T. M. and R. W. Kiefer. 2000. *Remote Sensing and Image Interpretation,* 4th edition. Hoboken, NJ: John Wiley and Sons.

Longley, P. A., M. F. Goodchild, D. J. Maguire and D. W. Rhind. 2001. *Geographic Information Systems and Science*. Baffins Lane, Chichester UK: John Wiley and Sons, Ltd.

Lotek Wireless Inc. 2003. *Lotek Wireless Fish and Wildlife Monitoring, http://www.lotek.com* (accessed July 31, 2004).

Monterey Bay Aquarium. 2002. *White sharks migrate thousands of miles across the sea, new study finds.* Press release January 2, 2002.

ReefBase. 2004. Interactive map server. ReefBase.org *http://reefgis.reefbase.org/mapper.asp* (accessed July 31, 2004).

Riordan, R. and K. Barker. 2003. Cultivating biodiversity in Napa. *Geospatial Solutions,* November, *http://www.geospatial-online.com/geospatialsolutions/article/articleDetail.jsp?id=76573* (accessed October 11, 2004).

Sietzen, F., Jr. 2004. High-growth jobs initiative. *Geospatial solutions.* Advanstar Communications. June 1, 2004. *http://www.geospatial-online.com/geospatialsolutions/article/articleDetail.jsp?id=96795* (accessed November 1, 2004).

Tomlinson, R. 2003. *Thinking about GIS: Geographic information system planning for managers.* Redlands, CA: ESRI Press.
TOPP. 2002. *Tagging of Pacific Pelagics: Program overview.* Press release February 4, 2002.
U. S. Coral Reef Task Force. 2004. *http://www.coralreef.gov* (accessed July 31, 2004).
Veisze, P. and K. Karpov. 2002. Geopositioning a remotely operated vehicle for marine species and habitat analysis. In *Undersea with GIS,* ed. D. J. Wright, 105–115. Redlands, CA: ESRI Press.

# *Time for a Sea Change*

Sylvia A. Earle

The *Defying Ocean's End (DOE)* Conference took place at a momentous time, when decisions—or lack of them—would have a magnified impact on the future of the sea, and in turn, on the fortunes and ultimate fate of humankind. Early in the 21st century, after years of widespread ignorance and indifference about the decline of ocean health, various events converged to create something close to a "tipping point," a time when enough people had become aware that the ocean is in trouble that a general shift in behavior seemed imminent—a sea change, if you will.

Holding a conference to assess global ocean issues, envision actions and develop a business plan to "defy ocean's end" was not intended as an end in itself, but rather a means to other ends—a catalytic meeting that would bring specialists from diverse fields together to define a clear, strategic way forward and then encourage combined efforts on the global plan, with checkpoints along the way to gauge progress for at least ten years. Those assembled seemed to be of one mind in believing that if perverse policies can be modified, if the public and decision-makers can be better informed and motivated, if resources can be mobilized—at least two billion dollars per year—current troubling trends can be turned toward more positive outcomes over the next decade. In 2003, turning the "ifs" into "whens" seemed achievable and the time frame reasonable.

Although convened by Conservation International, *DOE* involved representatives from twenty countries and more than seventy organizations. And implementation of the *DOE* vision and the ultimate outcomes will depend on the talents and efforts of many individuals and organizations, including many who were not a part of the original Conference. While there is broad concurrence on the goals—the need for improved governance of the global commons, fisheries reform, increased ocean research and exploration, expanded protection of ocean areas, and better communication of ocean issues to the public—individuals and organizations vary in their priorities, methods and abilities. Many *DOE* participants agreed that by joining forces we could accomplish more—faster and more effectively—than the business-as-usual approach of operating independently.

In Washington, D.C., many organizations participated in bi-monthly follow-on meetings to share information on their individual programs and discuss promising new coalitions and partnerships emerging from the *DOE* Conference. In June 2004

**Facing page**
Red Pygmy Seahorse.
© WOLCOTT HENRY

an anniversary session was held at the Cosmos Club to take stock of progress over the year and to plan future actions. Amid a continuing avalanche of bad news about the decline of marine life, ocean pollution and widespread apathy, the following actions appeared to be promising signs of progress.

## Ocean Governance

There is a convergence in the logic of policies needed to stem the flow of noxious substances into the sea, a recognition of the need to start at the tops of mountains, the headwaters of rivers, and continue through cities and towns, fields and farms, to ships in harbors and at sea, as well as the sources of what goes into the skies above, because the sea eventually receives it all. The U. S. State Department's "White Water to Blue Water" conference in 2003 highlighted the need for national and international cooperation to deal with shared problems concerning pollution and resource management.

Findings from the Pew Oceans Commission of 2003 and the U. S. National Ocean Commission of 2004—both focused on U.S. domestic issues—have much in common with each other and with resolutions put forth in recent years at various international meetings, by government agencies, private organizations and numerous concerned individuals. In 2003, the state of Australia's coral reefs was reviewed during meetings in Cairns, and early in 2004 Australia's ocean policies were considered at its Coast-to-Coast conference. Such meetings are examples of a growing worldwide interest in improving governance within national Exclusive Economic Zones. All of these are largely aligned with the recommendations of the independent, international, broadly focused *DOE* Conference.

A major concern at the *DOE* Conference focused on the immense heart of the ocean that lies beyond the jurisdiction of any nation—the 60% of the sea that aver-

STERLING ZUMBRUNN, CONSERVATION INTERNATIONAL

Coral polyps in the Galápagos Archipelago, Ecuador.

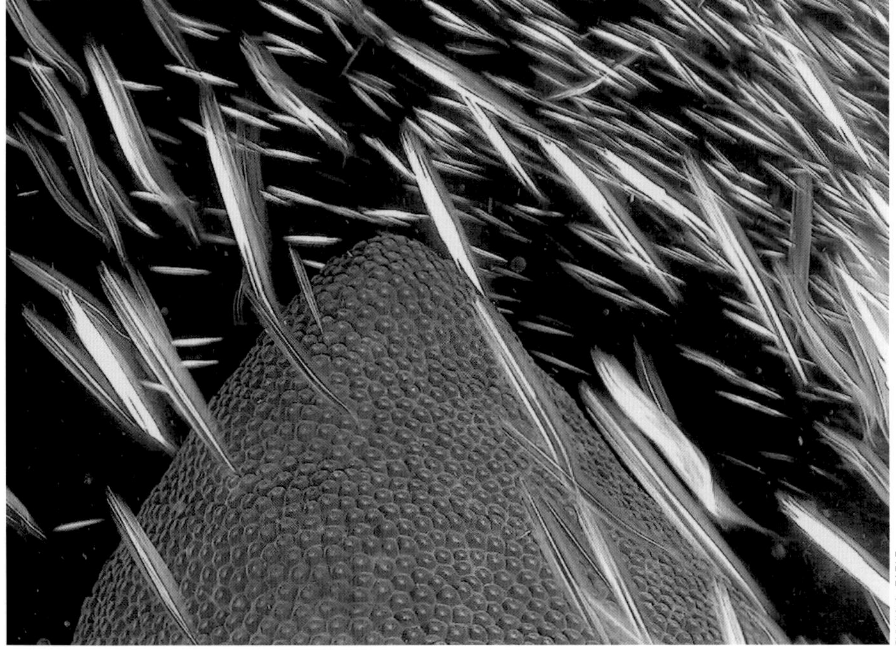
SYLVIA A. EARLE

Great star coral head surrounded by a passing school of fish in the Gulf of Mexico.

ages four kilometers in depth and encompasses broad, deep ocean plains, valleys, tens of thousands of unexplored submerged islands or seamounts and literally millions of not yet identified species of marine life including microbes that may outnumber all other species of life on Earth combined. Currently, a relatively small number of individuals from a few nations are destructively exploiting the ocean's "global commons." This increases the sense of urgency for establishing protective guidelines that would benefit all nations, including land-locked countries whose vital ocean "life support" benefits are at risk along with those of the rest of the world. Conservation International is supporting the preparation of law review articles to look at the concept of a "World Ocean Public Trust"—an international regime for overseeing activities on the high seas outside of national jurisdictions—and to discuss the feasibility and potential processes involved in its implementation.

The Natural Resources Defense Council (NRDC) led a new coalition that pulled together efforts to address destructive bottom-trawl fishing on seamounts in the high seas beyond national EEZs and secure a United Nations General Assembly resolution calling for a moratorium. Another group is urging the United Nations to ban the use of longlines on the high seas because of the direct, deadly impact this fishing technique is having on highly endangered leatherback sea turtles and other non-targeted species. There is precedent for such actions in the 1992 U.N. Resolution banning the use of driftnets in high seas fishing.

The Global Marine Program of IUCN–The World Conservation Union has initiated an international High Seas Marine Protected Areas working group that communicates often and convened experts in November 2004 at the triennial World Conservation Congress in Thailand to consider next steps concerning policies for protecting this vast, largely unknown part of the world ocean.

## Fisheries Reform/Aquaculture

Areas once protected by their remoteness, especially in the deepsea and open ocean, are no longer immune to disruption and destruction by modern technologies and international trade in marine wildlife "products." Economic motives drive the taking of wildlife from the sea, but the "profits" must be balanced against the existence of billions of dollars of "perverse subsidies." There is also a lack of credible accounting for the hidden costs to the environment that is ultimately the joint "property" of all humankind. Some maintain there is a need to capture ocean wildlife to satisfy food requirements for a growing human population, now exceeding six billion. With care, it may be possible to extract from healthy ecosystems enough ocean wildlife to feed local families and communities on a "sustainable" basis, but the major decline of most commercially exploited marine species strongly suggests that the prospect of sustainably feeding large numbers of people with ocean wildlife is no more realistic than sustainably supporting billions of people with wild-caught birds or land mammals.

In reality, wild fish supply a relatively small amount of the basic food needs of people worldwide—only about 5%. About a quarter of the fish caught globally goes to feed farmed carnivorous fish, such as salmon and trout, and farm animals—cows, pigs, chicken and other raised fish—that are basically herbivores (plant eaters) and normally would not consume fish at all. Much of the rest of the catch supports a high-end luxury market for tuna, swordfish and sharks, all high-on-the-food-chain carnivores (meat eaters) that represent a sizeable investment in food production. It takes many tons of plants—and tons of bigger fish eating smaller fish—to yield just one pound of a popular 10-year-old ocean carnivore species such as bluefin tuna. Also popular are Chilean sea bass, monkfish and orange roughy—all deep-water species

GREG FOSTER

Pink anemonefish hover above their home, a "magnificent anemone," in Palau.

that live as long as humans, don't reach reproductive age until their "teens" and whose populations are collapsing because they are being caught faster than they can reproduce. Neither of these fishery approaches are efficient or sustainable ways for humans to feed themselves.

Most of the calories that feed most people most of the time come from a small number of plants (rice, corn, wheat, soybeans) and plant-eating farm-raised animals. For centuries, cultivated fish in Asian countries have supplied significant amounts of protein. With new aquaculture methods, herbivorous carp, tilapia and catfish now account for millions of tons of energy-efficient protein, largely consumed in developing African and Asian countries. It takes only about two pounds of plants to yield a pound of such fish, about the same amount required for chickens. Cultivating carnivorous fish (tuna, salmon, trout, char) is far more costly and usually more difficult than raising plant-eaters, and future success will require finding food sources other than wild-caught animals. Showing the most promise are closed-system aquaculture methods that tightly control water quality and quantity ("more crop per drop"), rely on farm-raised brood stock and focus on herbivores, or at least low-on-the-food-chain organisms such as shrimp, oysters and clams.

While there is a long history of people trying to extract sustainable yields from wild populations of whales, fish, shrimp, lobsters and other marine wildlife, the biological history of the animals involved is much, much longer. Nothing in the millions of years these organisms have spent in developing survival strategies has prepared them for the recent arrival of a relentless, voracious terrestrial predator that not only insists on depleting hundreds of species by more than half and calling the resulting populations "sustainable," but also competes for food, destroys critical habitats, takes the largest, most productive adults, often at peak times of breeding, and alters the water chemistry with toxic compounds that interfere with their fine-tuned biological processes.

Fisheries reform is now widely endorsed as a high priority requirement, with many nations facing up to the reality that goals set for sustainable extraction of wildlife from the sea have been far too ambitious for far too long. There is also a growing awareness that fish and other ocean wildlife *alive* are important to humankind for reasons that transcend their value when killed and sold as commodities. The ecosystem role and value of living fish—sometimes likened to birds on the land—is increasingly being acknowledged by policymakers, as is the economic importance of live fish such as the Great Barrier Reef's famous potato cod, the Red Sea's Napoleon wrasses, the manta rays of Bora Bora and the hammerhead sharks of the Galápagos and Cocos islands. Such creatures bring a high price—once caught and sold to swim with lemon slices and butter—but when left swimming in the ocean they bring sustained high revenues over many years to local businesses who support divers who pay for the privilege of being with the big fish underwater.

"Business as usual" is no longer acceptable, especially when the destructive nature of bottom trawling and the enormous collateral damage inflicted by trawling and other fishing techniques on non-targeted fish, marine mammals, birds, sea turtles and countless invertebrates are taken into account. Collateral damage takes another form when the relationship between ecosystem health and loss of large carnivores and major grazers is taken into account. Recent studies show clear connections between

STERLING ZUMBRUNN, CONSERVATION INTERNATIONAL

Stone Gossard, *Pearl Jam* guitarist and ambassador for the ocean.

declines in kelp forests, coral reefs and seagrass meadows worldwide and the depletion of major herbivores such as parrotfish, sea turtles, manatees and sea urchins, and carnivores such as grouper, snapper, sharks and lobsters.

Reforms aimed at addressing these issues are beginning, but changes depend on changes in governance approaches, more focused research, increased public awareness and expanded policies relating to marine protected areas.

## Communications

Among the most serious issues relating to marine protection is the persistent belief that the ocean is so vast there is little humans can do to harm the way it works and, even when concerns about pollution and depletion of ocean wildlife are acknowledged, the relevance to our everyday lives is almost entirely unknown. Many people do not understand the profound impact the ocean has on all aspects of human life—the air we breathe, the water we drink, the benign climate, weather and temperatures we enjoy—the very underpinnings of our global physical, economic and social well-being. Nor do they know how human activities are altering the capacity of the ocean to provide the "life support" benefits they take for granted. A recent campaign to improve "ocean literacy" has been embraced by a number of organizations, including efforts to bring knowledge of the sea and its importance to the public through curricula, education programs in public aquariums, production of films and television programs, and internet communications.

Following the *DOE* Conference, The Ocean Conservancy and Conservation International, in collaboration with IUCN– The World Conservation Union, began a campaign to "Shatter the Myth" concerning ill-founded assertions about the resilience of marine animals and plants to excessive taking, pollution and habitat destruction. SeaWeb, a Washington-based non-governmental organization, stepped up its program of coordinating effective communications concerning ocean issues through electronic media and direct meetings with targeted audiences. SeaWeb and Conservation International are planning a *DOE*-based global ocean communications effort. And a new Deep Sea Conservation Coalition, working toward a moratorium for bottom-trawl fishing on the high seas, has established a cross-organizational communications group to support its efforts in reaching U.N. delegations from around the world. A process for facilitating communication and collaboration among national, regional and international organizations concerned about ocean issues is being developed through expansion of the *DOE* website and establishment of a *DOE* listserv.

While results are difficult to measure, investment in the communications of ocean issues is vital if positive changes in attitude are to occur. Even with knowing, people may not care, but they cannot care and will not act if they do not know.

## Marine Protected Areas

Concern about the loss of wild lands and terrestrial wildlife inspired U. S. President Theodore Roosevelt and other visionaries to take bold actions early in the 20th century to protect the nation's natural, cultural and historic heritage through a system of National Parks and wildlife refuges. Having National Parks, some say, is the "best idea North America ever had." The idea has been widely emulated worldwide and it has gradually been extended beyond the shore into the sea.

At the September 2003 World Parks Congress in Durban, South Africa, the need for accelerated protection for marine species and ecosystems was repeatedly acknowledged. Twelve percent of the land has been protected in the past century, while at the same time, only a fraction of 1% of the ocean has gained comparable status. Some proposed that by the time of the next World Parks Congress in ten years, at least 10% of the ocean should be protected in the manner of terrestrial national parks, while others argued that at least 30% under protection is needed to maintain the integrity of ocean systems.

Whatever the percentage, there was agreement at the *DOE* Conference that a scientific basis for establishing priorities is needed. Mirroring the approach taken by the *Defying Nature's End* Conference in 2001, *DOE* participants identified areas of high biodiversity, high risk, feeding, breeding and nursery areas, and vulnerable wilderness regions to be considered for inclusion in protected areas. It was also acknowledged, however, that timely opportunities to establish protected areas should not be missed because of "lack of sufficient knowledge." It is relatively easy to destroy the distillation of millions of years of natural development of any species or wild place, but once they're gone, gone too is the chance to protect them.

The *DOE* Conference also clearly endorsed the concept of large national and internationally managed multiple-use marine areas or "seascapes," with new emphasis on incorporating significantly large, fully protected reserves. New Zealand and Australia have recently added to their inventory of protected marine areas, including a number of offshore seamounts where fishing is now prohibited. In 2004, after years of monitoring the Great Barrier Reef and documenting serious decline, the government of Australia increased the areas of full protection from 4% to 33% of the reef.

© WOLCOTT HENRY

Coral reefs worldwide are in need of greater protection.

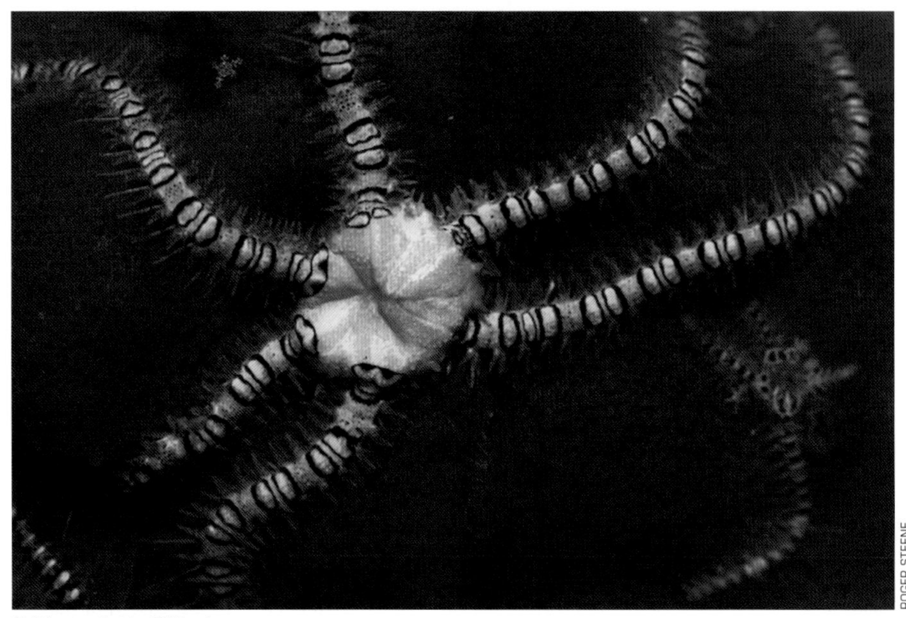
ROGER STEENE

Brittlestar in the Philippines.

In February 2004, Malaysia, Indonesia and the Philippines signed a Memorandum of Understanding adopting a conservation plan that promotes the sustainable development of the Sulu-Sulawesi Marine Ecoregion, an area of almost a million square kilometers representing a critical center of global marine biodiversity.

In April 2004, representatives from Costa Rica, Peru, Ecuador and Colombia signed the "San Jose Declaration," a plan for managing the Eastern Tropical Pacific Seascape, a mega-marine corridor encompassing numerous existing marine protected areas, two World Heritage Sites (the Galápagos and Cocos islands) and two others that are proposed at Coiba Island and Malpelo Island. Also in April the government of Costa Rica agreed to establish 25% of its entire EEZ for protection, and within the greater Eastern Tropical Pacific Seascape, to designate intensive protection for the Cocos—Las Baulas corridor to accommodate migrating dolphin, shark, tuna and turtle species. Also in 2004, Madagascar committed 20% of their coastal waters for conservation management, and the Bahamian Parliament passed a resolution to extend full no-take protection to 20% of its territorial sea.

At the *DOE* Conference, plans for the first international meeting focused entirely on marine protected areas was announced; it will take place in Geelong, Australia, in October 2005, and it will include a global assessment of areas protected, an evaluation of policies and practices, and consideration of new networks of multinational seascapes. In time, local, national and regional policies may be consciously connected in thoughtful, science-based ways into preserving an enduring place for ourselves on planet Earth—the ultimate "seascape."

## Global Science

Observations from satellites and ships collect global overviews of the sea surface, and ocean-oriented geographical information systems are yielding new insights into heretofore obscure scientific patterns and relationships, but access to the depths below remains woefully inadequate, as does our knowledge of the deep sea. Many scientists still use traditional nets, dredges and trawls to sample the ocean—in relatively ineffective and destructive ways—despite an increasingly available array of remotely operated vehicles, autonomous systems, manned submersibles and various sophisticated sensing, sampling and photo-documentation equipment.

Within the next five years, there will be several new manned submersibles capable of deep exploration and research built in the United States, Japan and China, and a new remotely operated system capable of reaching full ocean depth. These will greatly improve ocean access and help to implement the ambitious but vital research goals established at the *DOE* Conference. Research and conservation actions were recommended for four high-priority "unknown" habitats—seamounts, deep coral reefs, continental slopes and cold seeps, and marine caves. Research needs are recommended for microbial life and organisms in the water column, deepsea areas and polar waters. In the past year, a number of expeditions have been launched that will contribute vital new insights into life in these ocean areas and will help to establish priorities for protected areas and management policies.

Among the most promising developments is the Ocean Biogeographic Information System (OBIS) of the Census of Marine Life. New taxonomic authority groups have been started, and assessment of the status of marine species through the IUCN Red List is underway in cooperation with the marine division of the Species Survival Commission. The goal of building OBIS to 25 million records within ten years appears achievable but will require significant new resources and the commitment of many people worldwide.

## Next Steps

The vision of developing a global agenda for safeguarding the ocean—including specific recommendations and costs—is well underway, but there is an increasing sense of urgency as forces causing habitat destruction, biodiversity loss, pollution and other negative impacts from human activities continue to close our options.

Annual assessments are needed to update and refine the goals from the *DOE* Conference, and to assess efforts to fund them. There is an increasing need to demonstrate to the general public and government leaders the relevance of ocean degradation to our health, security, economic prosperity, social stability, quality of life, recreation, new knowledge and the viability of life itself. The network of people and organizations emanating from and linked to the *DOE* Conference is growing, and with it there is hope that when the five- and ten-year goals are assessed, satisfaction will outweigh despair, areas protected will exceed those lost, enduring policies will be in place to protect the health of the ocean, and an ethic of caring for the ocean will have been inspired that will both support and transcend the need for layers of laws.

Participants in the *Defying Ocean's End* Conference in Los Cabos, Mexico.

CHRISTINA MITTERMEIER

# Conference Participants

**Robin Abadia** *Conservation International*
**Julia Acuna** *Conservation International*
**John Adams** *Natural Resources Defense Council*
**Gerry Allen** *Conservation International*
**Chris Anderson** *Sapling Foundation*
**Ragnar Arnason** *University of Iceland*
**Farooq Azam** *Scripps Institution of Oceanography*
**Todd Baldwin** *Island Press*
**Ivan Barkhorn** *Meridian Strategy Group, LLC*
**Nancy Baron** *SeaWeb*
**Jamie Bechtel** *Conservation International*
**Charles Betlach**
**Juan Bezaury** *The Nature Conservancy*
**Dee Boersma** *University of Washington*
**Charlotte Boyd** *Conservation International*
**George Branch** *University of Cape Town*
**Thomas Brooks** *Conservation International*
**Lauretta Burke** *World Resources Institute*
**Scott Burns** *World Wildlife Fund–US*
**Rodrigo Bustamante** *CSIRO Marine Research & Charles Darwin Research Station*
**Kraig Butrum** *Conservation International*
**Louis Cabot** *Cabot-Wellington LLC*
**Claudio Campagna** *Wildlife Conservation Society*
**Jim Cannon** *Conservation International*
**Marielle Canter** *Conservation International*
**Maria de los Angeles Carvajal** *Conservation International*
**Ed Cassano** *Ocean Futures Society*
**Sarah Chasis** *Natural Resources Defense Council*
**Penny Chisholm** *Massachusetts Institute of Technology*
**Lew Coleman** *Moore Foundation*
**Jean-Michel Cousteau** *Ocean Futures Society*
**The Honorable Rohkmin Dahuri** *Minister for Marine Affairs and Fisheries—Indonesia*
**Neil Davies** *Richard B. Gump South Pacific Research Station*
**Lloyd Davis** *University of Otago, New Zealand*
**Susan Dawson** *Sapling Foundation*
**Brian Day** *International Institute for Environmental Communication*
**Guilherme Dutra** *Conservation International*
**Sylvia A. Earle** *Conservation International*
**Sara Everett** *DOE GIS Specialist*
**Exequiel Ezcurra** *Instituto Nacional de Ecologia*
**Enyuan Fan** *Chinese Academy of Fishery Sciences*
**Robert Fine** *DOE Info Technology Team*
**John Francis** *National Geographic*
**Rodney Fujita** *Environmental Defense*
**Karen Garrison** *Natural Resources Defense Council*
**Claude Gascon** *Conservation International*
**Joe George** *DOE Info Technology Team*

**Kristina Gjerde** *Environmental Solutions International*
**Linda Glover** *Gloverworks Consulting*
**Barry Gold** *Packard Foundation*
**Edgardo Gomez** *University of The Phillipines*
**Montserrat Gorina-Ysern** *American University*
**Fred Grassle** *Rutgers University*
**Dietmar Grimm** *Meridian Strategy Group, LLC*
**David Guggenheim** *The Ocean Conservancy*
**Lynne Hale** *The Nature Conservancy*
**Martin Hall** *Inter-American Tropical Tuna Commission*
**Graham Harris** *Foundation Patagonia Natural*
**Jim Hauslein** *Hauslein & Company, Inc.*
**Hugh Hough** *Green Team Advertising, Inc.*
**Tom Houston** *International Seakeepers Society*
**Jeremy Jackson** *Scripps Institution of Oceanography*
**Jennifer Jeffers** *DOE Conference Associate*
**Aristides Katoppo** *Sinar Harapan Daily*
**Les Kaufman** *Boston University*
**Graeme Kelleher** *DOE Conference Chair*
**Paul Kelly** *Rowan Companies*
**Richard Kenchington** *Consultant*
**Peter Knights** *WildAid*
**Nancy Knowlton** *Scripps Institution of Oceanography*
**Phil Kramer** *University of Miami*
**Scott Kyllo** *DOE Info Technology Team*
**Zelinda Leao** *Universidade Federal da Bahia*
**Larry Linden** *Goldman Sachs & Co*
**Ghislaine Llewellyn** *World Wildlife Fund–US*
**Gaston Luken** *Former Chairman, GE Capital Mexico*
**Carl Lundin** *IUCN-World Conservation Union*
**Larry Madin** *Woods Hole Oceanographic Institution*
**Francois Martel** *Conservation International*
**Dan Martin** *Moore Foundation*
**Rod Mast** *Conservation International*
**John McCosker** *California Academy of Sciences*
**Anne McEnany** *International Community Foundation*
**Sheila McKenna** *Conservation International*
**Edmund McManus** *UNEP World Conservation Monitoring Center*
**Roger McManus** *Species Survival Commission*
**Cristina Mittermeier**
**Russell Mittermeier** *Conservation International*
**Gordon and Betty Moore** *Moore Foundation*
**Ken Moore** *Moore Foundation*
**Lance Morgan** *Marine Conservation Biology Institute*
**Yaya Mulyana** *Director of Marine Protected Areas–Indonesia*
**Robin Murphy** *Conservation International*
**Ransom A. Myers** *Dalhousie University*
**David Obura** *CORDIO*
**John Ogden** *Florida Institute of Oceanography*
**Stephen Olsen** *University of Rhode Island*
**Michael Orbach** *Duke University*
**Julie Packard** *Packard Foundation*
**Arthur Paterson** *NOAA*
**Maria Louisa Pena** *DOE Info Technology Team*
**Ed Penhoet** *Moore Foundation*
**Charles Peterson** *University of North Carolina*
**Brad Phillips** *Conservation International*
**Ellen Pikitch** *The Wildlife Conservation Society*
**Steve Polasky** *University of Minnesota*
**Marjorie Reaka-Kudla** *University of Maryland*
**Callum Roberts** *University of York*
**Alejandro Robles** *Conservation International*
**Patricio Robles Gil** *Agrupación Sierra Madre, Unidos para la Conservación*
**Martha Rosales** *Conservation International*
**Roger Rufe** *The Ocean Conservancy*
**Enric Sala** *Scripps Institution of Oceanography*
**Rod Salm** *The Nature Conservancy*
**David Sandalow** *World Wildlife Fund–US*
**Gabriel Larrea Santana** *Gobernador de Baja California Sur Representative*
**Mick Seidl** *Moore Foundation*
**Peter Seligmann** *Conservation International*
**Myriam Sibuet** *IFREMER*
**Mike Smith** *Conservation International*
**Anne Solbraekke** *Moore Foundation*
**Andrew Solow** *Woods Hole Oceanographic Institution*
**Mark Spalding** *Consultant*
**Karen Stocks** *Scripps Institution of Oceanography*
**Greg Stone** *New England Aquarium*
**Jatna Supriatna** *Conservation International*
**Lee M. Talbot** *George Mason University*
**Martha Talbot**
**Phil Trathan** *British Antarctic Survey*
**John Wesley Tunnell** *Texas A&M*
**Tom Tusher**
**Ben Vitale** *Conservation International*
**Anne Walls** *BP p.l.c.*
**Sue Wells**
**Tim Werner** *Conservation International*
**James Wilen** *University of California, Davis*
**Captain Bill Wright** *Royal Caribbean Cruise Lines*
**Jim Wyss** *Conservation International*
**Dan Zimble** *ESRI*

Affiliations listed are those at the time of the *DOE* Conference, May/June 2003.

**Facing page**
The Galápagos sealion, one of the most numerous marine mammals in the Archipelago.
STERLING ZUMBRUNN, CONSERVATION INTERNATIONAL

# Index

*Italic* page numbers indicate tables and figures

## C

## D



## E

A Caribbean reef squid moves beneath moon-speckled surface waters in the Hol Chan Marine Reserve, Belize.



## F

## G



## H



## I

Footprints on the beach in Los Cabos, Mexico.



## J



## K



## L



## M

## N



## O



## P

# R



# S

## T